GEOMICROBIOLOGY

GEOMICROBIOLOGY

Editors

Sudhir K. Jain

School of Studies in Microbiology
Vikram University
Ujjain, Madhya Pradesh
India

Abdul Arif Khan

College of Pharmacy
King Saud University
Riyadh
Saudi Arabia

Mahendra K. Rai

Biotechnology Department
SGB Amravati University
Maharashtra
India

CRC Press
Taylor & Francis Group
Boca Raton London New York

CRC Press is an imprint of the
Taylor & Francis Group, an **informa** business

CRC Press
Taylor & Francis Group
6000 Broken Sound Parkway NW, Suite 300
Boca Raton, FL 33487-2742

First issued in paperback 2019

© 2010 by Taylor & Francis Group, LLC
CRC Press is an imprint of Taylor & Francis Group, an Informa business

No claim to original U.S. Government works

ISBN-13: 978-1-57808-665-8 (hbk)
ISBN-13: 978-0-367-38426-5 (pbk)

Library of Congress Cataloging-in-Publication Data

Geomicrobiology/editors, Sudhir K. Jain, Abdul Arif Khan, Mahendra K. Rai.
      p. cm.
   Includes bibliographical references and index.
   ISBN 978-1-57808-665-8 (hardcover)
1. Geomicrobiology. I. Jain, Sudhir K. II. Khan, Abdul Arif. III. Rai, Mahendra.
   QR103G455   2010
   579175   -- dc22

2009039707

Visit the Taylor & Francis Web site at
http://www.taylorandfrancis.com

and the CRC Press Web site at
http://www.crcpress.com

Preface

Geomicrobiology is a combination of geology and microbiology, which includes the study of interaction of microorganisms with their environment, such as in sedimentary rocks. This is a new and rapidly-developing field that has led in the past decade to a radically-revised view of the diversity and activity of microbial life on Earth. Geomicrobiology examines the role that microbes have played in the past and are currently playing in a number of fundamental geological processes.

Microorganisms are very important for the environment. Biogeochemical cycles performed by these organisms are necessary to maintain normal life on Earth. These microorganisms are very important in the petroleum and pharmaceutical industries from the economic point of view. Beside which, increased agricultural productivity through soil microorganisms, natural products and a wide range of industrially important substances isolated from these microorganisms, have popularized this field as a thrust area for present and future researchers. The curiosity to know about life on other planets is also evident in present-day researchers. The Earth's earliest biological sediments can solve the enigma about the origin of life on Earth and give an indication about the presence of life on other planets. The future of this field is definitely bright, people are dependent on these microorganisms not only for their food but also for maintaining a proper healthy environment.

Inspite of the bright future of this field, quality literature related to geomicrobiology is not sufficient to fulfil the requirements of students and researchers. There is an immense need to gather quality literature for scientists working in this thrust area.

The present book is of great importance for researchers working in the field of microbiology, biotechnology, geology and environmental science. It can be a major reference book for students as well as researchers.

Sudhir K. Jain
Abdul Arif Khan
M.K. Rai

Contents

List of Contributors

A. Esteve-Núñez

Centro de Astrobiología, CSIC/INTA, Associated to the NASA Astrobiology Institute, Ctra de Ajalvir. Km. 4, 28850 Torrejón de Ardoz, Madrid, Spain.

E-mail: estevena@inta.es

Abdul Arif Khan

College of Pharmacy, King Saud University, Riyadh, Saudi Arabia

E-mail: abdularifkhan@gmail.com

Arabinda Ray

Department of Chemistry, Sardar Patel University, Vallabh Vidyanagar - 388120, Gujarat, India.

E-mail: arabinda24@yahoo.co.in

Ata Akcil

BIOMIN Group, Mineral Processing Division, Department of Mining Engineering, Suleyman Demirel University, Isparta TR 32260, Turkey.

E-mail: ata@mmf.sdu.edu.tr

Brendan P. Burns

School of Biotechnology and Biomolecular Sciences and Australian Centre for Astrobiology, The University of New South Wales, 2052 Australia.

Brett A. Neilan

School of Biotechnology and Biomolecular Sciences and Australian Centre for Astrobiology, The University of New South Wales, 2052 Australia.
E-mail: b.neilan@unsw.edu.au

Chiaki Kato

Extremobiosphere Research Center, Japan Agency for Marine Earth Science and Technology (JAMSTEC), 2-15 Natsushima-cho, Yokosuka 237-0061, Japan.
E-mail: kato_chi@jamstec.go.jp

Diana E. Northup

Biology Department, University of New Mexico, Albuquerque, NM 87131-0001 USA.
E-mail: dnorthup@unm.edu

E. Lázaro

Centro de Astrobiología, CSIC/INTA, Associated to the NASA Astrobiology Institute, Ctra de Ajalvir. Km. 4, 28850 Torrejón de Ardoz, Madrid, Spain.
E-mail: lazarolm@inta.es

Elijah Ohimain

Biological Sciences Department, Faculty of Science, Niger Delta University, Wilberforce Island, Amassoma, Bayelsa State, Nigeria.
E-mail: eohimain@yahoo.com

Falicia Goh

School of Biotechnology and Biomolecular Sciences and Australian Centre for Astrobiology, The University of New South Wales, 2052 Australia.

Haci Deveci

Mineral Processing Division, Department of Mining Engineering, Karadeniz Technical University, Trabzon TR 61080, Turkey.
E-mail: hdeveci@ktu.edu.tr

Hazel A. Barton

Department of Biological Sciences, Northern Kentucky University, Highland Heights, KY 41099, USA.
E-mail: bartonh@nku.edu

Hiren Doshi

Department of Chemistry, Sardar Patel University, Vallabh Vidyanagar - 388120, Gujarat, India.

Present Address: Ashok & Rita Patel Institute of Integrated Study and Research in Biotechnology and Allied Sciences (ARIBAS), New V.V. Nagar - 388121, Gujarat, India.

E-mail: drhirendoshi@yahoo.co.in

I.L. Kothari

Department of Bioscience, Sardar Patel University, Vallabh Vidyanagar - 388120, Gujarat, India.

E-mail: ilkothari@yahoo.com

J. Martínez-Frías

Centro de Astrobiología, CSIC/INTA, Associated to the NASA Astrobiology Institute, Ctra de Ajalvir. Km. 4, 28850 Torrejón de Ardoz, Madrid, Spain.

E-mail: martinezfj@inta.es

Jiasong Fang

College of Natural and Computational Sciences, Hawaii Pacific University, 45-045 Kamehameha Highway, Kaneohe, HI 96744, USA.

E-mail: jfang@hpu.edu

Kathleen H. Lavoie

Arts and Sciences, State University of New York, College of Plattsburgh, Plattsburgh, NY 12901-2681 USA.

E-mail: lavoiekh@plattsburgh.edu

M.K. Rai

Department of Biotechnology, SGB Amravati University, Amravati 444 602, Maharashtra, India.

E-mail: mkrai123@rediffmail.com

Michelle A. Allen

School of Biotechnology and Biomolecular Sciences and Australian Centre for Astrobiology, The University of New South Wales, 2052 Australia.

Preeti Bhatnagar

Department of Microbiology, College of Life Sciences, Cancer Hospital & Research Institute, Gwalior (M.P.), India.

E-mail: bhatnagarpreeti01@gmail.com

Rachael Shi

School of Biotechnology and Biomolecular Sciences and Australian Centre for Astrobiology, The University of New South Wales, 2052 Australia.

Sudhir K. Jain

School of Studies in Microbiology, Vikram University, Ujjain (M.P.), India.

E-mail: sudhirkjain1@rediffmail.com

Todd J. Harvey

Vice President, Base Metal Technologies, GeoBiotics, LLC, 12345 W. Alameda Pkwy #310, Lakewood, CO, USA 80228.

E-mail: tharvey@geobiotics.com

CHAPTER 1

Microbe–Mineral Interactions: Cave Geomicrobiology

Kathleen H. Lavoie[1*], Diana E. Northup[2] and Hazel A. Barton[3]

INTRODUCTION

Geologists tend to attribute observable mineral changes to abiotic processes while biologists try to explain everything in terms of biotic mechanisms (Barton et al. 2001), the actual mechanisms are often a combination of abiotic and biotic. Geomicrobiology recognizes that microorganisms are important active and passive promoters of redox reactions that can influence geological formation (Ehrlich 1996). The temporal and spatial scales of geomicrobiological processes vary greatly, from minutes to eons and microniche to global, yet the contributions made by microbes to most processes are unknown and unseen due to the very small size of a typical microbial cell. Microbes are crucial to nutrient cycling at local and global scales, able to discriminate among the stable isotopes of H, C, O, N, and S, resulting in fractionation and enrichment of lighter isotopes which are used as a marker of biological activity. Microbial activity also plays an important role in industry, being largely responsible for the production and accumulation of fossil fuels, extensive iron mineral deposits, sulfur domes, marine manganese nodules, and potentially the accumulation of uranium and gold (Ehrlich 1996, 1998).

[1]Arts and Sciences, State University of New York, College of Plattsburgh, Plattsburgh, NY 12901-2681 USA, E-mail: lavoiekh@plattsburgh.edu
[2]Biology Department, University of New Mexico, Albuquerque, NM 87131-0001 USA, E-mail: dnorthup@unm.edu
[3]Department of Biological Sciences, Northern Kentucky University, Highland Heights, KY 41099 USA, E-mail: bartonh@nku.edu
*Corresponding author

Scientists have determined that despite abundant life on the surface, most of the biomass of life on Earth is below the ground and primarily microbial (Whitman et al. 1998). Microbes form metabolic products and selectively reduce or oxidize mineral constituents of rocks and soils. All sedimentary and igneous rock is susceptible to microbially-mediated dissolution and diagenesis (Ehrlich 1998). It is also believed that if life exists on other planetary bodies in the solar system, such as on Mars, and the moons, Europa and Enceladus, life will be found in subterranean environments due to their challenging surface conditions. Caves therefore serve as terrestrial analogs for subterranean microbial ecosystems, allowing entry points to study such life below the surface without the necessity of invasive techniques.

Caves are ideal experimental study systems, not because they are strange, but because they are simple. Caves are often locally abundant, allowing for replicate studies (Frey 1963, Culver 1982). Nonetheless, studies of microorganisms in caves have been predominantly descriptive with only a few experimental studies reported. This situation is changing; during the past decade there has been extensive research into microbial interactions with minerals within cave environments (see papers in Sasowsky and Palmer 1994, Northup et al. 1997, Northup and Lavoie 2001, Barton and Northup 2007, Engel 2007). We review these efforts in the context of dissolution and precipitation reactions of caves and their speleothems (secondary mineral formations) involving carbonates, moonmilk, silicates, clays, iron and manganese, sulfur, saltpeter, and the formation of biokarst and phytokarst. Caves may also serve as models in Astrobiology. Suggestions for further study are also included.

CAVES AND SPELEOTHEMS

A cave is any natural void below the surface that is accessible to humans (Gillieson 1996).

Such voids below the Earth's surface range in size from meters to >500 kilometers, and, theoretically, most have no natural human-accessible entrance (Curl 1966). Caves can be classified in several ways, particularly by the type of rock and method of formation (Palmer 1991). The most common types of caves are those formed by dissolution in limestone, gypsum, and other calcareous rocks, and as lava tube caves in basaltic rock. Other types of caves are usually limited in extent and include those in granite, talus, quartzite, ice, and sandstone.

There are three primary mechanisms for the formation of caves.

- Dissolution by carbonic acid. Classical limestone caves such as the Mammoth Cave, Kentucky, and Altamira Cave, Spain, are formed as water passes through the litter and soil zone and absorbs biogenic CO_2, forming a dilute solution of carbonic acid. Water can also be acidified by microbes living in the carbonate rock (Schwabe et al. 2001). If this weakly acidic water comes into contact with limestone ($CaCO_3$), the calcium carbonate is dissolved (Gillieson 1996). Continued dissolution leads to cavity and cave formation. Opening of the cave to the surface releases supersaturated carbon dioxide and allows the formation of secondary carbonate speleothems.

- Sulfuric acid-driven speleogenesis creates some limestone caves when hydrogen sulfide gas rises along fissures until it encounters oxygenated water and is oxidized to sulfuric acid, which dissolves the limestone (Hill 1987, 1990, 1995, 1996, 2000, Jagnow et al. 2000). Biogenic production of sulfuric acid directly by microbes also contributes to such cave formation (Engel et al. 2004, Barton and Luiszer 2005). Examples include Carlsbad Cavern and Lechuguilla Cave in New Mexico, Movile Cave, Romania, and Cueva de Villa Luz, Mexico.

- Lava tube caves are formed by actively flowing lava, as seen on Hawaii's Kilahua volcano. As molten lava flows out of a volcano, the surface lava cools more quickly, providing a layer of insulation for the molten lava flowing beneath. When the eruption stops, the rapidly flowing lava may drain, leaving an empty tubular conduit behind (Palmer 1991). Microbes are not involved in formation of lava tube caves, but do influence subsequent weathering and breakdown.

The bedrock in which the cave is formed and the mechanism of speleogenesis can have a profound effect on the secondary deposits found there (Hill and Forti 1997). Spelothems, such as stalactites and stalagmites, are secondary mineral deposits formed by a physiochemical reaction from a primary mineral in a cave (Moore 1952). Hill and Forti (1997) recognize 38 speleothem types (Fig. 1.1), with numerous subtypes and varieties, and describe over 250 minerals found in caves. A particular speleothem can be composed of any of a number of different minerals.

MICROBIAL ECOLOGY OF CAVES

Caves are considered to be extreme environments for life (Howarth 1993), and are often severely resource-limited due to the absence of light that precludes primary production of organic material (reviewed in Poulson and Lavoie 2000). Physical parameters, however, tend to be temperate, predictable, and constant. On entering a cave, one goes

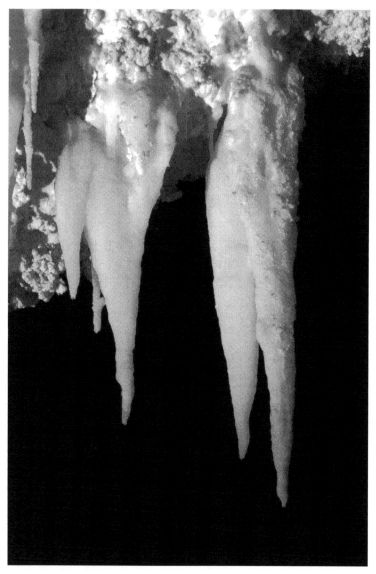

FIG. 1.1 Speleothems like these calcite stalactites are secondary mineral deposits common in caves. (Image courtesy and copyright of Diana Northup and Kenneth Ingham.)

through a series of zones, beginning with an entrance zone that is strongly impacted by surface conditions of light and temperature. Deeper in is a twilight zone where limited light penetrates and surface conditions are moderated by deep cave conditions. In the deep cave there is an absence of light, high humidity (~99%), and temperatures at or near the mean annual surface temperature (MAST) for the area. While microbial growth under nutrient limitation in not unique, the level of organic carbon available in cave systems may be 1,000-fold lower that in comparative surface environments (Morita 1997, Barton et al. 2007). To overcome such nutrient limitation, cave microbial ecosystems may develop oligotrophic energy acquisition systems to scavenge the tiny amount of allochthonous energy input percolating into such systems (Barton and Jurado 2007). Alternatively, cave systems have been identified that do not rely on any allochthonous energy input, instead these systems rely on exclusively chemoautotrophic primary production as in the case of sulfidic cave systems (reviewed by Engel 2007). In return, these sulfidic systems offer new environmental stresses for organisms, including high levels of toxic gases and extremely acidic conditions, which have to be overcome for ecosystem function (Spear et al. 2007).

A wide range of microbial species has been identified through molecular techniques in both terrestrial and aquatic cave environments around the world. Novel resident microbes have not been isolated in cultivation studies. In one of the earliest publications on cultivated cave microbes, Høeg (1946) described microbes on cave walls in Norway. Early reviews of cultivated microorganisms from caves include Caumartin (1963), Vandel (1965), Dyson and James (1973), Dickson and Kirk (1976), Jones and Motyka (1987), and Rutherford and Huang (1994). Many cultured microbes identified from deep caves are identical to surface forms, opportunistic, and active only under favorable growth conditions (Dickson and Kirk 1976, Jones and Motyka 1987, James 1994). However, these enrichment-based and cultural studies have focused on typical heterotrophic microbes known from medical and surface studies. Such techniques relying on nutrient-rich media grow less than 1% of microbes present (Amann et al. 1995, Pace 1997). In cave environments there is rarely an overlap between species identified through cultivation versus molecular identification. Nonetheless, pure cultures are necessary to test for biochemical and physiological characteristics of these species that have adapted to subsist under low levels of available organic carbon.

Microscopy has also proved useful in cave studies, despite the low biomass and cell numbers typically found in caves (Barton et al. 2006). Light microscopy used with stains such as Gram's Stain, or dyes such as acridine orange or DAPI, are used to visualize microorganisms and for enumeration. Staining with metabolic dyes such as INT can help to show microbial activity. Electron microscopy (e.g. scanning SEM or transmission TEM) allows greater magnification and examination of microbial interactions with their environment (Fig. 1.2). However,

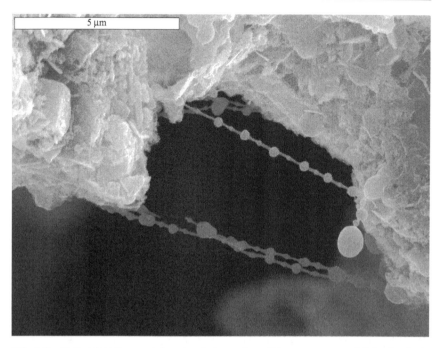

FIG. 1.2 Unique beads on a string morphology of possible manganese-oxidizing bacteria from ferromanganese deposits from Snowing Passage in Lechuguilla Cave. EDX analysis shows that the mineral matrix is manganese oxide. (Scale bar 5 mm.) (SEM courtesy of Diana Northup and Mike Spilde.)

Southam and Donald (1999) caution that SEM cannot be used to differentiate between geochemical and geomicrobiological precipitates. Confocal microscopy allows for imaging of live cells over time and in three dimensions, although this interesting technique has not yet been utilized in cave studies.

Newer approaches to microbial identification do not require culture, allowing one to describe microbial populations *in situ*. One no longer needs to model the microbial contribution to ecosystems or processes as a mysterious black box (Macalady and Banfield, 2002). Instead, 16S ribosomal RNA gene sequence and molecular phylogenetics allow us to better understand the link between biological and geochemical processes at the molecular level (Newman and Banfield 2002). Such culture-independent, molecular phylogenetic techniques have shown that many novel organisms can be found in caves which, while broadening our understanding of the diversity of such systems, highlight the current paucity of such data on cave systems (Angert et al. 1998, Vlasceanu et al.

2000, Barton et al 2004, Chelius and Moore 2004, Macalady et al. 2006, Spear et al. 2007).

ENERGY FLOW AND REDOX REACTIONS IN CAVES

Microbial life has been able to colonize every habitat on Earth where organic molecules can function and a thermodynamically favorable energy couple exists. Prior to the 1990s, much of the work on microorganisms in caves concentrated on descriptions of organisms, and documented, to a limited extent, the relationship of organisms to redox environments (Barton and Northup 2007). Caves have been viewed as highly oxidized environments with little in the way of reduced compounds to support the growth of microorganisms, with growth limited to the entry of flowing, dripping and seeping water bringing allochthonous energy, albeit in a geologically and hydrologically controlled manner (Brooke 1996). Nonetheless studies have shown that microorganisms in caves exploit a variety of environments with redox gradients (see papers in Sasowsky and Palmer 1994 and in the special issue of *Geomicrobiology Journal*, August 2001, on the Geomicrobiology of Caves).

Microorganisms living at the interface between the host rock and cave passages can utilize reduced compounds in the host rock for energy production; limestones and basalts and often rich in reduced sulfur, iron, and manganese. The primary productivity of chemolithotrophic microorganisms using iron, sulfur, and manganese, have traditionally been considered to be of limited importance in caves (Caumartin 1963, Barr 1968, Jones and Motyka 1987), but recent studies (Sarbu et al. 1996, Hose et al. 2000) show that their contributions to the food base in some caves can be considerable. With the chemical heterogeneity of the geologic environment, redox relationships are often complicated and interrelated, with many microbes found at boundaries or redox interfaces where gradients of compounds occur. To dissect the energetics of such systems, stable isotopic ratios are very useful in determining the contribution of microbes to the cave food chain and their role in the production of secondary mineral deposits.

For subaqeous systems, the Edwards Aquifer of Texas provides an excellent example of the role of microbial species at subterranean redox boundaries (Schindel et al. 2000), including freshwater and saline interfaces with redox boundaries of sulfate, petroleum, and chlorides. Fungi growing in shallower artesian wells in the aquifer were presumed to utilize the energy from hydrocarbons at an oxygen boundary interface (Kuehn and Koehn 1988), while current studies are investigating the bacterial community and its role in carbonate dissolution in the Edwards Aquifer (Randall et al. 2005, Randall 2006).

In another subaqueous system, anchialine caves are water-filled limestone caves with inland and oceanic entrances, allowing subterranean connections of marine and freshwaters. Anchialine caves are characterized by a halocline with a colder, freshwater lens flowing over a warmer, denser saline layer (Bottrell et al. 1991, Moore et al. 1992). Around the halocline layer are gradients of dissolved nitrates, oxygen, salinity, and temperature, that support high densities of microbial species stratified in the water column (Stoessell et al. 1993, Wilson and Morris 1994, Pohlman et al. 1997, Socki et al. 2002). Enrichment in the light isotope of sulfur in the sulfides found in such systems by up to – 63.2‰ demonstrates the very active role of bacterial metabolism in the energetics of such systems (Socki et al. 2002).

MICROBE–MINERAL INTERACTIONS IN CAVES

Microbes contribute to the formation of caves through dissolution. Microbially influenced dissolution or corrosion of mineral surfaces can occur through mechanical attack, secretion of exoenzymes, organic and mineral acids (e.g. sulfuric acid), and a variety of other proposed mechanisms (Sand 1997, Jones 2001, Bullen et al. 2008). Of particular interest in cave dissolution processes are reactions involving sulfur-, iron-, and manganese-oxidizing bacteria. These microbially mediated reactions can generate considerable acidity that can dissolve cave walls or speleothems. Laboratory investigations show that the ability to precipitate calcium carbonate is common among bacteria (Boquet et al. 1973), and bacteria isolated from caves are even more effective in precipitation (Danielli and Edington 1983, Engel et al. 2001). Nonetheless, some microbial structures, such as lipids and phospholipids, and humic acids from soils that enter caves from drip waters (Saiz-Jimenez and Hermosin 1999), inhibit calcite precipitation.

In addition to dissolution, various minerals can precipitate onto bacterial cell surfaces through biologically controlled mineralization (direct or enzymatic) or biologically induced mineralization (indirect or passive) (Lowenstam and Weiner 1989, Banfield and Nealson 1997, Konhauser 1997, 1998, Woods et al. 1999, Neuweiler et al. 2000, Bosak and Newman 2005). In biologically induced mineralization, amphoteric functional groups (carboxyl, phosphoryl, and amino constituents) on negatively-charged cell wall surfaces, sheaths, or capsules, sorb dissolved metals. Bound metals (such as iron) provide sites for chemical interactions, reduce activation energy barriers and serve as nucleation sites for crystal growth. Additionally, metabolic activity of microorganisms can also change the local pH and lead to precipitation of various minerals. In caves, such biologically controlled and induced mineralization is implicated in the formation of numerous minerals, including carbonates (including moonmilk), silicates, clays, iron and manganese oxides, sulfur, and nitrates (saltpeter).

Carbonates

Calcium carbonate speleothems predominate in most limestone caves and investigators have examined such structures, including stalactites, stalagmites, helictites, moonmilk, pool fingers, terrestrial oncoids, cave pisoliths, and cave pearls, for a microbial role in their formation. Riding (2000) provides an excellent review of microbial carbonates in general, including some cave deposits, and Jones (2000) described his extensive long-term investigations of microbial sediments in the Cayman Islands. Like surface travertine deposits, carbonate speleothem deposition and structure are influenced by local geochemistry, temperature, microbial activity, CO_2 concentrations and out-gassing (Jones and Kahle 1993, Palmer 1996, Fouke et al. 2000, Frisia et al. 2002: Sanchez-Moral et al. 2003, Blyth and Frisia 2008).

Fungi, algae, and bacteria have all been implicated in the precipitation of carbonate dripstone in both non-cave, carbonate, and travertine literature (e.g. Ehrlich 1996) and in caves (Went 1969, Danielli and Edington 1983, Cox et al. 1989a and b, 1995, Engel et al. 2001), where microorganisms are commonly found fossilized within carbonate speleothems (e.g., Jones and Motyka 1987, Polyak and Cokendolpher 1992, Jones and Kahle 1995, Melim et al. 2001). Carbonate pool fingers (Fig. 1.3) in Lechuguilla Cave were first identified by Davis (2000) as potentially biogenic in origin, which is supported by the abundance of fossil filaments and a small shift in the carbon isotope ratios of micritic layers but not in clear calcite layers within these formations (Melim et al. 2001). Similar results were found by Baskar et al. (2006) from microcrystalline calcite within stalactites. Terrestrial oncoids from sinkholes in dolostones from Grand Cayman and Cayman Brac showed a diverse microbial component (Jones 1991). The oncoids grew when calcifying filaments and spores trapped and bound detritus within the associated mucus, allowing Jones (1991) to provide criteria for differentiating abiotic from biotic coatings on grains. Terrestrial oncoids resemble cave pearls (Fig. 1.4), a speleothem that may have a microbial association. Gradzínski (1999) found a strong microbial component to irregularly shaped, but not symmetrical, cave pearls.

It is known that bacteria and fungi can precipitate calcium carbonate extracellularly through a variety of processes that include ammonification, denitrification, sulfate reduction, and anaerobic sulfide

FIG. 1.3 Small u-loops connecting pendant pool fingers in Hidden Cave, New Mexico, resemble acidic biofilms (snottites). (Image courtesy and copyright of Kenneth Ingham.)

A.

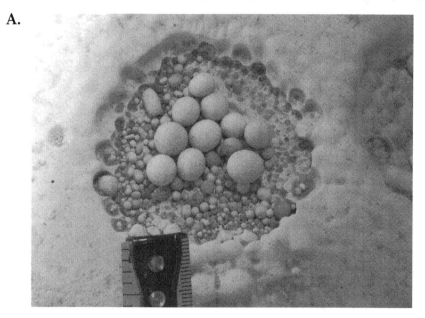

Fig. 1.4 Contd.

Fig. 1.4 Contd.

B.

FIG. 1.4 **A.** Cave pearls from Carlsbad Cavern, NM. (Image courtesy and copyright of Kenneth Ingham.) **B.** Scanning electron micrograph of the interior of a cave pearl from a cave in Tabasco, Mexico. Note the filamentous structures. (SEM by Mike Spilde and Diana Northup.)

oxidation (Simkiss and Wilbur 1989, Ehrlich 1996, Castanier et al. 1999, Riding 2000). These processes change the local saturation index of the solution and increase the local concentration of HCO_3^-, leading to increased alkalinity for calcium carbonate precipitation, subsequent $CaCO_3$ precipitation may be purely inorganic. Described as biologically controlled, organic-matrix, or boundary-organized precipitation; cellular substrates such as extracellular polymeric substances (EPS) are often critical to the creation of microbial carbonates, trapping sediments necessary in biomineralization (Simkiss and Wilbur 1989, Riding 2000, Bazylinski and Frankel 2003). The cells that participate in this calcite precipitation are trapped within the growing deposit and die (Barton et al. 2001), although the advantage to the microbial cell may be detoxification of Ca^{2+} ions. In culture, non-photosynthetic bacterial species isolated from caves are capable of precipitating significant quantities of $CaCO_3$ when provided with an organic calcium salt, such as calcium acetate or calcium succinate (Le Métayer-Levrel et al. 1997). Calcium carbonate precipitation in cultures is only seen in live cultures and not in killed controls, suggesting that biologically induced mineralization (BIM) is less important in carbonate precipitation. Bacterial metabolism and removal of inhibitors of precipitation, such as Mg^{2+}, can ameliorate such precipitation, while the S layer of certain bacterial species can aid in the coordination of molecules to increase the likelihood of mineralization. Nonetheless, we have identified species actively producing organic acids that can simultaneously dissolve and re-crystallize $CaCO_3$ within the same colony (Banks et al. 2008). Recent work by Sanchez-Moral et al. (2003) on the physiochemistry of bacterial carbonate precipitation in caves suggest that CO_2 is actively taken up by bacterial cells in this process, although the presence of specific carbonic anhydrase enzymes to mediate such uptake is not known. These observations are supported by the work of Cacchio et al. (2004), who showed that the $CaCO_3$ of speleothems is isotopically light.

In 1997, Le Métayer-Levrel et al. published a study investigating helictites from Clamouse Cave, France, containing bacteria incorporated into the crystalline structure of the formation, acting as nucleation axes for piled calcite rhombohedra. Carbonate production in Clamouse was controlled by the type of bacteria present, temperature, salinity, the nature and amount of nutrients available, and time. *In vitro* experiments of carbonatogenic yield suggested that bacterial carbonatogenesis plays a major role in limestone deposition. Cacchio et al. (2004) went on to show a selective enrichment of microbes from speleothems in Cervo Cave, Italy, that were able to deposit $CaCO_3$ at a rate much higher than isolates from non-karst environments. Models for carbonate precipitation

FIG. 1.5 Extensive deposits of moonmilk are found in Spider Cave, NM. The moonmilk is discolored in places from flooding of the cave. (Image courtesy and copyright of Kenneth Ingham.)

based on both autotrophic and heterotrophic metabolic processes have been proposed (Castanier et al. 1999, Hammes and Verstraete 2002, Cacchio et al. 2004).

Electron microscopy and XRD analysis of secondary deposits have been important tools in the study of microbial carbonate deposition in caves. Enrichment culture studies have demonstrated a higher incidence of the phenotype for carbonate precipitation among cave species than their terrestrial counterparts (Cacchio et al. 2003, 2004). Whether such activities provide an evolutionary advantage to cave adapted microbial species, such as the ability to detoxify Ca^{2+} ions within a calcium-rich environment (Simkiss 1986), will require the integration of techniques in isotopic analysis and molecular biology.

Moonmilk

Moonmilk (Fig. 1.5) is a common speleothem in many caves, composed of calcite or aragonite in limestone caves or hydromagnesite in dolomite caves (Hill and Forti 1997) and has been used since ancient times for its putative medicinal value (Shaw 1997). Moonmilk, also known as Mondmilch and by a wide variety of other names (Bernasconi 1981, Reinbacher 1994), describes a range of soft, moist, microcrystalline mineral aggregates with a physical appearance ranging from a paste to cottage cheese. Differences in the texture of moonmilk mainly relate to the amount of water present in the crystalline matrix and the mineral complexity (Gradzínski et al. 1997).

Formation of moonmilk is not limited to a defined location within a cave or to a specific chemical reaction, but to physiochemical conditions. Given the variability of minerals that form moonmilk, it is not surprising that several mechanisms, biotic and abiotic, have been proposed for its formation. Høeg (1946) reported identifying microorganisms in moonmilk and suggested that microbial activity was the cause of such deposition, an idea later supported by Davies and Moore (1957). Williams (1959) and Bernasconi (1976) provide early reviews of these mechanisms with more recent reviews by Hill and Forti (1997) and Northup et al. (1997) that include the effects of freezing, disintegration of bedrock or speleothems, precipitation from groundwater, and microorganisms. Separating the contributions of these mechanisms to the formation of moonmilk is complicated by the structural disruptions of crystalline microstructure caused by the collection and transport of such delicate samples (Jones 2001).

Gradzínski et al. (1997) evaluated moonmilk from several caves in southern Poland and concluded that deposits were the result of either microbially mediated precipitation of autochthonous carbonates or of microbial degradation of the host rock. They propose stages in the progressive formation of moonmilk where cells and an organic matrix provide a structural framework, active bacterial cells are then calcified and the extracellular organic matrix fills the remaining space with calcite through epitaxial growth. These authors went on to suggest that the growth of hydrogen-oxidizing (knallgas) bacteria facilitates calcite precipitation by creating an alkaline microenvironment. An extensive survey of moonmilk deposits from high altitude caves in the Italian Alps by Borsato et al. (2000) showed no evidence of microbial involvement in calcite precipitation, although the majority of samples were from fossil deposits. Nonetheless, a wide range of microbes including fungi, algae, protozoa, and particularly bacteria and *Streptomycetes* species can be cultured from moonmilk, often in very high densities (Northup et al. 2000). Putative microbial cells and an organic matrix can also be frequently seen in thin sections or with SEM of moonmilk, but not in all cases.

Cañaveras et al. (1999) investigated the possible role of microbes, predominantly *Streptomycetes*, in the formation of hydromagnesite moonmilk from Altamira Cave, Spain. These investigators concluded that microbes could be involved in hydromagnesite moonmilk formation, but were unable to rule out a possible inorganic origin. A later study by Cañaveras et al. (2006) on the role of filamentous microbes in the formation of moonmilk, used cultures, molecular phylogenetics, and microscopic examination of crystal structure concluded that fungal filaments were not present, but filamentous *Proteobacteria* were able to precipitate calcite in culture. Other researchers found *Crenarchaeota* present in moonmilk, but were unable to establish a role for them in moonmilk formation (Gonzalez et al. 2006). Cañaveras et al. (2006) now propose a model in which moonmilk formation begins with microbial colonization of host rock. Subsequent microbial growth leads to the local dissolution of $CaCO_3$ at the rock surface followed by re-crystallization by species demonstrating a slightly different metabolic profile above the host-rock/community interface. Over time, moisture accumulation in the colony allows inorganic epitaxial growth of the bacterially precipitated $CaCO_3$ crystals. This process continues, with the sustained dissolution of the host rock and build-up of crystals, allowing moonmilk deposits to reach the 10+ cm thickness that has been identified within cave deposits. A direct metabolic causality in the formation of such deposits has been confirmed by Blyth and Frisia (2008), supporting the findings of Barabesi et al. (2007) who demonstrated a genetic link to fatty acid metabolism in the precipitation of calcite by *Bacillus subtilis*. Given the variety of mineral types identified in moonmilk and the range of physicochemical conditions under which it is observed, it is likely that a great deal remains to be understood regarding the formation of this ubiquitous secondary deposit.

Sulfur Compounds

Microbes have been involved in transformations of sulfur since early in the history of life on Earth, with isotopic evidence of microbial sulfate reduction at 3.47 bya (Shen et al. 2001). The important role of sulfur cycling in caves is documented by the discovery of sulfur deposits of various oxidation states in caves and cenotes (Hill 1987, 1990, Marcella et al. 1994, Martin and Brigmon 1994, Wilson and Morris 1994, Brigmon et al. 1994a and b, Hill 1995, Hill and Forti 1997, Hill 2000, Socki et al. 2002, Engel 2007). To establish a biogenic origin for such sulfur in caves, investigators have identified the presence of gypsum and sulfur deposits enriched in the light isotope of sulfur (Hill 1987, 1990, Mylroie et al. 1994, Pisarowicz 1994, Hill 1996, 2000, Hose et al. 2000). Based on sulfur isotope analysis of deposits in Lechuguilla Cave, Hill (1995) supported the role of microbes in the reduction of sulfate to hydrogen sulfide and the probable role of microbes in gypsum depositions, but concluded that deposits of elemental sulfur in caves are probably abiotic.

Lowe and Gunn (1995) suggest that all subsurface carbonate porosity is from sulfuric acid, and the Palmers (Palmer 1991, Palmer and Palmer 1994) attribute this process to the formation of petroleum reservoirs in limestone. Egemeier (1973, 1981) first suggested that the hydrogen sulfide in the Kane Caves, Wyoming, was biogenic in origin, producing the H_2SO_4 that contributed to the formation of these caves. In contrast, in Lower Kane Cave, Engel et al. (2004) demonstrated that the low levels of hydrogen sulfide in the water were actively consumed by sulfate-oxidizing bacteria before sulfuric acid could be produced, rather, *Epsilonproteobacteria* species found in the springs were producing sulfuric acid that locally dissolved solution pockets in the limestone, leading to speleogenesis. Molecular phylogenetic studies of acidic biofilms in Cueva de Villa Luz (Fig. 1.6) and caves of the Frasassi Gorge, Italy, demonstrate the presence of *Thiobacillus* and *Acidithiobacillus* spp. (Hose et al. 2000, Vlasceanu et al. 2000, Macalady et al. 2006). These thiobacilli gain energy from the oxidation of sulfur or sulfide to sulfuric acid and can contribute to dissolution of carbonate in these caves. Such sulfuric acid-driven speleogenesis is implicated in the formation of numerous caves throughout the world (Principi 1931, Davis 1980, Egemeier 1981, Hill 1987, 1990, Galdenzi 1990, Korshunov and Semikolennyh 1994, Galdenzi and Menichetti 1995, Hill 1995, Galdenzi et al. 1997, Hill 2000, Macalady et al. 2006, Engel 2007). Hose et al. (2000) reported observing

FIG. 1.6 Acid biofilms (snottites) in Cueva de Villa Luz, Tabasco, Mexico. The water drips at the end of the soft, mucoid structures have a pH of 0-2. (Image courtesy and copyright of Kenneth Ingham.)

gypsum falling off the walls in Cueva de Villa Luz as major contributions to enlargement in that active sulfur cave.

In addition to examining mineral deposits, sulfur-cycling bacterial communities within active caves have also been investigated (Angert et al. 1998, Hose and Pisarowicz 1999, Hose et al. 2000, Sarbu et al. 2000, Vlasceanu et al. 2000, Barton and Luiszer, 2005, reviewed by Engel 2007) using culture-independent molecular techniques. These studies have highlighted the contributions of sulfur utilizing microbes in the process of sulfuric acid speleogenesis (Spirakis and Cunningham 1992, Cunninghan et al. 1993, 1994, Jagnow et al. 2000, Boston et al. 2006), with microbial species and their metabolic activities demonstrating similarities in sulfidic karst systems across caves and continents (Engel 2007). Investigations of Movile Cave, Romania, by Sarbu et al. (1996) identified thick microbial mats of fungi and bacteria floating in partially flooded galleries of the cave. These mats contained sulfide-oxidizing species of *Thiobacillus* and *Beggiatoa,* utilizing hydrogen sulfide levels approaching 1300 µM in the system for energy production. Living in coexistence with these organisms were sulfate-reducing species such as *Desulfovibrio* (Sarbu et al. 1994, 1996, Vlasceanu et al. 1997), with metabolic activities leading to the deposition of elemental sulfur within the mat. These floating microbial communities appear to receive no organic input from the surface, relying instead on chemolithotrophic energy generation to support a unique and diverse biota, including numerous macroscopic species (Sarbu and Popa 1992, Sarbu and Kane 1995, Sarbu et al. 1996). Molecular phylogenetic analysis of the filamentous bacterial communities in Sulphur River of Parker Cave, Kentucky, also demonstrated the presence of probable sulfide-oxidizing bacteria. These filaments precipitated elemental sulfur and produced sulfuric acid, as evident by the elemental sulfur and gypsum deposits on artificial substrates placed in the cave (Olson and Thompson 1988, Thompson and Olson 1988, Angert et al. 1998). Other caves which have come under investigation for such sulfur-cycling include the caves of Frasassi Gorge (Vlasceanu et al. 2000, Macalady et al. 2006), Cueva de Villa Luz (Hose and Pisarowicz 1999, Hose et al. 2000), Glenwood Canyon, Colorado (Barton and Luiszer 2005, Spear et al. 2007), and submarine caves in Cape Palinuro, Italy (Southward et al. 1996, Mattison et al. 1998).

The discovery of caves around the world that contain microbial communities utilizing sulfur compounds for chemoautotrophic energy production has greatly expanded our opportunities to study the sulfur cycle. In contrast to previously examined terrestrial systems, these sulfidic cave springs are temperate (rather that classic hot springs) and are not impacted by seasonal or diurnal variations. Indeed, microbial mats in such hypogean environments do not have the stratification or high levels of insect predation seen in terrestrial sulfidic spring systems (Cohen 1989), allowing a direct analysis of the effects of chemolithotrophy on such systems in caves. Without such ecological pressures and grazing, these microbial systems can form unique mineral-biofabrics not observed elsewhere (Spear et al. 2007). The important role that microbial sulfur-cycling plays in speleogenesis, as evidenced by the development of huge cave systems such as the 170+ km Lechuguilla Cave, New Mexico, suggests that the true geological extent of such metabolic activities are only beginning to be elucidated (Engel 2007). As a result, microbially mediated biogenic sulfidic-speleogenesis promises to remain one of the most active fields within cave geomicrobiology for some time.

Iron and Manganese Mineralization

Iron

Historically, circumstantial evidence for iron biomineralization in caves comes largely from observational studies (Crabtree 1962, Caumartin 1963, Caldwell and Caldwell 1980, Dyson and James 1981, Luiszer 1992, Klimchouk 1994, Maltsev 1997). In one of the more detailed studies of possible biogenic structures, Jones and Motyka (1987) found spherical bodies with high concentrations of iron or manganese in stalactites from Grand Cayman Island, British West Indies. The most unusual iron oxide deposits found in caves are stalactites from Lechuguilla Cave, New Mexico, that resemble the rusticles found on the hull of the Titanic. These formations consist of iron oxides coating organic filaments that may be fossilized *Clonothrix*, or other iron-oxidizing bacteria (Davis et al. 1990, Provencio and Polyak 2001).

Several descriptive studies have established the association of bacteria with iron and manganese deposits in caves, but experimental evidence for the metabolic role of microbial species in the formation of these deposits in caves is lacking (Barton and Northup 2007). Microbial iron mineral formation has been documented in the formation of ferric hydroxide (e.g. ferrihydrite), iron oxides, magnetite, iron sulfates and sulfides, iron silicates, iron phosphates, and carbonates (siderite, $FeCO_3$) in a wide variety of aqueous and some terrestrial habitats (reviewed in Konhauser 1997, 1998). Generally, abiotic formation of iron oxides occurs above pH 6, while biologically produced iron oxides form at low pH, however, recent evidence suggests that biological iron oxidation can occur at circumneutral pH, despite the very low solubility of iron minerals (Emerson and Moyer 1997, Straub et al., 2001, Sobolev and Roden, 2002). Sources of reduced iron for metabolism in caves include trace components in limestone (Melim 1991) and infiltrating waters. Microbially oxidized iron can in turn precipitate onto a wide range of bacterial cell surfaces through enzymatic or passive mineralization (Lowenstam and Weiner 1989, Konhauser 1997, 1998). Hydrated iron oxide is the initial form of iron precipitated in caves, where crystalline forms such as goethite or hematite then form; goethite (FeOOH) is commonly found in cave sediments and may contribute to the formation of boxwork, flowstone, or helictites; hematite (Fe_2O_3) occurs as crystal inclusions in other minerals (Hill and Forti 1997).

Such precipitated iron oxides and hydroxides are most often observed in caves as coatings or crusts, powder in clastic cave fills or a pigmentation component of speleothems such as stalactites. Kasama and Murakami (2001) showed that microbes increased ferrihydrate precipitated on stalactites by up to four orders of magnitude at neutral pH. While the study of biogenic fractionation of iron isotopes has the potential of providing new data in investigating the role of iron and manganese oxidizing-bacteria in caves (Beard et al. 1999), cultivation and metabolic phenotyping may prove the most critical in elucidating the role of Fe- and Mn-based energy generation systems in these starved environments.

Manganese

Several studies have proposed biological oxidation in the formation of cave manganese deposits (Crabtree 1962, Broughton 1971, White 1976, Laverty and Crabtree 1978, Moore and Sullivan 1978, Hill 1982, Peck 1986, Cílek and Fábry 1989, Jones 1992, Gradzínski et al. 1995, Onac et al. 1997a and b, Northup et al. 2000), which can speed up the rate of oxidation by up to five orders of magnitude (Tebo et al. 1997). Manganese compounds are found as soft deposits that may occur in clastic sediments (Cílek and Fábry 1989), as coatings on walls or speleothems (Moore and Sullivan 1978, Gascoine 1982, Hill 1982, Rogers and Williams 1982, Kashima 1983, Moore and Sullivan 1997), or as consolidated crusts (Moore 1981, Hill 1982, Peck 1986, Jones 1992). Biotic oxidation of manganese can occur indirectly or directly; indirectly from the release of oxidants, acids, or bases into the environment surrounding the microbial cell, changing the redox chemistry surrounding the cell (reviewed in Tebo et al. 1997), directly by binding nonspecific negatively charged substances on the bacterial cell surface, or through the action of intra and extracellular Mn(II)-binding proteins (Ghiorse 1984). Such Mn oxidation is accomplished by a phylogenetically diverse set of bacteria that fall within subdivisions of the *Proteobacteria*, the Gram-positive bacteria and fungi and is mediated by oxygen, temperature, pH and metal ion concentrations in the environment.

The most common manganese mineral found in caves is birnessite (Hill and Forti 1997), while poorly crystalline manganese oxides and hydroxides, such as pyrolusite, romanechite, todorokite and rhodochrosite, are also found (Onac et al. 1997a, b). Moore (1981) attributed manganese-oxidizing bacteria such as *Leptothrix,* in a stream in Matts Black Cave, West Virginia, to the formation of birnessite in this cave, via precipitation of manganese around sheaths of bacteria. The presence of rods, sheets, strands, and smooth spheroid morphologies in fossil remains of manganese precipitates in stalactites, karst breccia, and calcrete crusts on plant roots in the Grand Caymen caves led Jones (1992) to conclude that many of these manganese precipitates were also biogenic.

Irregularly shaped crusts of manganese flowstone (2–20 mm thick) in Jaskinia Czarna Cave, Poland, contained filaments and globular bodies, which were interpreted as bacterial or fungal cells. Due to the three-dimensional morphology, the amorphic character and the high Mn/Fe ratio (72.1:1), it is thought that microbial species participated in the formation of such flowstones (Gradzínski et al. 1995). The submerged caves of the Mediterranean also contain black coatings that have been attributed to microbial utilization of manganese in seawater, deposited by microbial biofilm communities (Allouc and Harmelin 2001), however, in these subaqueous systems, the Mn/Fe ratio is negatively correlated with the level of nutrients available to these biofilm communities.

Ferromanganese

Within caves, it is not uncommon to see both iron and manganese oxides precipitated together. A study by Peck (1986) examining such deposits reported on the presence of *Gallionella ferruginea* and *Leptothrix* sp. from cave pools, sumps, moist Fe and Mn wall crusts and formations in Level Crevice Cave, Iowa. Cultured *Gallionella ferruginea* produced iron hydroxide precipitations and *Leptothrix* sp. showed iron-impregnated sheaths, while sterile controls showed no such precipitation. While this study demonstrated the presence of iron-oxidizing microbial species in cave sediments, it did not link such activity with energy generation *in situ.* Nonetheless, evidence for such activity *in situ* can be found in the extensive ferromanganese deposits originally described as corrosion residues (Fig. 1.7) of Lechuguilla and Spider Caves, New Mexico (Cunningham 1991, Cunningham et al. 1995). These ferromanganese deposits are varied in color and in chemical composition containing variable amounts of clay and aluminum oxides (Spilde et al. 2005). Initially they were thought to be produced by corrosive air attacking limestone bedrock (Cunningham 1991, Queen 1994). Bulk chemistry and XRD studies demonstrated that corrosion residues are not simply dissolution products, but are highly enriched in certain elements such as Fe and Mn, possibly by microbial processes (Dotson et al. 1999). Davis (2000) described these corrosion residues as biogenic in origin. Utilizing

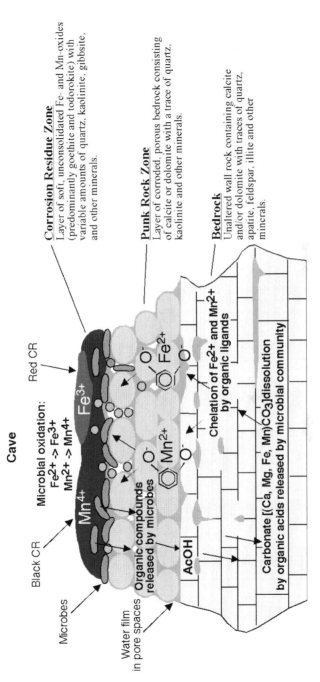

FIG. 1.7 Proposed formation mechanism for cave ferromanganese deposits (a.k.a. corrosion residues). (Image courtesy of Michael Spilde, modified from Larry Mallory, unpublished.)

molecular phylogenetic techniques and enrichment cultures targeting iron and manganese oxidizers, Northup et al. (2000, 2003) showed a diverse microbial community within these deposits related to known iron- and manganese-oxidizing bacteria. Metabolically active bacteria were identified in the punk rock between the residue and the bedrock, representing a transition zone that was enriched up to four times in reduced manganese. The mineral structures in the transition zone could be produced by cultured bacterial in a reduced metal medium (Spilde et al. 2005). These culture-based studies also documented the progression from amorphous to highly crystalline structures of manganese and iron oxides, but not in controls killed after the formation of amorphous deposits. When microbes selectively mobilize minerals, the undissolved residue becomes enriched in the non-mobilized components (Ehrlich 1998) and these ferromanganese deposits are also often enriched in rare earth-phosphate minerals. Similar ferromanganese deposits, described as ochres are found in Ochintá Aragonite Cave, Slovakia (Bosák et al. 2002). Paoleofills in Wind Cave, South Dakota, were also shown to contain microbial communities similar to those from Lechuguilla Cave (Northup et al. 2003; Chelius and Moore 2004).

Correlation of the location and activity of microorganisms to changes in mineral composition will assist one in determining the extent of microbial involvement in the formation of iron and manganese minerals in caves (Spilde et al. 2000). It may be possible to analyze deposits for the presence of enzymes and proteins involved in iron and manganese oxidation providing a simple way of field-screening samples. Microcosm studies *in situ* could establish rates of production of those minerals in caves.

Saltpeter

Nitrocalcite (calcium nitrate) is the saltpeter commonly found in dry cave sediments throughout the world, which was mined from caves in the United States for the niter component of gunpowder during the Civil War and the War of 1812 (Shaw 1997). Hill et al. (1983) used 'action archaeology' to recreate the mining process in order to understand the steps and chemistry involved in the historical manufacture of gunpowder. Water is used to lixiviate the calcium nitrate (Fig. 1.8) from cave saltpeter soils. Concentrated potash lye (5.4% K) added to concentrated calcium nitrate leachate precipitated calcium and magnesium as hydroxides and produced niter as KNO_3. A unique feature of saltpeter mining is that leached saltpeter earth will regenerate itself in three to ten years if placed in contact with cave walls and floor. Experiments in Mammoth Cave by Olson and Krapac (1995, 1997) show

FIG. 1.8 Saltpeter vats from Er Wang Dong, Jielong, China. Cave soil was transported to the vat and leached with water. The lixiviated liquor was concentrated, and calcium nitrate was converted to potassium nitrate by the addition of potash, similar to the process used at Mammoth Cave and described by Hill et al. (1983). (Photo by Hazel Barton.)

that longer regeneration times may be needed. While much remains to be learned about the biogeochemical interactions of calcium nitrate deposition, many studies support the role of nitrifying bacteria in the origin of saltpeter earth (Faust 1949, 1967, 1968, Hill 1981a and b, Hill et al. 1983). Fliermans and Schmidt (1977) isolated *Nitrobacter* from saltpeter deposits in Mammoth Cave, where they reported cave bacterial densities a hundred times higher than in surface soils, although different species predominated. Stable isotope analysis also demonstrated that saltpeter is selectively enriched in the lighter isotope of nitrogen, indicating a biogenic source (Jameson et al. 1994).

One question that still remains relates to the original source of the nitrogen (Hill 1992, Lewis 1992, Moore 1994). Hess (1900) first proposed a seeping groundwater hypothesis, wherein bacterial decomposition of organic matter above the cave released nitrate ions that were transported into the cave. Evaporation of water in dry passages would then result in a buildup of nitrate in the saltpeter earth. Pace (1971) and Hill (1981a and b) proposed modified seeping groundwater mechanisms, with ammonia or ammonium ions carried in from surface soils. This hypothesis was supported with 30 cm deep drill cores of surface limestone, which showed very low levels of nitrate, while similar cores from caves had high concentrations of nitrate (a few ppm vs. hundreds to thousands of ppm). Potential sources of nitrogen in caves also include bat guano (Hill 1987), ammonium–urea from amberat made of cave rat feces and urine (Moore and Sullivan 1978), bacterial nitrogen fixation (Faust 1967, Lewis 1992), fertilizers and sewage, volcanic rocks and forest litter (Hess 1900, Hill 1981a and b, Moore 1994). Hill (1981a and b) made a strong case for highly organic surface soil as the primary source for nitrogen compounds entering the cave. As the nitrogen content in surface soils varies with climate, vegetation type, topography, soil type, and porosity, so does the correlation with observed saltpeter caves (Bartholomew and Clark 1965, Hill 1981a and b).

Microbial growth in subterranean environments is tightly coupled to the availability of essential nutrients, such as phosphate and nitrogen, and unraveling the source of nitrates in these systems may prove critical to a broader understanding of subterranean microbial processes.

Silicates and Clays

Silicate minerals make up more than 95% of the Earth's crust, with silica in groundwater mostly derived from weathering processes. Despite the presence of silicates in groundwater, silicate speleothems are not as abundant as carbonate or sulfate minerals. Within caves, these minerals occur in any of three groups (Hill and Forti 1997): 1) framework-structures, including silica minerals such as quartz, 2) sheet-structures, primarily as clay minerals, and 3) ore silicate minerals that tend to be rare and localized. Quartz usually has a high-temperature origin and can occur in caves as spar crystals, or as a dense variety known as calcedony (Hill and Forti 1997). Microbes are known to cause the dissolution of quartz; Hiebert and Bennett (1992) and Feldmann et al. (1997) observed basidiomycete fungal hyphae boring into quartz crystals by dissolving SiO_2, the resultant hyphae were rimmed with iron encrustations that may have formed from reactions of calcium oxalate with Fe^{3+}.

The most common non-clay silicate mineral in caves is opal, an amorphous precipitate commonly found in lava tubes or interlayered with calcite in limestone caves (Hill and Forti 1997). A variety of microbial forms have been observed associated with opal in caves (Urbani 1976, 1977, Kunicka-Goldfinger 1982, Urbani 1996a, b, Léveillé et al. 2000, Wray 1999, Aubrecht et al. 2008) and in opal-sulfur corralloids, flowstone, conulites, and crusts in Santa Ninfa Cave, Sicily (Forti and Rossi 1987, Forti 1989, 1994). Other descriptions include silicaceous algal diatoms (*Meolosira*) concentrated in layers and covering silica coralloids in Togawa Sakaidanipdo Cave, Japan, and cyanobacterial-deposited microbalites in basaltic Hawaiian sea caves, which are frequently oriented toward the cave entrance (Kashima 1986, Kashima et al. 1989, Urbani 1996a, Léveillé 1999). While Urbani (1996a and b) and Aubrecht et al. (2008) have described a possible role for microbial activity in the development of opal formation and Willems et al. (1998) suggest a bacterial role in siliceous karst genesis in Niger, a mechanism for these activities has yet to be described.

A wide range of clay minerals are found in caves, but most are detrital in origin and washed into caves, and thus not true cave minerals (Hill and Forti 1997). Nonetheless, clays constitute part of soils, corrosion residues, and wall vermiculations in caves, and may enhance microbial colonization through a number of mechanisms: clays are rich in iron, can contain anaerobic pockets, and can concentrate and/or neutralize metabolic waste products. Konhauser and Urrutia (1999) showed that microbial surfaces can initiate biomineralization of metals at low concentrations, which then became autocatalytic in formation of iron and aluminum silicates in natural and experimental studies. Such enhancement may be seen in the ferromanganese deposits of Lechuguilla and Spider Caves, New Mexico. These microbe-rich deposits are also rich in clays, with the makeup of these clays changing substantially from the host rock outward (Northup et al. 2000) and substantially reduced in silica. The extent to which the microbial community present is responsible for this transformation is currently under investigation.

Vermiculations are thin, irregular, discontinuous deposits of mud and clays ranging in size from 1 mm to 1 cm or more, often surrounded by a light colored halo (Parenzan 1961). Vermiculations can form from a variety of mechanisms, including biological, on many different cave surfaces (Hill and Forti 1997). Anelli and Graniti (1967) suggested that the halo surrounding vermiculations is caused by acids and other organic substances secreted by fungi, while Urbani (1996b) described vermiculations composed of clay, gypsum, and green algae. Vermiculations in Cueva de Villa Luz (Fig. 1.9) contain a diverse population of microorganisms and macroorganisms and are described as biovermiculations (Hose et al. 2000). The biovermicualtions rapidly reform in areas scraped clean of them (Hose and Northup 2004).

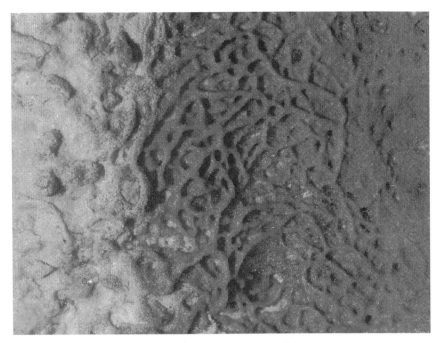

FIG. 1.9 Vermiculations in Ragu Passage, Cueva de Villa Luz, Tabasco, Mexico. (Image courtesy and copyright of Kenneth Ingham.)

The involvement of microbes in the formation and weathering of silicate speleothems is an area open to study. Opal and quartz are widely distributed in caves and experimental studies of their mechanism of formation, or a confirmed biogenic origin, remain to be elucidated. While clays in caves have been cultured for microorganisms, no studies have conclusively shown a role for microorganisms in their formation or alteration.

Biokarst and Phytokarst: Landforms Shaped by Microorganisms

Karst is defined as large areas of limestone with distinctive surfacial and underground geomorphology. With biokarst, scale effects move from corrosion and deposition to erosion and consolidation, and dominance of organic influences on specific process–form relationships on a landscape scale (Sweeting 1973, Viles 1984). Within karst, biokarst is often recognized as small-scale features of localized importance, and in caves usually involve stalactites, moonmilk deposits, and sinter crusts. These features were considered to form from primarily hydrological mechanisms, but biotic action (soil and vegetation) influenced by climate, lithology, and human impacts can modify limestone.

Viles (1984) reviewed the major forms of biological modifications of karst, including bioerosion (Neuman 1966) and phytokarst (Folk et al. 1973, Jones 1989) as well as the organisms involved, ranging from bacteria and cyanobacteria up to trees and animals. Folk et al. (1973) studied spongy phytokarst pinnacles in Hell on Grand Caymen Island, which was later attributed by Viles and Spencer (1986) to a suite of weathering processes as well as phytokarstic algal boring and erosion. Bull and Laverty (1982) described phytokarst pinnacles in cave entrances in Sarawak, Malaysia, that were oriented to light and formed as erosional structures from boring and solution by red algal and cyanobacterial growth. This study characterized four distinct forms of directed phytokarst, with formation controlled by both structure and aspect, but specifically excluding lithological control. Cox et al. (1989a and b) reported on speleothems with a characteristic morphology resembling crustaceans from two caves in New South Wales, Australia. Locally described as craybacks or lobsters, the structures are up to 4 m long by 3 m high. Environmental factors, including location within the twilight zone of these caves, provided strong control on formation. The authors describe these speleothems as cyanobacterial subaerial stromatolites. Formation is both biologically controlled and biologically induced, with

FIG. 1.10 Rillenkarin show sharp edges from biogenic sulfuric acid dissolution, such as these from Cueva de Villa Luz, MX. Note cave fish in water with suspended sulfur. (Image courtesy and copyright of Kenneth Ingham.)

calcite precipitation and aeolian sediment trapping important in deposition.

Other biokarst features include rillenkarren, first identified in Lechuguilla Cave and thought to be biogenic in origin (Davis 2000) and were also identified as prominent features (Figure 1.10), in Cueva de Villa Luz (Hose et al. 2000). In both cases, subaerial dissolution of limestone through condensation of H_2S saturated water can produce locally highly aggressive sulfuric acid, capable of rapidly forming such striking features (Barton and Luiszer 2005).

As changes in human activities lead to more rapid changes in surface vegetation and soil profiles, biokarst modification of limestone may become more pronounced. In order to better understand such processes more work is needed. Viles (1984) urged more work on biokarst in three critical areas, 1) observation and quantitative description, 2) establishing biotic and abiotic mechanisms and rates of formation, 3) establishment of process-form links, where the biotic contribution is the dominant force in development of a particular biokarst feature.

CONCLUSIONS

John Muir wrote "when you tug at a single thing in nature, you will find it connected to the rest of the universe" [*My First Summer in the Sierra* (1911)]. This statement is no truer than in cave geomicrobiology, where molecular transformations carried out by individual species within the environment can cause scale effects up to the level of cave formation itself. In turn, geomicrobiology studies in caves may provide insights into biomineralization in general and subterranean geochemical interactions at a global scale. Nonetheless, we still know very little regarding the full potential of microbial metabolic reactions, their interrelations, and regulation in cave systems. In this chapter we have shown that studies of geomicrobiology of caves are largely still qualitative in nature. Most of the culpability for the few comparative studies can be attributed to the nature of the cave environment itself, caves are completely dark, may contain deep pits, constricted passageways, exposed traverses, and slippery slopes, all of which collectively make traveling through them with research equipment and working in them a logistical and technical challenge. Caves are also often muddy, dusty, or contain large amounts of bat guano, yet routinely have limited biomass (Barton et al. 2006, Barton et al. 2007), making conditions miserable for the working microbiologist. But despite these inhospitable conditions, researchers are beginning to make significant discoveries within the realm of cave geomicrobiology (Engel et al. 2004, Macalady et al. 2006, Macalady et al. 2007, Bullen et al. 2008).

While qualitative studies serve as the critical base for future studies, we encourage researchers to consider more quantitative, comparative and experimental studies in cave geomicrobiology. To determine the microbial contribution to mineral formation in caves we need to draw on the wealth of rigorous research done on surface systems: one should expand the use of molecular phylogenetic techniques to study the makeup of microbial communities in caves, taking advantage of emerging technologies in environmental metagenomics, metabolomics, and microarray technology (Sogin et al., 2006, Xu 2006, Suttle 2007). Comparative genomics of cultivated species can provide information on how different mineral environments drive metabolic flux, energetics and gene transfer (Xu 2006). Techniques in microscale geosciences, such as atomic force microscopy and x-ray powder diffractometry, may allow a better understanding of the mechanism of energy generation from microbe-mineral interactions under extreme starvation (Lower et al. 2001). Techniques in materials chemistry, such as attenuated total reflectance Fourier transformed infrared spectroscopy and x-ray adsorption fine-structure spectroscopy, may help elucidate the role that reduced metal ions play in microbial metabolism and secondary mineral deposition (Sparks 2005). Finally, the incorporation of stable isotope techniques can provide information on microbial contribution to mineral formation (Hose et al. 2000) and ecosystem bioenergetics (Sarbu et al. 1996).

A systematic analysis of patterns of microbial metabolic activity in caves may provide strong empirical evidence for the biogeochemical interaction in subterranean, starved, and aphotic systems that support life. Such evidence may provide vital clues for the recognition of biomarkers for subsurface life on Earth and other planetary bodies in the emerging field of Astrobiology. (Boston et al. 1992, McKay et al. 1994, Cunningham et al. 1995, Boston et al. 2001, Boston et al. 2006). While no single feature can be taken as conclusive evidence for life, a consortium of biomarkers could provide strong circumstantial evidence of extinct, or even extant, life (Allen et al. 2000, Boston et al. 2001), allowing one to understand the important contributions that subterranean environments may have had to the development of biological systems.

ACKNOWLEDGEMENTS

We thank Rick Olsen at Mammoth Cave National Park, and Paul Burger, Stan Allison, Tom Bemis, and Dale Pate at Carlsbad Caverns National Park for their help and support of our research.

HAB is supported by the Kentucky EPSCoR Program (NSF# 0447479), the Kentucky Space Grant Consortium, the Merck/AAAS Undergraduate Science Research Program and the Center for Integrative Natural Science and Mathematics (CINSAM) at NKU. Infrastructure support was provided, in part, by the National Institutes of Health KY INBRE program (5P20RR016481-05).

DEN is supported by a National Science Foundation grant from the Biogeosciences program of the Geosciences Directorate (EAR0311932) and support from the University of New Mexico's Molecular Biology Facility which is supported by NIH Grant Number 1P20RR18754 from the Institute Development Award (IDeA) Program of the National Center for Research Resources.

REFERENCES

Allen, C.C., F.G. Albert, H.S. Chafetz, C. Combie, C.R. Graham, T.L. Kieft, S.J. Kivett, D.S. McKay, A. Steele, A.E. Taunton, M.R. Taylor, K.L. Thomas-Keprta, and F. Westall. 2000. Microscopic physical biomarkers in carbonate hot springs: Implications in the search for life on Mars. Icarus 147: 49-67.

Allouc, J. and J.-G. Harmelin. 2001. Les dépôs d'enduits manganoferrifères en environment marin littoral. L'example de grottes sous-marines en Méditerranée nord-occidentale. Bull Société Géologique de France 171: 765-778.

Amann, R.I., W. Ludwig, and K.-H. Schleifer. 1995. Phylogenetic identification and *in situ* detection of individual microbial cells without cultivation. Microbiol Rev 59: 143-169.

Anelli, F. and A. Graniti. 1967. Aspetti microbiologici nella genesi delle vermicolazioni argillose delle Grotte di Castellana (Murge di Bari). Grotte d'Italia, ser 41: 131-140.

Angert, E.R., D.E. Northup, A.-L. Reysenbach, A.S. Peek, B.M. Goebelm, and N.R. Pace. 1998. Molecular phylogenetic analysis of a bacterial community in Sulphur River, Parker Cave, Kentucky. Am Mineral 83: 1583-1592.

Aubrecht, R., C. Brewer-Carias, B. Šmida, M. Audy, and L'. Kováčik. 2008. Anatomy of biologically mediated opal speleothems in the world's largest sandstone cave: Cueva Charles Brewer, Chimantá Plateau, Venezuela. Sedimen Geol 203: 181-195.

Banfield, J.F. and K.H. Nealson. 1997. Geomicrobiology: Interactions between microbes and minerals. *In:* P.H. Ribbe [ed.]. Rev Mineralogy, 35: 448.

Banks, E.D., N.M. Taylor, B.R. Lubbers, and H.A. Barton. 2008. A physiological role for the bacterial precipitation of calcium carbonate minerals. Abstr Proc Am Society Microbiol 108th Gen Mtg N-327.

Barabesi, C., A. Galizzi, G. Mastromei, M. Rossi, E. Tamburini, and B. Perito. 2007. *Bacillus subtilis* gene cluster involved in calcium carbonate biomineralization. J Bacteriol 189: 228-235.

Barr, T.C. Jr. 1968. Cave ecology and the evolution of troglobites. Evol Biol 2: 35-102.

Bartholomew, W.V. and F.E. Clark (eds.). 1965. Soil Nitrogen. Madison, WI: Am Soc Agronomy.

Barton, H.A. and V. Jurado. 2007. What's up down there: Microbial diversity in starved cave environments. Microbe 2: 132-138.

Barton, H.A. and F. Luiszer. 2005. Microbial metabolic structure in a sulfidic cave hot spring: Potential mechanisms of biospeleogenesis. J Cave Karst Stu 67: 28-38.

Barton, H.A. and D.E. Northup. 2007. Geomicrobiology in cave environments: Past, current, and future perspectives. 65th Anniversary Issue. J Cave Karst Stu 69(1): 163-178.

Barton, H.A., J.R. Spear, and N.R. Pace. 2001. Microbial life in the underworld: Biogenecity in secondary mineral formations. Geomicrobiology J 18: 359-368.

Barton, H.A., M.R. Taylor, and N.R. Pace. 2004. Molecular phylogenetic analysis of a bacterial community in an oligotrophic cave environment. Geomicrobiology J 21: 11-20.

Barton, H.A., N.M. Taylor, B.R. Lubbers, and A.C. Pemberton. 2006. DNA extraction from low-biomass carbonate rock: An improved method with reduced contamination and the low-biomass contaminant database. J Microbiol Methods 66: 21-31.

Barton, H.A., N.M. Taylor, M.P. Kreate, A.C. Springer, S.A. Oehrle, and J.L. Bertog. 2007. The impact of host rock geochemistry on bacterial community structure in oligotrophic cave environments. Int J Speleo 36: 93-104.

Baskar, S., R. Baskar, L. Mauclaire, and J.A. McKenzie. 2006. Microbially induced calcite precipitation in culture experiments: Possible origin for stalactites in Sahastradhara Caves, Dehradun, India. Cur Sci 90: 58-64.

Bazylinski, R.B. and D.A. Frankel. 2003. Biologically induced mineralization by bacteria. Rev Mineralogy and Geochem 54(1): 95-114.

Beard, B.L., C.M. Johnson, L. Cox, H. Sun, K.H. Nealson, and C. Aquilar. 1999. Iron isotope biosignatures. Science 285: 1889-1892.

Bernasconi, R. 1976. The physio-chemical evolution of moonmilk. Cave Geol 1: 63-88. Translated from French by Mansker, W.L. Original publication: 1961. Memoir V Rassegna Speleologica Italiana, Proc Int Congr Speleol, Varenna, Italy, 1960. 1: 75-100.

Bernasconi, R. 1981. Mondmilch (moonmilk): Two questions of terminology. Proc 8th Int Congr Speleol 1: 113-116.

Blyth, A.J. and S. Frisia. 2008. Molecular evidence for bacterial mediation of calcite formation in cold high-altitude caves. Geomicrobiology J 25: 101-111.

Boquet, E., A. Bornate, and A. Ramos-Cormenzana. 1973. Production of calcite (calcium carbonate) crystals by soil bacteria as a general phenomenon. Nature 246: 527-529.

Borsato, A., S. Frisia, B. Jones, and K. van der Borg. 2000. Calcite moonmilk: crystal morphology and environment of formation in caves in the Italian Alps. J Sedimen Res 70: 1179-1190.

Bosák, P., P. Bella, V. Cílek, D.C. Ford, H. Hercman, J. Kadlec, A. Osborne, and P. Pruner. 2002. Ochtína aragonite cave (western Carpathians, Slovakia): Morphology, mineralogy of the fill and genesis. Geologica Carpathica 53: 399-410.

Bosak, T. and D.K. Newman. 2005. Microbial kinetic controls on calcite morphology in supersaturated solutions. J Sed Res 75: 190-199.

Boston, P.J., M.V. Ivanov, and C.P. McKay. 1992. On the possibility of chemosynthetic ecosystems in subsurface habitats on Mars. Icarus 95: 300-308.

Boston, P.J., M.N. Spilde, D.E. Northup, L.A. Melim, D. Soroka, L. Kleina, K. Lavoie, L. Hose, and L. Mallory. 2001. Cave biosignature suites: Microbes, minerals and Mars. Astrobiol J 1: 57-70.

Boston, P.J., L.D. Hose, D.E. Northup, and M.N. Spilde. 2006. The microbial communities of sulfur caves, a newly appreciated geologically driven system on Earth and potential model for Mars. Spec Paper GSA 404: 331-344.

Bottrell, S.H., P.L. Smart, F. Whitaker, and R. Raiswell. 1991. Geochemistry and isotope systematics of sulphur in the mixing zone of Bahamian blue holes. Appl Geochem 6: 97-103.

Brigmon, R.L., G. Bitton, S.G. Zam, H.W. Martin, and B. O'Brien. 1994a. Identification, enrichment, and isolation of *Thiothrix* spp. from environmental samples. Curr Microbiol 28: 243-246.

Brigmon, R.L., H.W. Martin, T.L. Morris, G. Bitton, and S.G. Zam. 1994b. Biogeochemical ecology of *Thiothrix* spp. in underwater limestone caves. Geomicrobiol J 12: 141-159.

Broughton, P.L. 1971. Origin and distribution of mineral species in limestone caves. Earth Sci J 5: 36-43.

Brooke, M. 1996. Infiltration pathways at Carlsbad Caverns National Park determined by hydrogeologic and hydrochemical characterization and analysis. Master's Thesis. Colorado School of Mines. Golden, Colorado.

Bull, P.A. and M. Laverty. 1982. Observations on phytokarst. Z Geomorph NF 26: 437-457.

Bullen, H.A., S.A. Oehrle, A.F. Bennett, N.M. Taylor, and H.A. Barton. 2008. The use of attenuated total reflectance fourier transformed infrared (ATR-FTIR) spectroscopy to identify microbial metabolic products on carbonate mineral surfaces. Appl Environ Microbiol 74(14): 4553-4559.

Cacchio, P., C. Ercole, G. Cappuccio, and A. Lepidi. 2003. $CaCO_3$ precipitation by bacterial strains isolated from a limestone cave and from a loamy soil. Geomicrobiol J 20: 85-98.

Cacchio, P., R. Contento, C. Ercole, G. Cappuccio, M.P. Martinez, and A. Lepidi. 2004. Involvement of microorganisms in the formation of carbonate speleothems in the Cervo Cave (L'Aquila – Italy). Geomicrobiol J 21: 497-509.

Caldwell, D.E. and S.J. Caldwell. 1980. Fine structure of *in situ* microbial iron deposits. Geomicrobiol J 2: 39-53.

Cañaveras, J.C., M. Hoyos, S. Sanchez-Moral, E. Sanz-Rubio, J. Bedoya, V. Soler, I. Groth, P. Schumann, L. Laiz, I. Gonzalez, and C. Saiz-Jimenez. 1999. Microbial communities associated with hydromagnesite and needle-fiber aragonite deposits in a karstic cave (Altamira, northern Spain). Geomicrobiol J 16: 9-25.

Cañaveras, J.C., S. Cuezva, S. Sanchez-Moral, J. Lario, L. Laiz, J.M. Gonzalez, and C. Saiz-Jimenez. 2006. On the origin of fiber calcite crystals in moonmilk deposits. Naturwissenschaften 93: 27-32.

Castanier, S., G. Le Métayer-Levrel, and J.P. Perthuisot. 1999. Ca-carbonates precipitation and limestone genesis—the microbiogeologist point of view. Sediment Geol 126: 9-23.

Caumartin, V. 1963. Review of the microbiology of underground environments. Bull Natl Speleol Soc 25: 1-14.

Chelius, M.K. and J.C. Moore. 2004. Molecular phylogenetic analysis of Archaea and Bacteria in Wind Cave, South Dakota. Geomicrobiol J 21: 123-134.

Cílek, V. and J. Fábry. 1989. Epigenetické, manganem bohaté polohy v krasóvych vyplních Zlatého koně v Ceském krasu: Ceskonolov Kras 40: 37-55.

Cohen, Y. 1989. Microbial Mats: Physiological Ecology of Benthic Microbial Communities. ASM Press, Washington, DC.

Cox, G., J.M. James, R.A.L. Osborne, and K.E.A. Leggett. 1989a. Stromatolitic crayfish-like stalagmites. Proc Univ Bristol Speleol Soc 18: 339-358.

Cox, G., J.M. James, K.E.A. Leggett, and R.A.L. Osborne. 1989b. Cyanobacterially deposited speleothems: Subaerial stromatolites. Geomicrobiol J 7: 245-252.

Cox, G., A. Salih, J. James, and B. Allaway. 1995. Confocal microscopy of cyanobacteria in calcite speleothems. Zool Studies 34: 5-6.

Crabtree, P.W. 1962. Bog ore from Black Reef Cave. Cave Sci 4: 360-361.

Culver, D.C. 1982. Cave Life: Evolution and Ecology. Harvard University Press, Cambridge, MA.

Cunningham, K.I. 1991. Organic and inorganic composition of colored corrosion residues: Lechuguilla Cave: Preliminary report. Natl Speleol Soc News 49: 252, 254.

Cunningham, K.I., H.R. DuChene, and C.S. Spirakis. 1993. Elemental sulfur in caves of the Guadalupe Mountains, New Mexico. New Mexico Geological Society Guidebook 44th Field Conference, Carlsbad regions, New Mexico and west Texas: Socorro, NM. New Mexico Geol Soc, pp. 129-136.

Cunningham, K.I., H.R. DuChene, C.S. Spirakis, and J.S. McLean. 1994. Elemental sulfur in caves of the Guadalupe Mountains, New Mexico. *In:* I.D. Sasowsky and M.V. Palmer [eds.]. Breakthroughs in karst geomicrobiology and redox geochemistry: Abstracts and field-trip guide for the symposium held February 16 through 19, 1994 Colorado Springs, CO. Spec Pub 1. Karst Waters Institute, Inc Charles Town, WV. pp. 11-12.

Cunningham, K.I., D.E. Northup, R.M. Pollastro, W.G. Wright, and E.J. Larock. 1995. Bacteria, fungi and biokarst in Lechuguilla Cave, Carlsbad Caverns National Park, New Mexico. Environ Geol 25: 2-8.

Curl, R.L. 1966. Caves as a measure of karst. J Geol 74: 798-830.

Danielli, H.M.C. and M.A. Edington. 1983. Bacterial calcification in limestone caves. Geomicrobiol J 3: 1-16.

Davis, D.G. 1980. Cave development in the Guadalupe Mountains: A critical review of recent hypotheses. Bull Natl Speleol Soc 42: 42-48.

Davis, D.G. 2000. Extraordinary features of Lechuguilla Cave, Guadalupe Mountains, New Mexico. J Cave Karst Stu 62: 147-157.

Davies, W.E. and G.W. Moore. 1957. Endellite and hydromagnesite from Carlsbad Caverns. Bull Natl Speleol Soc 19: 24-27.

Davis, D.G., M.V. Palmer, and A.N. Palmer. 1990. Extraordinary subaqueous speleothems in Lechuguilla Cave, New Mexico. Bull Natl Speleol Soc 52: 70-86.

Dickson, G.W. and P.W. Kirk. 1976. Distribution of heterotrophic microorganisms in relation to detritivores in Virginia caves (with supplemental Bibliography on Cave Mycology and Microbiology). *In:* B.C. Parker and M.K. Roane [eds.]. The distributional history of the biota of the Southern Appalachians. Part IV. Algae and fungi. University Press of Virginia. Charlottesville, VA, pp. 205-226.

Dotson, K.E., R.T. Schelble, M.N. Spilde, L.J. Crossey, and D.E. Northup. 1999. Geochemistry and mineralogy of corrosion residues, Lechuguilla and Spider Caves, Carlsbad Caverns National Park, NM: Biogeochemical processes in an extreme environment. Geol Soc Am Abstr Progr 31: A154.

Dyson, H.J. and J.M. James. 1973. A preliminary study of cave bacteria. J Sydney Speleol Soc 17: 221-230.

Dyson, H.J. and J.M. James. 1981. The incidence of iron bacteria in an Australian cave. *In:* Proc 8th Int Congr Speleol 1: 79-81.

Egemeier, S.J. 1973. Cavern development by thermal waters with a possible bearing on ore deposition. PhD dissertation. Stanford University. Palo Alto, CA.

Egemeier, S.J. 1981. Cavern development by thermal waters. Bull Natl Speleol Soc 43: 31-51.

Ehrlich, H.L. 1996. Geomicrobiology. 3rd ed. Marcel Dekker, Inc., New York.

Ehrlich, H.L. 1998. Geomicrobiology: Its significance for geology. Earth-Sci Rev 45(1): 45-60.

Emerson, D. and C. Moyer. 1997. Isolation and characterization of novel iron-oxidizing bacteria that grow at circumneutral pH. Appl Environ Micro 63: 4784-4792.

Engel, A.S. 2007. Biodiversity of sulfidic karst habitats. 65th Anniversary Issue. J Cave Karst Stu 69(1): 187-206.

Engel, A.S., M.L. Porter, B.K. Kinkle, and T.C. Kane. 2001. Ecological assessment and geological significance of microbial communities from Cesspool Cave, Virginia. Geomicrobiol J 18: 259-274.

Engel, A.S., L.A. Stern, and P.C. Bennett. 2004. Microbial contributions to cave formation: New insights into sulfuric acid speleogenesis. Geol 32: 369-372.

Faust, B. 1949. The formation of saltpetre in caves. Bull Natl Speleol Soc 11: 17-23.

Faust, B. 1967. Saltpetre mining in Mammoth Cave, Kentucky. Filson Club Hist Quart 42: 1-96.

Faust, B. 1968. Notes on the subterranean accumulation of saltpetre. J Spelean Hist 1: 3-11.

Feldmann, M., J. Neher, W. Jung, and F. Graf. 1997. Fungal quartz weathering and iron crystallite formation in an Alpine environment, Piz Alv, Switzerland. Ecologae Geol Helv 90: 541-556.

Fliermans, C.B. and E.L. Schmidt. 1977. Nitrobacter in Mammoth Cave. Int J Speleol 9: 1-19.

Folk, R.L., H.H. Roberts, and C.H. Moore. 1973. Black phytokarst from Hell, Cayman Islands, British West Indies. Geol Soc Am Bull 84: 2351-2360.

Forti, P. 1989. Le concrezioni e le mineralizzazioni delle grotte in gesso di Santa Ninfa, Trapani. Mem Inst Ital Speleol Ser 23: 137-154.

Forti, P. 1994. Los depósitos químicos de la Sima Aondo Superior y de otras cavidades del Auyántepui, Venezuela. Bol Soc Venez Espéleol 28: 1-4.

Forti, P. and A. Rossi. 1987. Le concrezioni poliminerali della Grotta di Santa Ninfa: un esempio evidente dell'influenza degli equilibri solfuri-solfati sulla minerogenesi carsica. Atti Mem Com Grotta "E Boegan" 26: 47-64.

Fouke, B.W., J.D. Farmer, D.J. Des Marais, J.R. Pratt, N.C. Sturchio, P.C. Burns, and M.K. Discipulo. 2000. Depositional facies and aqueous-solid geochemistry of travertine-depositing hot springs (Angel Terrance, Mammoth Hot Springs, Yellowstone National Park, USA). J Sed Petrol 70: 565-585.

Frey, D.G. 1963. Limnology in North America. University of Wisconsin Press, Madison, WI.

Frisia, A., A. Borsato, I.J. Fairchild, F. McDermott, and E.M. Selmo. 2002. Aragonite-calcite relationships in speleothems (Grotte de Clamouse, France): Environment, fabrics, and carbonate geochemistry. J Sed Res 72(5): 687-699.

Galdenzi, S. 1990. Un modello genetico per la Grotta Grande del Vento. Mem Ist Ital Speleol Ser 24: 123-142.

Galdenzi, S. and M. Menichetti. 1995. Occurrence of hypogenic caves in a karst region: Examples from central Italy. Environ Geol 26: 39-47.

Galdenzi, S., M. Menichetti, and P. Forti. 1997. La corrosione di placchette calcaree and opera di acque sulfuree: dati sperimentali in abiente ipogeo. Proc 12th Int Congr Speleol 1: 187-190.

Gascoine, W. 1982. The formation of black deposits in some caves of south east Wales. Cave Sci 9: 165-175.

Ghiorse, W.C. 1984. Biology of iron- and manganese-depositing bacteria. Ann Rev Microbiol 38: 515-550.

Gillieson, D. 1996. Caves: Processes, Development, and Management. Blackwell Publishers, Ltd., Oxford.

Gonzalez, J.M., M.C. Portillo, and C. Saiz-Jimenez. 2006. Metabolically active Crenarchaeota in Altamira Cave. Naturwissenschaften 93: 42-45.

Gradzínski, M. 1999. Role of micro-organisms in cave pearls formation. J Conf Abs. 11th Bathurst Meeting July 13th-15th, 1999 Cambridge, UK 4: 924.

Gradzínski, M., M. Banas, and A. Uchman. 1995. Biogenic origin of manganese flowstones from Jaskinia Czarna Cave, Tatra Mts., Western Carpathians. Ann Soc Geologorom Poloniae 65: 19-27.

Gradzínski, M., J. Szulc, and B. Smyk. 1997. Microbial agents of moonmilk calcification. Proc 12th Int Congr Speleol 1: 275-278.

Hammes, F. and W. Verstraete. 2002. Key roles of pH and calcium metabolism in microbial carbonate precipitation. Rev Environ Sci Biotechnol 1: 3-7.

Hess, W.H. 1900. The origin of nitrates in cavern earths. J Geol 8: 129-134.

Hiebert, F.K. and P.C. Bennett. 1992. Microbial control of silicate weathering in organic-rich ground water. Science 258: 278-281.

Hill, C.A. 1981a. Origin of cave saltpeter. Bull Natl Speleol Soc 43: 110-126.

Hill, C.A. 1981b. Origin of cave saltpeter. J Geol 89: 252-259.

Hill, C.A. 1982. Origin of black deposits in caves. Bull Natl Speleol Soc 44: 15-19.

Hill, C.A. 1987. Geology of Carlsbad Cavern and other caves in the Guadalupe Mountains, New Mexico and Texas. New Mexico Bur Mines Mineral Res Bull 117. New Mexico Bureau of Mines & Mineral Resources. Socorro, NM.

Hill, C.A. 1990. Sulfuric acid speleogenesis of Carlsbad Cavern and its relationship to hydrocarbons, Delaware Basin, New Mexico and Texas. Am Assoc Petrol Geol Bull 74: 1685-1694.

Hill, C.A. 1992. On the origin of cave saltpeter: A second opinion — reply. Bull Natl Speleol Soc 54: 31-32.

Hill, C.A. 1995. Sulfur redox reactions: hydrocarbons, native sulfur, Mississippi Valley-type deposits, and sulfuric acid karst in the Delaware Basin, New Mexico and Texas. Environ Geol 25: 16-23.

Hill, C.A. 1996. Geology of the Delaware Basin, Guadalupe, Apache and Glass Mountains, New Mexico and West Texas. Society of Economic Paleontologists & Mineralogists. Permian Basin Section. Pub No 96-39. Permian Basin Section-SEPM. Midland, TX.

Hill, C.A. 2000. Overview of the geologic history of cave development in the Guadalupe Mountains, New Mexico. J Cave Karst Stu 62(2): 60-71.

Hill, C.A. and P. Forti. 1997. Cave Minerals of the World. 2nd ed. National Speleological Society. Huntsville, AL.

Hill, C.A, P.G. Eller, C.B. Fliermans, and P.M. Hauer, 1983. Saltpeter conversion and the origin of cave nitrates. Nat Geogr Soc Res Report 15: 295-309.

Høeg, O.A. 1946. Cyanophyceae and bacteria in calcareous sediments in the interior of limestone caves in Nord-Rana, Norway. Nytt Magasin for Natuvidenspene 85: 99-104.

Hose, L.D. and J.A. Pisarowicz. 1999. Cueva de Villa Luz, Tabasco, Mexico: Reconnaissance study of an active sulfur spring cave and ecosystem. J Cave Karst Stu 61: 13-21.

Hose, L.D. and D.E. Northup. 2004. Biovermiculations: Living vermiculations in Cueva de Villa Luz, Mexico. (Abstract). J Cave Karst Stu 66(3): 112.

Hose, L.D., A.N. Palmer, M.V. Palmer, D.E. Northup, P.J. Boston, and H.R. DuChene. 2000. Microbiology and geochemistry in a hydrogen-sulphide-rich karst environment. Chem Geol 169: 399-423.

Howarth, F.G. 1993. High-stress subterranean habitats and evolutionary change in cave-inhabiting arthropods. Am Nat 142: S65-S77.

Jagnow, D.H., C.A. Hill, D.G. Davis, H.R. Duchene, K.I. Cunningham, D.E. Northup, and J.M. Queen. 2000. History of the sulfuric acid theory of speleogenesis in the Guadalupe Mountains. J Cave Karst Stu. New Mexico. 62: 54-59.

James, J.M. 1994. Microorganisms in Australian caves and their influence on speleogenesis. *In:* I.D. Sasowsky and M.V. Palmer [eds.]. Breakthroughs in karst geomicrobiology and redox geochemistry: Abstracts and field-trip guide for the symposium held February 16 through 19, 1994. Colorado Springs, CO. Spec Pub 1. Karst Waters Institute, Inc. Charles Town, WV. pp. 31-34.

Jameson, R.A., D.G. Boyer, and E.C. Alexander Jr. 1994. Nitrogen isotope analysis of high-nitrate and other karst waters and leached sediments at Friar's Hole Cave, West Virginia. *In:* I.D. Sasowsky and M.V. Palmer [eds.]. Breakthroughs in karst geomicrobiology and redox geochemistry: Abstracts and field-trip guide for the symposium held February 16 through 19, 1994. Colorado Springs, CO. Spec Pub 1. Karst Waters Institute, Inc. Charles Town, WV. pp. 36-37.

Jones, B. 1989. The role of microorganisms in phytokarst development on dolostones and limestones, Grand Cayman, British West Indies. Can J Earth Sci 26: 2204-2213.

Jones, B. 1991. Genesis of terrestrial oncoids, Cayman Islands, British West Indies. Can J Earth Sci 28: 382-397.

Jones, B. 1992. Manganese precipitates in the karst terrain of Grand Cayman, British West Indies. Can J Earth Sci 29: 1125-1139.

Jones, B. 2000. Microbial sediments in tropical karst terrains: A model based on the Cayman Islands. *In:* R.E. Riding and S.M. Awramik [eds.]. Microbial sediments. Springer-Verlag, Berlin. pp. 171-178.

Jones, B. 2001. Microbial activity in caves: A geological perspective. Geomicrobiol J 18: 345-357.

Jones, B. and A. Motyka. 1987. Biogenic structures and micrite in stalactites from Grand Cayman Island, British West Indies. Can J Earth Sci 24: 1402-1411.

Jones, B. and C.F. Kahle. 1993. Morphology, relationship, and origin of fiber and dendritic calcite crystals. J Sediment Petrology 63: 1018-1031.

Jones, B. and C.F. Kahle. 1995. Origin of endogenetic micrite in karst terrains: A case study from the Cayman Islands. J Sediment Res A65: 283-293.

Kasama, T. and T. Murakami. 2001. The effect of microorganisms in Fe precipitation rates at neutral pH. Chem Geol 180: 117-128.

Kashima, N. 1983. On the wad minerals from the cavern environment. Int J Speleol 13: 67-72.

Kashima, N. 1986. Cave formations from noncalcareous caves in Kyushu, Japan. Proc 9th Int Congr Speleol 2: 41-43.

Kashima, N., T. Ogawa, and S.H. Hong. 1989. Volcanogenic speleo-minerals in Cheju Island, Korea. J Speleol Soc Japan 14: 32-39.

Klimchouk, A. 1994. Speleogenesis in gypsum and geomicrobiological processes in the Miocene sequence of the pre-Carpathian Region. *In:* I.D. Sasowsky and M.V. Palmer [eds.]. Breakthroughs in karst geomicrobiology and redox geochemistry: Abstracts and field-trip guide for the symposium held February 16 through 19, 1994. Colorado Springs, CO. Spec Pub 1. Karst Waters Institute, Inc. Charles Town, WV. pp. 40-42.

Konhauser, K.O. 1997. Bacterial iron mineralization in nature. FEMS Microbiol Rev 20: 315-326.

Konhauser, K.O. 1998. Diversity of bacterial iron mineralization. Earth-Sci Rev 43: 91-121.

Konhauser, K.O. and M.M. Urrutia. 1999. Bacterial clay authigenesis: A common biogeochemical process. Chem Geol 161: 399-413.

Korshunov, V. and A. Semikolennyh. 1994. A model of speleogenic processes connected with bacterial redox in sulfur cycles in the caves of Kugitangtou Ridge, Turkmenia. *In:* I.D. Sasowsky and M.V. Palmer [eds.]. Breakthroughs in karst geomicrobiology and redox geochemistry: Abstracts and field-trip guide for the symposium held February 16 through 19, 1994.Colorado Springs, CO. Spec Pub 1. Karst Waters Institute, Inc. Charles Town, WV. pp. 43-44.

Kuehn, K.A. and R.D. Koehn. 1988. A mycofloral survey of an artesian community within the Edwards Aquifer of central Texas. Mycologia 80: 646-652.

Kunicka-Goldfinger, W. 1982. Preliminary observations on the microbiology of karst caves of the Sarisariñama plateau in Venezuela. Bol Soc Venez Espeleol 10: 133-136.

Laverty, M. and S. Crabtree. 1978. Ranciéite and mirabilite: Some preliminary results on cave mineralogy. Trans Brit Cave Res Assoc 5: 135-142.

LeMétayer-Levrel, G., S. Castanier, J.-F. Loubière, and J.P. Perthuisot. 1997. Bacterial carbonatogenesis in caves. SEM study of an helictite from Clamouse, Herault, France. C R Acad Sci Ser IIA Sci Terre Planets 325: 179-184.

Léveillé, R.J., W.S. Fyfe, and F.J. Longstaffe. 2000. Geomicrobiology of carbonate-silicate microbialites from Hawaiian basaltic sea caves. Chem Geol 169: 339-355.

Lewis, W.C. 1992. On the origin of cave saltpeter: A second opinion. Bull Natl Speleol Soc 54: 28-30.

Lowe, D. and J. Gunn. 1995. The role of strong acid in speleo-inception and subsequent cavern development. *In:* I. Barany-Kevei and L. Mucsi [eds.]. Special Issue Acta Geographic (Szeged), 24: 33-60.

Lowenstam, H.A. and S. Weiner. 1989. On Biomineralization. Oxford University Press, New York.

Lower, S.K., M.F. Hochella, and T.J. Beveridge. 2001. Bacterial recognition of mineral surfaces: Nanoscale interactions between Shewanella and α-FeOOH. Science 292: 1360-1363.

Luiszer, F.G. 1992. Chemolithoautotrophic iron sediments at Cave of the Winds and the Iron Springs of Manitou Springs, Colorado. Bull Natl Speleol Soc 54: 92.

Macalady, J. and J.F. Banfield. 2002. Molecular geomicrobiology: Genes and geochemical cycling. Earth Planetary Sci Letters 209: 1-17.

Macalady, J.L., E.H. Lyon, B. Koffman, L.K. Albertson, K. Meyer, S. Galdenzi, and S. Mariani. 2006. Dominant microbial populations in limestone-corroding stream biofilms, Frasassi Cave system, Italy. Appl Environ Microbiol 72: 5596-5609.

Macalady, J.L., D.S. Jones, and E.H. Lyon. 2007. Extremely acidic, pendulous cave wall biofilms from the Frasassi Cave system, Italy. Environ Microbiol 9: 1402-1414.

Maltsev, V. 1997. Cupp-Coutunn Cave, Turkmenistan. *In:* C.A. Hill and P. Forti [eds.]. Cave Minerals of the World. 2nd ed. Nat Speleol Soc. Huntsville, AL. pp. 323-328.

Marcella, L., E. Heydari, R.K. Stoessell, and M. Schoonen. 1994. Sulfur geochemistry and limestone dissolution in coastal Yucatan cenotes. *In:* I.D. Sasowsky and M.V. Palmer [eds.]. Breakthroughs in karst geomicrobiology and redox geochemistry: Abstracts and field-trip guide for the symposium held February 16 through 19, 1994. Colorado Springs, CO. Spec Pub 1. Karst Waters Institute, Inc. Charles Town, WV. p. 48.

Martin, H.W. and R.L. Brigmon. 1994. Biogeochemistry of sulfide oxidizing bacteria in phreatic karst. *In:* I.D. Sasowsky and M.V. Palmer [eds.]. Breakthroughs in karst geomicrobiology and redox geochemistry: Abstracts and field-trip guide for the symposium held February 16 through 19, 1994. Colorado Springs, CO. Spec Pub 1. Karst Waters Institute, Inc. Charles Town, WV. pp. 49-51.

Mattison, R.G., M. Abbiati, P.R. Dando, M.F. Fitzsimons, S.M. Pratt, A.J. Southward, and E.C. Southward. 1998. Chemoautotrophic microbial mats in submarine caves with hydrothermal sulphidic springs at Cape Palinuro, Italy. Microb Ecol 35: 58-71.

McKay, C.P., M. Ivanov, and P.J. Boston. 1994. Considering the improbable: Life underground on Mars. Planetary Report XIV: 13-15.

Melim, L.A. 1991. The origin of dolomite in the Permian (Guadalupian) Capitan Formation, Delaware Basin, West Texas and New Mexico: Implications for dolomitization models. Ph.D. Dissertation. Southern Methodist University. Dallas, TX.

Melim, L.A., K.M. Shinglman, P.J. Boston, D.E. Northup, M.N. Spilde, and J.M. Queen. 2001. Evidence for microbial involvement in pool finger precipitation, Hidden Cave, New Mexico. Geomicrobiol J 18: 311-329.

Moore, G.W. 1952. Speleothem—a new cave term. Natl Speleol Soc News 10: 2.

Moore, G.W. 1981. Manganese deposition in limestone caves. *In:* Proc 8th Intl Congr Speleol II: 642-645.

Moore, G.W. 1994. When will we have an accepted explanation for cave nitrate deposits? *In:* I.D. Sasowsky and M.V. Palmer [eds.]. Breakthroughs in karst geomicrobiology and redox geochemistry: Abstracts and field-trip guide for the symposium held February 16 through 19, 1994. Colorado Springs, CO. Spec Pub 1. Karst Waters Institute, Inc. Charles Town, WV. pp. 53-54.

Moore, G.W. and G.N. Sullivan. 1978. Speleology: The Study of Caves. Rev 2nd ed. Cave Books, Inc. St. Louis, MO.

Moore, G.W. and G.N. Sullivan. 1997. Speleology: Caves and the Cave Environment. Rev 3rd ed. Cave Books, Inc. St. Louis, MO.

Moore, Y.H., R.K. Stoessell, and D.H. Easley. 1992. Fresh-water/sea-water relationship within a ground-water flow system, Northeastern coast of the Yucatan peninsula. Ground Water 30: 343-350.

Morita, R.Y. 1997. Bacteria in Oligotrophic Environments: Starvation-Survival Lifestyle. Chapman and Hall, London.

Mylroie, J.E., J.L. Carew, S.H. Bottrell, and W.J. Balcerzak. 1994. Microbial processes and cave development in the freshwater lens of a Quaternary carbonate island, San Salvador, the Bahamas. *In:* I.D. Sasowsky and M.V. Palmer [eds.]. Breakthroughs in karst geomicrobiology and redox geochemistry: Abstracts and field-trip guide for the symposium held February 16 through 19, 1994 Colorado Springs, CO. Spec Pub 1. Karst Waters Institute, Inc. Charles Town, WV. pp. 54-55.

Neuman, A.C. 1966. Observations on coastal erosion in Bermuda and measurements of the boring rate of the sponge *Cliona lampa*. Limnol Oceanogr 11: 92-108.

Neuweiler, F., M. Rutsch, G. Geipel, A. Reimer, and K.-H. Heise. 2000. Soluble humic substances from *in situ* precipitated microcrystalline calcium carbonate, internal sediment, and spar cement in a Cretaceous carbonate mud-mound. Geol 28(9): 851-854.

Newman, D.K. and J.F. Banfield. 2002. Geomicrobiology: How molecular-scale interactions underpin geochemical systems. Science 296: 1071-1077.

Northup, D.E and K.H. Lavoie. 2001. Geomicrobiology of caves: A review. Geomicrobiol J 18(3): 199-222.

Northup, D.E., A.-L. Reysenbach, and N.R. Pace. 1997. Microorganisms and speleothems. *In:* C.A. Hill and P. Forti [eds.]. Cave Minerals of the World. 2nd ed. Nat Speleo Soc. Huntsville, AL. pp. 261-266.

Northup, D.E., C.N. Dahm, L.A. Melim, M.N. Spilde, L.J. Crossey, K.H. Lavoie, L.M. Mallory, P.J. Boston, K.I. Cunningham, and S.M. Barns. 2000. Evidence for geomicrobiological interactions in Guadalupe caves. J Cave Karst Stud 62: 80-90.

Northup, D.E., S.M. Barns, L.E. Yu, M.N. Spilde, R.T. Schelble, K.E. Dano, L.J. Crossey, C.A. Connolly, P.J. Boston, and C.N. Dahm. 2003. Diverse microbial communities inhabiting ferromanganese deposits in Lechuguilla and Spider Caves: Environ Microbiol 5: 1071-1086.

Olson, R. and I.G. Krapac. 1995. Regeneration of nitrates in Mammoth Cave sediment: A mid-term report. *In:* Proceedings of Mammoth Cave National

Park's fourth science conference, Mammoth Cave National Park, July 6-7, 1995. pp. 109-118.

Olson, R, and I. Krapac. 1997. Origin and regeneration of nitrates in Mammoth Cave sediment. Karst-O-Rama, The Electric Caver 34: 51-60.

Olson, R.A. and D.B. Thompson. 1988. Scanning electron microscopy and energy dispersive x-ray analysis of artificial and natural substrates from the Phantom flowstone of Sulphur River in Parker Cave, Kentucky, USA. Bull Natl Speleol Soc 50: 47-53.

Onac, B.P., R.B. Pedersen, and M. Tysseland. 1997a. Presence of rare-earth elements in black ferromanganese coatings from Vântului Cave (Romania). J Cave Karst Stud 59: 128-131.

Onac, B.P., M. Tysseland, M. Bengeanu, and A. Hofenpradli. 1997b. Deposition of black manganese and iron-rich sediments in Vântului Cave (Romania). Proc 12th Int Congr Speleol 1: 235-238.

Pace, N.R. 1971. Caves and saltpeter: A novel hypothesis for saltpeter formation. Caving in the Rockies 13: 7-9.

Pace, N.R. 1997. A molecular view of microbial diversity and the biosphere. Science 276: 734-740.

Palmer, A.N. 1991. Origin and morphology of limestone caves. Geol Soc Am Bull 103: 1-21.

Palmer, M.V. 1996. Influence of carbon dioxide outgassing rates and accessory ions on $CaCO_3$ crystal shapes. Geol Soc Am, 28th Ann Mtg, Abstracts with Programs 28: 48.

Palmer, M.V. and A.N. Palmer. 1994. Sulfate-induced diagenesis in the Madison Aquifer of South Dakota. *In:* I.D. Sasowsky and M.V. Palmer [eds.]. Breakthroughs in karst geomicrobiology and redox geochemistry: Abstracts and field-trip guide for the symposium held February 16 through 19, 1994 Colorado Springs, CO. Spec Pub 1. Karst Waters Institute, Inc. Charles Town, WV. pp. 57-58.

Parenzan, P. 1961. Sulle formazioni argillose-limose dette vermicolari. Atti Int Symp, Varenna 1: 120-125.

Peck, S.B. 1986. Bacterial deposition of iron and manganese oxides in North American caves. Bull Natl Speleol Soc 48: 26-30.

Pisarowicz, J.A. 1994. Cueva de Villa Luz—an active case of H_2S speleogenesis. *In:* I.D. Sasowsky and M.V. Palmer [eds.]. Breakthroughs in karst geomicrobiology and redox geochemistry: Abstracts and field-trip guide for the symposium held February 16 through 19, 1994 Colorado Springs, CO. Spec Pub 1. Karst Waters Institute, Inc Charles Town, WV. pp. 60-61.

Pohlman, J.W., T.M. Iliffe, and L.A. Cifuentes. 1997. A stable isotope study of organic cycling and the ecology of an anchialine cave ecosystem. Marine Ecol Progr Ser 155: 17-27.

Polyak, V.J. and J.C. Cokendolpher. 1992. Recovery of microfossils from carbonate speleothems. Bull Natl Speleol Soc 54: 66-68.

Poulson, T.L. and K.H. Lavoie. 2000. The trophic basis of subsurface ecosystems. *In:* H. Wilkens, D.C. Culver, and J.W. Humphreys [eds.]. Ecosystems of the World 30: Subterranean Ecosystems. Elsevier, Amsterdam. pp. 231-249.

Principi, P. 1931. Fenomeni di idrologia sotterranea nei dintorni di Triponzo (Umbria). Grotte d'Ital 5: 1-4.

Provencio, P.P. and Polyak, V.J. 2001. Iron oxide-rich filaments: Possible fossil bacteria in Lechuguilla Cave, New Mexico. Geomicrobiol J 18: 297-309.

Queen, J.M. 1994. Influence of thermal atmospheric convection on the nature and distribution of microbiota in cave environments. *In:* I.D. Sasowsky and M.V. Palmer [eds.]. Breakthroughs in karst geomicrobiology and redox geochemistry: Abstracts and field-trip guide for the symposium held February 16 through 19, 1994 Colorado Springs, CO. Spec Pub 1. Karst Waters Institute, Inc. Charles Town, WV. pp. 62-64.

Randall, K.W. 2006. Assessing the potential impact of microbes in the Edwards and Trinity Aquifers of central Texas. M.S. Thesis. Louisiana State University, Louisiana.

Randall, K.W., L. Johnson, and A.S. Engel. 2005. Assessing the potential impact of microbes in the Edwards and Trinity aquifers of central Texas. GSA Abs. Prog. 37: 217.

Reinbacher, W.R. 1994. Is it gnome, is it berg, is it mont, is it mond? An updated view of the origin and etymology of moonmilk. Bull Natl Speleol Soc 56: 1-13.

Riding, R. 2000. Microbial carbonates; the geological record of calcified bacterial-algal mats and biofilms. Sedimentology 47: 179-214.

Rogers, B.W. and K.M. Williams. 1982. Mineralogy of Lilburn Cave, Kings Canyon National Park, California. Bull Natl Speleol Soc 44: 23-31.

Rutherford, J.M. and L.H. Huang. 1994. A study of fungi of remote sediments in West Virginia caves and a comparison with reported species in the literature. Bull Natl Speleol Soc 56: 38-45.

Saiz-Jimenez, C. and B. Hermosin. 1999. Thermally assisted hydrolysis and methylation of dissolved organic matter in dripping water from the Altimara Cave. J Analyt Appl Physics 49(1): 337-347.

Sanchez-Moral, S., J. Bedoya, L. Luque, J.C. Cañaveras, V. Jurado, L. Laiz, and C. Saiz-Jimenez. 2003. Biomineralization of different crystalline phases by bacteria isolated from catacombs. *In:* C. Saiz-Jimenez [ed.]. Molecular Biology and Cultural Heritage, Balkema, Rotterdam, Netherlands. pp. 179-185.

Sanchez-Moral, S., J.C. Cañaveras, L. Laiz, C. Saiz-Jimenez, J. Bedoya, and L. Luque. 2003. Biomediated precipitation of calcium carbonate metastable phases in hypogean environments: A short review. Geomicrobiol J 20(5): 491-500.

Sand, W. 1997. Microbial mechanisms of deterioration of inorganic substrates — A general mechanistic review. Int Biodeter Biodegrad 40: 183-190.

Sarbu, S.M. and R. Popa. 1992. A unique chemoautotrophically based cave ecosystem. *In:* A.I. Camacho [ed.]. The natural history of biospeleology. Museo Nacional de Ciencias Naturales. Consejo Superior de Investigaciones Científicas. Madrid. pp. 637-666.

Sarbu, S.M. and T.C. Kane. 1995. A subterranean chemoautotrophically based ecosystem. Bull Natl Speleol Soc 57: 91-98.

Sarbu, S.M., B.K. Kinkle, L. Vlasceanu, T.C. Kane, and R. Popa. 1994. Microbiological characterization of a sulfide-rich groundwater ecosystem. Geomicrobiol J 12: 175-182.

Sarbu, S.M., T.C. Kane, and B.K. Kinkle. 1996. A chemoautotrophically based cave ecosystem. Science 272: 1953-1954.

Sarbu, S.M., S. Galdenzi, M. Menichetti, and D. Gentile. 2000. Geology and biology of Grotte di Frasassi (Frasassi Caves) in Central Italy, an ecological multi-disciplinary study of a hypogenic underground karst system. *In:* H. Wilkens, D.C. Culver, and W.F. Humphreys [eds.]. Subterranean ecosystems. Ecosystems of the World, Vol. 30. Elsevier Science, Amsterdam. pp. 359-378.

Sasowsky I.D. and M.V. Palmer [eds.]. 1994. Breakthroughs in karst geomicrobiology and redox geochemistry: Abstracts and field-trip guide for the symposium held February 16 through 19, 1994, Colorado Springs, Colorado Spec Pub 1. Karst Waters Institute, Inc. Charles Town, WV.

Schindel, G.M., S.R.H. Worthington, E.C. Alexander Jr, and G. Veni. 2000. An overview of the San Antonio segment of the Edwards (Balcones Fault Zone) Aquifer in south-central Texas. J Cave Karst Stud 62: 197.

Schwabe, S.J., R.A. Herbert, and J.L. Carew. 2001. A hypothesis for biogenic cave formation: A study conducted in the Bahamas. *In:* Proceedings of the 13th Symposium on the Geology of the Bahamas and other Carbonate Regions, pp. 141-152.

Shaw, T.R. 1997. Historical introduction. *In:* C.A. Hill and P. Forti [eds.]. Cave Minerals of the World. 2nd ed. National Speleological Society. Huntsville, AL. pp. 27-43.

Shen, Y., R. Buick, and D.E. Canfield. 2001. Isotopic evidence for microbial sulfate reduction in the early Archaean era. Nature 410: 77-81.

Simkiss, K. 1986. The process of biomineralization in lower plants and animals — An overview. *In:* B.S.C. Leadbeater and R. Riding [eds.]. Biomineralization in lower plants and animals. The Systematics Assoc, Spec Vol. 30: 19-37, Clarendon Press, Oxford.

Simkiss, K. and K.M. Wilbur. 1989. Biomineralization: Cell Biology and Mineral Deposition. Academic Press, San Diego.

Sobolev, D. and E.F. Roden. 2002. Evidence for rapid microscale bacterial redox cycling of iron in the circumneutral environments. Antonie Van Leeuw 81: 587-597.

Socki, R.A., E.C. Perry, and C.S. Romanek. 2002. Stable isotope systematics of two cenotes from Northern Yucatan, Mexico. Limnol Oceanogr 47: 1808-1818.

Sogin, M.L., H.G. Morrison, J.A. Huber, D.M. Welch, S.M. Huse, P.R. Neal, J.M. Arrieta, and G.J. Herndl. 2006. Microbial diversity in the deep and underexplored 'rare biosphere'. PNAS. 103: 12115-12120.

Southam, G. and R. Donald. 1999. A structural comparison of bacterial microfossils vs. 'nanobacteria' and nanofossils. Earth-Science Rev 48: 251-264.

Southward, A.J., M.C. Kennicutt II, J. Alcalà-Herrera, M. Abbiati, L. Airoldi, F. Cinelli, C.N. Bianchi, C. Morri, and E.C. Southward. 1996. On the biology of submarine caves with sulphur springs: Appraisal of $^{13}C/^{12}C$ ratios as a guide to trophic relations. J Mar Biol Assoc UK 76: 265-285.

Sparks, D.L. 2005. Toxic metals in the environment: The role of surfaces. Elements 1: 193-197.

Spear, J.R., H.A. Barton, K.J. Roberts, C.A. Francis, and N.R. Pace. 2007. Microbial biofabrics in a geothermal mine adit. Appl Environ Microbiol 73: 6172-6180.

Spilde, M.N., L.J. Crossey, K.E. Dotson, R.T. Schelble, D.E. Northup, S.M. Barns, and C.N. Dahm. 2000. Biogenic influence on mineral corrosion and deposition at Lechuguilla and Spider Caves, New Mexico. Geol Soc Am Abstr Progr, 32, A256.

Spilde, M.N., D.E. Northup, P.J. Boston, R.T. Schelble, C.A. Dano, and N.R. Pace. 2005. Geomicrobiology of cave ferromanganese deposits: A field and laboratory investigation. Geomicrobiol J 22: 99-116.

Spirakis, C. and K.I. Cunningham. 1992. Genesis of sulfur deposits in Lechguilla Cave, Carlsbad Caverns National Park, New Mexico. *In:* G. Wessel and B. Wimberley [eds.]. Native sulfur—Developments in geology and explorations. Am Inst Mining, Metallurgic, Petrol Engr (AIME) p. 139-145.

Stoessell, R.K., Y.H. Moore, and J.G. Coke. 1993. The occurrence and effects of sulfate reduction and sulfide oxidation on coastal limestone dissolution in Yucatan Cenotes. Ground Water 31: 566-575.

Straub, K.L., M. Benz, and B. Schink. 2001. Iron metabolism in anoxic environments at near neutral pH. FEMS Microbiol Ecol 34: 181-186.

Suttle, C.A. 2007. Marine viruses—major players in the global ecosystem. Nat Rev Microbiol 5: 801-812.

Sweeting M.M. 1973. Karst Landforms. Columbia University Press, New York.

Tebo, B.M., W.C. Ghiorse, L.G. van Waasbergen, P.L. Siering, and R. Caspi. 1997. Bacterially mediated mineral formation: Insights into manganese(II) oxidation from molecular genetic and biochemical studies. Rev Mineral 35: 225-266.

Thompson, D.B. and R. Olson. 1988. A preliminary survey of the protozoa and bacteria from Sulphur River in Parkers Cave, Kentucky, USA. Bull Natl Speleol Soc 50: 42-46.

Urbani, P.F. 1976. Ópalo, calcedonia y calcita en la cueva del Cerro Autana (Am. 11), Territorio Federal Amazonas, Venezuela. Bol Soc Venez Espeleol 7: 129-145.

Urbani, P.F. 1977. Novedades sobre estudios realizados en las formas cársicas y pseudocársicas del Escudo de Guayana. Bol Soc Venez Espeleol 8: 175-197.

Urbani, P.F. 1996a. Espeleothems de opaló de origen biogénico en cavidades desarrollades en rocas silíceas. Bolívar y Apure (abs): 46th Conv Ann Aso VAC, Barquisimeto, Acta Clinet Venezez 47: supl 1.

Urbani, F. 1996b. Venezuelan cave minerals: A review. Bol Soc Venez Espeleol 30: 1-13.

Vandel, A. 1965. Biospeleology: The Biology of Cavernicolous Animals. (Translated from the 1964 French edition by B.E. Freeman.). Pergamon Press, Oxford.

Viles, H.A. 1984. Biokarst: Review and prospect. Prog Physical Geogr 8: 523-542.

Viles, H.A. and T. Spencer. 1986. 'Phytokarst', blue-green algae and limestone weathering. *In:* K. Paterson and M.M. Sweeting [eds.]. New directions in karst. Proc Anglo-French Karst Symp, Sept 1983. Geo Books. Norwich. pp. 115-140.

Vlasceanu, L., R. Popa, and B.K. Kinkle. 1997. Characterization of *Thiobacillus thioparus* LV43 and its distribution in a chemoautotrophically based groundwater ecosystem. Appl Environ Microbiol 63: 3123-3127.

Vlasceanu, L., S.M. Sarbu, A.S. Engel, and B.K. Kinkle. 2000. Acidic cave-wall biofilms located in the Frasassi Gorge, Italy. Geomicrobiol J 17: 125-139.

Went, F.W. 1969. Fungi associated with stalactite growth. Science 166: 385-386.

White, W.B. 1976. Cave minerals and speleothems. *In:* T.D. Ford and C.H. Cullingford [eds.]. The Science of Speleology. Academic Press, London. pp. 267-327.

Whitman, W.B., D.C. Coleman, and W.J. Wieble. 1998. Prokaryotes: The unseen majority. Proc Natl Acad Sci USA 95: 6578-6583.

Willems, L., P. Compere, and B. Sponholz. 1998. Study of siliceous karst genesis in eastern Niger: Microscopy and X-ray microanalysis of speleothems. Z Geomorph 42: 129-142.

Williams, A.M. 1959. The formation and deposition of moonmilk. Transact Cave Res Group 5: 135-138.

Wilson, W.L. and T.L. Morris. 1994. Cenote Verde: A meromictic karst pond, Quintana Roo, Mexico *In:* I.D. Sasowsky and M.V. Palmer [eds.]. Breakthroughs in karst geomicrobiology and redox geochemistry: Abstracts and field-trip guide for the symposium held February 16 through 19, 1994, Colorado Springs, CO. Spec Pub 1. Karst Waters Institute, Inc. Charles Town, WV. pp. 77-79.

Woods, A.D., D.J. Bottjer, M. Mutti, and J. Morrison. 1999. Lower Triassic large sea-floor carbonate cements; their origin and a mechanism for the prolonged biotic recovery from the end-Permian mass extinction. Geol 27: 645-648.

Wray, R.A.L. 1999. Opal and chalcedony speleothems on quartz sandstones in the Sydney region, southeastern Australia. Australian J Earth Sci 46: 623-632.

Xu, J. 2006. Microbial ecology in the age of genomics and metagenomics: Concepts, tools and recent advances. Mol Ecol 15: 1713-1731.

CHAPTER 2

Deep-sea Piezophilic Bacteria: Geomicrobiology and Biotechnology

Jiasong Fang[1*] and Chiaki Kato[2]

INTRODUCTION

Geomicrobiology is a sub-discipline arising from interdisciplinary studies of microbiology, geology, chemistry and other disciplines. It is the study of the interactions between microorganisms and their inorganic environments in the Geosphere (Fang and Bazylinski 2008). The rapid development and growth of geomicrobiology can partially be attributed to studies and discoveries in the last several decades of extremophiles present in many different environments that play key roles in biogeochemical cycles which occur in these environments. The rapidly-developing studies of extremophiles and geomicrobiology have revolutionized our views of the origin of life, the effects that microbial life has had on the Earth, the effect that the Earth has had on the development and evolution of life, and the interaction between microbes and the environment through time. Recently, a number of publications have been devoted to geomicrobiology in a variety of environments (Banfield and Nealson 1997, Ehrlich 1998, Newman and Banfield 2002, Macalady and

[1]College of Natural and Computational Sciences, Hawaii Pacific University, Kaneohe, HI 96744, USA, E-mail: jfang@hpu.edu
[2]Extremobiosphere Research Center, Japan Agency for Marine Earth Science and Technology (JAMSTEC), 2-15 Natsushima-cho, Yokosuka 237-0061, Japan, E-mail: kato_chi@jamstec.go.jp
*Corresponding author

Banfield 2003, Burton and Lappin-Scott 2005, Edwards et al. 2005, Fang and Bazylinski 2008). Here we present a brief review of recent advances in deep-sea geomicrobiology and biotechnology.

Piezophilic bacteria are microorganisms that display optimal growth at pressures greater than 0.1 MPa or that showing a requirement for increased pressure for growth (Yayanos 1995). Based on their response to pressure (optimal growth pressure), piezophiles can be classified as piezotolerant (growth from 0.1–10 MPa), piezophilic (10–70 MPa), and hyperpiezophilic (extremely piezophilic) bacteria (no growth at pressure lower than 50 MPa) (Fig. 2.1, Fang and Bazylinski 2008). Piezotolerant bacteria are those which are capable of growth at high pressure as well as atmospheric pressure and are easily distinguished from piezophilic and hyperpiezophilic bacteria that require elevated pressures for optimal growth (Kato et al. 1998, Abe and Horikoshi 2001). Hyperpiezophilic bacteria only grow at very high pressures (>70 MPa) (Kato et al. 1998, Bartlett 2002). Representative bacterial strains of these groups are listed in (Table 2.1).

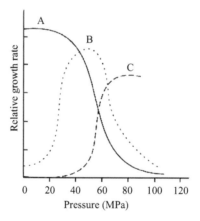

FIG. 2.1 Piezophilic bacteria grouped into piezotolerant (0.1–10 MPa), piezophilic (10–70 MPa), and hyperpiezophilic (no growth at less than 50 MPa) bacteria. A: piezotolerant; B: piezophilic; C: hyperpiezophilic. See text for further discussion.

TAXONOMY AND DIVERSITY OF DEEP-SEA PIEZOPHILIC BACTERIA

Most known cultured piezophilic bacteria belong to the γ-subdivision of the Proteobacteria based on 5S rRNA gene or 16S rRNA gene sequencing and lipid analysis (Fig. 2.2). DeLong et al. (1997) reported that 11

TABLE 2.1 Piezophilic microorganisms isolated from various sources

Organism	Source (depth)	Optimal growth conditions	Investigations	Reference
Piezotolerant bacteria (0.1–10 MPa)				
Sporosarcina sp. DSK25	Japan Trench 6,500 m	0.1 MPa, 35°C	Taxonomy	1
Piezophilic bacteria (10–70 MPa)				
Moritella japonica DSK1	Japan Trench 6,353 m	50 MPa, 15°C	Physiology, membrane lipids	2–4
			Carbon isotope fractionation	4
Shewanella SC2A	E. Pacific Ocean 1,957 m	14 MPa, 20°C	Physiology	5
Shewanella benthica WHB46	Weddell Sea 4,995 m	40 MPa, 5°C		6
Shewanella benthica F1A	N. Atlantic	41 MPa, 8°C	Phylogenetics	7
Photobacterium profundum DSJ4	Ryukyu Trench 5,110 m	10 MPa, 10°C	Taxonomy	8
Photobacterium profundum SS9	Sulu Sea 2,551 m	28 MPa, 15°C	Gene expression Membrane protein Membrane lipids Genome sequence	9–15
Shewanella violacea DSS12	Ryukyu Trench 5,110 m	30 MPa, 8°C	Taxonomy Gene expression Respiratory system Lipid biosynthesis	2, 4, 25 16–23 38, 41
Moritella PE36	E. Pacific Ocean 3,584 m	41 MPa, 15°C	Phylogenetics	7

Table 2.1 Contd.

Table 2.1 Contd.

Shewanella benthica DB5501	Suruga Bay 2,485 m	50 MPa, 10°C	Taxonomy Phylogenetic characterization	2, 24, 25
Shewanella benthica DB6101	Ryukyu Trench 5,110 m	50 MPa, 10°C	Taxonomy Phylogenetic characterization	2, 24, 25
Shewanella benthica DB6705	Japan Trench 6,356 m	50 MPa, 10°C	Taxonomy Phylogenetic characterization	2, 23, 24 25
Shewanella benthica DB6906	Japan Trench 6,269 m	50 MPa, 10°C	Taxonomy Phylogenetics	2, 24, 25
Psychromonas strain CNPT3	N. Pacific 5,700 m	52 MPa, 8°C	Physiology Membrane lipids Phylogenetics	7, 26–28 28
Shewanella benthica PT48	Philippine Trench 6,163 m	62 MPa, 8°C	Phylogenetics	7
Shewanella benthica DB172F	Izu-Bonin Trench 6,499 m	70 MPa, 10°C	Taxonomy Phylogenetics Respiratory	16, 24, 25 39,40
Shewanella benthica DB172R	Izu-Bonin Trench 6,499 m	70 MPa, 10°C	Taxonomy Phylogenetics	16, 24, 25
Shewanella benthica PT99	Philippine Trench 8,600 m	62 MPa, 8°C	Membrane lipids Phylogenetics	7, 28
Moritella profunda	Off West Africa Coast 2,815 m	20-24 MPa,6°C	Taxonomy	34
Moritella abyssi	Off West Africa Coast 2,815 m	29-30 MPa, 10°C	Taxonomy	34
Psychromonas profunda	Atlantic Ocean 2,770 m	25 MPa, 10°C	Taxonomy	35
Psychromonas kaikoae	Japan Trench 7,434 m	50 MPa, 10°C	Taxonomy	36
Colwellia piezophila	Japan Trench	60 MPa, 10°C	Taxonomy	37

Table 2.1 Contd.

Table 2.1 Contd.

Hyperpiezophilic bacteria (no growth less than 50 MPa)

Shewanella benthica DB21MT-2	Mariana Trench 10,898 m	70 MPa, 10°C	Physiology, taxonomy, Lipid biosynthesis	4, 23, 29
Moritella yayanosii DB21MT-5	Mariana Trench 10,898 m	70 MPa, 10°C	Physiology, taxonomy, Lipid biosynthesis	4, 23, 29, 30
Colwellia hadaliensis BNL1	Puerto Rico Trench 7,410 m	93 MPa, 10°C	Physiology	31
Strain PT64	N. Pacific	90 MPa, 9°C	Physiology	32
Strain MT199	N. Pacific	90 MPa, 13°C	Physiology	32
Colwellia MT41	Mariana Trench	103 MPa, 8°C	Physiology	7, 28, 33

(1) Kato et al., 1995b; (2) Kato et al., 1995a; (3) Nogi et al., 1998c; (4) Fang et al., 2003; (5) Yayanos and Diets, 1982; (6) Liesack et al., 1991; (7) DeLong et al., 1997; (8) Nogi et al., 1998a; (9) Bartlett et al., 1989; (10) Chi and Bartlett, 1995; (11) Welch and Bartlett, 1996; (12) Welch and Bartlett, 1998; (13) Allen and Bartlett, 1999; (14) Allen and Bartlett, 2000; (15) Vezzi et al., 2005; (16) Kato et al., 1996b; (17) Kato et al., 1997a; (18) Tamegai et al., 1998; (19) Nakasone et al., 1998; (20) Yamada et al., 2000; (21) Ikegami et al., 2000; (22) Kato and Nogi, 2001); (23) Fang et al., 2004; (24) Li et al., 1998; (25) Nogi et al., 1998b; (26) Yayanos et al., 1979; (27) DeLong and Yayanos, 1985; (28) DeLong and Yayanos, 1986; (29) Kato et al., 1998; (30) Nogi and Kato, 1999; (31) Deming et al., 1988; (32) Yayanos, 1986; (33) Yayanos et al., 1981; (34) Xu et al., 2003a; (35) Xu et al., 2003b; (36) Nogi et al., 2002; (37) Nogi et al., 2004; (38) Nakasone et al., 2002; (39) Qureshi et al., 1998a; (40) Qureshi et al., 1998b; (41) Tamegai et al., 2005.

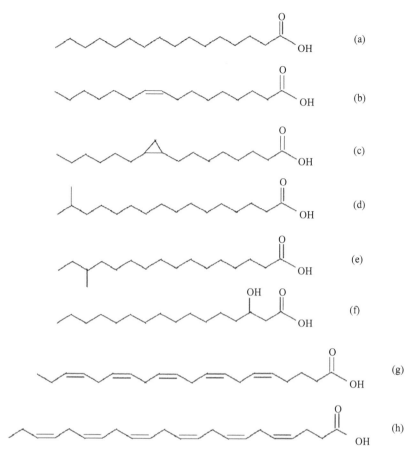

FIG. 2.2 Fatty acid biomarkers commonly detected in piezophilic bacteria. (a) a saturated fatty acid; (b) a monounsaturated fatty acid; (c) a cyclopropane fatty acid; (d) an *iso*-branched fatty acid; (e) an *anteiso*-branched fatty acid; (f) a β-hydroxy fatty acid; (g) eicosapentaenoic acid (EPA); (h) docosahexaenoic acid (DHA).

cultivated psychrophilic and piezophilic deep-sea bacteria were affiliated with one of the following five genera of the γ-subdivision of the Proteobacteria: *Shewanella, Photobacterium, Colwellia, Moritella,* and a genus containing strain CNPT3, recently identified as *Psychromonas* (Nogi et al. 2002). The most commonly found deep-sea piezophilic bacterial species are in the following five genera: *Shewanella* including *S. benthica* and *S. violacea* (Deming et al. 1984, MacDonell and Colwell 1985, Nogi et al. 1998b, Kato and Nogi 2001), *Photobacterium* including *P. profundum* (Nogi et al. 1998a), *Colwellia* including *C. hadaliensis* and *C. piezophila* (Deming et al. 1988, Nogi et al. 2004), *Moritella* including

M. japonica, M. yayanosii, M. profunda and *M. abyssi* (Nogi et al. 1998c, Nogi and Kato, 1999, Xu et al. 2003a), and *Psychromonas* including *P. kaikoi* and *P. profunda* (Nogi et al. 2002, Xu et al. 2003b). The phylogenetic relations of those piezophiles in γ-Proteobacteria are shown in Fig. 2.3. Overall, it seems that there is a limited diversity of the cultured piezophilic bacteria which may be an artifact of the use of rich, heterotrophic growth media. It is likely that diversity of piezophilic bacteria will increase as the use of different types of growth media increases (DeLong et al. 1997, Deming and Baross 2000).

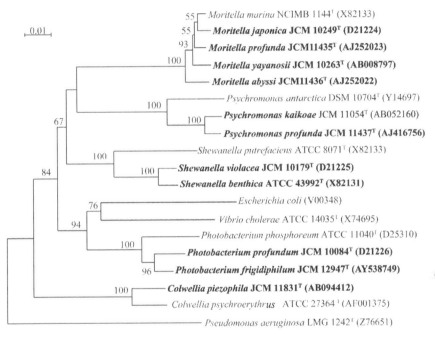

FIG. 2.3 Phylogenetic tree showing the relationships between isolated deep-sea piezophilic bacteria (in bold) within the γ-Proteobacteria subgroup determined by comparing 16S rRNA gene sequences using the neighbor-joining method (Saitou and Nei, 1987). The scale represents the average number of nucleotide substitutions per site. Bootstrap values (%) are shown for frequencies above the threshold of 50%.

Piezophilic Members of the Genus *Shewanella*

Piezophilic species of *Shewanella* and *Moritella* appear to dominate deep-sea piezophilic bacterial communities (e.g., Li and Kato 1999). The *Shewanella* species are particularly widely distributed in the marine environment (Venkateswaran et al. 1999) in the deep-sea, polar and other areas and many show piezophily and/or psychrophily (Nichols et al.

1997, Kato and Nogi, 2001). Most piezophilic bacteria cultured to date are the *Shewanella* species with a broad range of optimal growth pressure conditions from piezophilic to hyperpiezophilic (Table 2.1). Representative species include *S. violacea* strain DSS12, *S. benthica* strain PT99, and strains DB1172F, DB1172R, DB21MT-2, DB5501, DB6101, DB6705, DB6906, WHB46, F1A, PT48 (Table 2.1). *S. violacea* DSS12 is a moderately piezophilic bacterium and was isolated from sediment of the Ryukyu Trench at 5,110 m (Kato et al. 1995a, Nogi et al. 1998b). It is able to grow at hydrostatic pressures of up to 70 MPa (Kato et al. 1995a) with a relatively constant doubling time in a wide pressure range (0.1 MPa to 70 MPa), whereas the doubling times of most of the piezophilic *S. benthica* strains change substantially with increasing pressure (Kato et al. 1995a, 1996a, 1998). Because there are few differences in the growth characteristics of strain DSS12 under contrasting pressure conditions, this organism is an ideal model bacterium for studying lipid compositions and mechanisms of adaptation to high-pressure environments. Studies using this strain include analyses of the pressure-regulation of gene expression (Kato et al. 1997a,b), the role of *d*-type cytochromes in the growth of cells under high pressure (Kato et al. 1996b, Tamegai et al. 1998, Tamegai et al. 2005), and the biosynthesis and dietary uptake of fatty acids (Fang et al., 2004). Piezophilic strains DB6906, DB172F, DB172R and DSS12 are closely related phylogenetically based on their 16S rRNA gene sequences (Kato and Nogi 2001). These *Shewanella* species can be categorized taxonomically into two groups based on 16S rRNA gene sequences. Members of the *Shewanella* group 1 are high pressure, cold-adapted species and are either piezophilic (*S. benthica* and *S. violacea*) or piezotolerant (*S. gelidimarina* and *S. hanedai*), whereas those in group 2 are mesophilic and pressure-sensitive species and characterized by piezosensitive growth (no growth at pressure above 50 MPa) (Kato and Nogi 2001).

Hyperpiezophilic bacterium *Shewanella benthica* strain DB21MT-2 was isolated from the Mariana Trench sediment at 10,898 m (Kato et al. 1998). This strain grew well at pressures of 70–100 MPa (Kato et al. 1998). DB21MT-2 is a strain of *S. benthica* based on 16S rRNA gene sequence comparisons and DNA-DNA hybridization analysis (Kato et al. 1998). Other piezophilic bacteria in the *Shewanella* group include strain DB172R (Kato et al. 1996a, 2000), *Shewanella* strains PT99 (DeLong and Yayanos, 1986, DeLong et al. 1997), PT48 (DeLong and Yayanos 1986), and DB172F (Kato et al. 1996a, 2000) show optimal growth at pressures of 69, 62, and 70 MPa, respectively. However, these strains are able to grow at pressures of less than 50 MPa, thus, they are not hyperpiezophilic (Kato et al. 1998).

Piezophilic Members of the Genus *Moritella*

Two piezophilic bacterial isolates belong to the *Moritella* group, *Moritella japonica* DSK1 and *Moritella yayanosii* DB21MT-5 (Table 2.1, Fig. 2.3). Strain DSK1 is moderately piezophilic and was isolated from the Japan Trench sediment at 6,356 m (Kato et al. 1995a). This is the first piezophilic and psychrophilic species identified in the genus *Moritella* (Nogi et al. 1998c). Its optimal growth conditions are 10°C and 10-50 MPa (Kato et al. 1995a, Nogi et al. 1998c). Strain DB21MT-5 was isolated from sediment of the Mariana Trench at 11,000 m (Kato et al. 1998). Its optimal growth pressure was 80 MPa and no growth was observed at pressure less than 50 MPa. A facultative piezophilic strain of the genus *Moritella*, designated PE36, was isolated from the North Pacific Ocean at 3,586 m (Yayanos, 1986). All these strains produce DHA, one of the characteristic properties of this genus (DeLong et al. 1997, Kato et al. 1998, Nogi et al. 1998c, Fang et al. 2003).

Recently, Xu et al. (2003a) reported the isolation of two new piezophilic strains belonging to the genus *Moritella*. These strains, designated as *M. profunda* and *M. abyssi*, were isolated from surface sediments of the eastern tropical Atlantic at a depth of 2,815 m. These strains are facultative anaerobic, chemoorganotrophs. Like other *Moritella* species, they contain Q-8 as a major isoprenoid quinone and DHA as a major fatty acid. The maximum growth pressure of these strains is 20–24 MPa at 6°C. Thus, these strains are intermediate between *M. marina* which is not piezophilic and *M. yayanosii* which is hyperpiezophilic, with regard to their piezophilic characteristics (Xu et al. 2003a).

PIEZOPHILIC MEMBERS OF THE GENUS *COLWELLIA*

The first strict piezophilic strain isolated from the Mariana Trench was strain MT41 isolated from a dead amphipod, *Hirondellea gigas*, collected at a depth of 10,476 m (Yayanos et al. 1981). Phylogenetic analysis showed that MT41 is closely related to the genus *Colwellia* (DeLong et al. 1997). This hyperpiezophilic strain has an optimal pressure for growth of 100 MPa and is unable to grow at pressures less than 50 MPa (Yayanos, 1986). Another hyperpiezophilic bacterium in this genus is strain BNL-1 which was isolated from a seawater sample collected in the Puerto Rico Trench at a depth of 7,410 m (Deming et al. 1988). The optimal pressures for growth of strain BNL-1 were 93 MPa at 10°C and 74 MPa at 2°C (Deming et al. 1988). Strain BNL-1 was named *C. hadaliensis*. Recently two novel piezophilic strains were isolated from a surface sediment sample collected at 6,278 m in the Japan Trench (Nogi et al. 2004). 16S rRNA gene sequence analysis suggests that the strains represent novel, obligately piezophilic *Colwellia* species and are named as *C. piezophilia* (Nogi et al. 2004). These strains are different from other *Colwellia* piezophilic species in that they do not contain PUFA (EPA or DHA).

Piezophilic Members of the Genus *Photobacterium*

Two moderately piezophilic related strains, designated DSJ4 and SS9, were identified as *P. profundum* (Nogi et al. 1998a). *P. profundum* strain SS9 is one of the most extensively studied piezophilic bacteria and was isolated from the Sulu Sea at 2,551 m (Bartlett et al. 1989, Nogi et al. 1998a). It has been used as a model organism to study the molecular mechanisms of pressure-regulation (Bartlett et al. 1989, 1996, Welch and Bartlett 1998), membrane lipids and pressure adaptations (Allen et al. 1999, Allen and Bartlett 2000), phylogenetic characterization (DeLong et al. 1997), and the genomics (Vezzi et al. 2005). Another *P. profundum* isolate, strain DSJ4, was isolated from a sediment sample collected from the Ryukyu Trench at 5,110 m (Nogi et al. 1998a). *P. profundum* is the only species within the genus *Photobacterium* known to display piezophily, and the only one known to produce EPA.

Piezophilic Members of the Genus *Psychromonas*

Two novel obligately piezophilic bacteria, designated strains JT7301 and JT7304, were isolated from the Japan Trench at a depth of 7,434 m and were identified as *Psychromonas* species (Nogi et al. 2002). DNA-DNA hybridization between these strains and *P. antarctica* showed these new strains represented a new species of *Psychromonas*. The optimal growth temperature and pressure of these isolates were 10°C and 50 MPa, and they produced both EPA and DHA in the membrane layer (Nogi et al. 2002). The name *Psychromonas kaikoi* was proposed for the strain JT7304 which is the type strain of the species (Nogi et al. 2002). Another psychropiezophilic *Psychromonas* species, *P. profunda*, was isolated from the bottom sediment of the Atlantic Ocean at 2,770 m (Xu et al. 2003b). This strain grows from 0.1 to 25 MPa (at 6°C) and has an optimal growth pressure of 15–20 MPa. Strain CNPT3 appears to represent another *Psychromonas* species and was isolated from the North Pacific (Yayanos et al. 1979). However, the strain is comparatively related closely to *Psychromonas kaikoi* (Nogi et al. 2002).

MICROBIAL METABOLISM AND DEGRADATION OF ORGANIC MATTER IN THE OCEAN

The deep sea in general is considered to be an oligotrophic environment and the rates of bacterial growth and metabolism are low (Deming and

Baross 2000). The energy for deep-sea microorganisms is dependent on the "rain" of detritus of organic matter from the surface ocean. And most of the organic matter (>95%) produced photosynthetically in the surface waters are recycled in the upper 100–300 m (Jannasch and Taylor 1984, Wakeham and Lee 1993). Only a small proportion of the surface-produced organic matter (0.1-1%) reaches the deep sea (Wakeham and Lee 1993). This fact, on the other hand, suggests the high efficiency of the bacterial degradation processes (Azam et al. 1983, Harvey and Macko 1997a, Laureillard et al. 1997, Patching and Eardly 1997). Azam and Long (2001) suggest that marine bacteria can respond to the increased input of organic particles in the ocean and create hotspots of bacterial growth and carbon cycling. A number of studies of sediment trap materials and deep-sea sediments have demonstrated the extensive bacterial degradation of organic matter during vertical transport and the disappearance of the biomarker imprint of surface plankton in marine sediments (De Baar et al. 1983, Matsueda and Handa 1986, Wakeham and Canuel 1988, Laureillard et al. 1997). As such, microbial degradation of photosynthetically-derived organic matter results in the alteration of lipid and carbon isotope signature of organic matter preserved in the deep sea (e.g., Fang et al. 2006). Laboratory microbial degradation experiments showed that more than 99% of algal fatty acids were lost in a 77-day incubation period (Harvey and Macko 1997b). De Baar et al. (1983) showed that PUFA (including EPA and DHA) and total fatty acids decreased from 7.8 mg and 39 mg at 389 m to 10.4 µg and 930 µg at 5,068 m in sediment trap samples collected from the equatorial North Atlantic. The estimated decay rate constants with depth were 0.84 m^{-1} and 1.38 m^{-1} for total fatty acids and PUFA, respectively (De Baar et al. 1983). Thus, the decoupling of lipid signatures exists between surface plankton and deep-sea sediment. The work of Pinturier-Geiss et al. (2001) showed that sedimentary phospholipid fatty acids from the Western Crozet Basin (depth 1,720–4,750 m) were derived from bacteria, not from surface microplankton, even in areas where diatom cysts were still living in the fluff and where vertical transfer of particles occurred fast. Similarly, Matsueda and Handa (1986) and Laureillard et al. (1997) observed that imprints of surface plankton input of sinking particulates are lost in samples collected in deep-water column and seafloor-sediment samples. On the contrary, a good correlation of fatty acid composition was observed between surface water particulates and sediment in shallow marine environments (Smith et al. 1986, 1989). These studies highlight the importance and consequence of bacterial degradation of organic matter during vertical transit before reaching the seafloor. That is, microbial degradation of organic matter in the water column and at the sediment-water interface results in the incorporation of microbial lipids into sedimentary organic matter (Perry et al. 1979) and thus, alters the biomarker signature of fatty acids that can be markedly different from that of the material originally biosynthesized (Wakeham and Lee 1993, Laureillard et al. 1997). For example, the *iso-* and *anteiso-*C_{15} and C_{17} branched fatty acids isolated by De Baar et al. (1983) in deep traps of the North Atlantic Ocean might be from a viable bacterial community colonizing the particles (Wakeham and Lee 1993). The high levels of PUFA detected in organic matter collected at 1,500 m using filters of 1 μm pore size (Wakeham and Canuel 1988) were probably from piezophilic bacteria (Wakeham and Lee 1993).

In the deep sea, microbial mediated redox reactions are expected to take place in a sequential series based on the free energy yield of the reactions: NO_3^- reduction and denitrification, dissimilatory Mn(IV)- and Fe(III)-reduction, sulfate reduction, and methanogenesis (autotrophic, fermentation and acetoclastic). Since temperature is relatively stable and uniform in the deep sea (except the hydrothermal vent area), pressure may exert more important influence on microbial metabolic reactions (Fang and Bazylinski 2008). The electron towers constructed based on electron activity at the biological standard state ($p\varepsilon^{o\prime}$) at 25°C and 0.1 MPa (Fig. 2.4a) and 2°C and 40 MPa (Fig. 2.4b) show that the additive effect of temperature and pressure has raised the $p\varepsilon^{o\prime}$ values of the first four reactions commonly taking place at more positive redox potentials and lowered the $p\varepsilon^{o\prime}$ values for other five reactions that take place at more reducing conditions (Fig. 2.4b). Thus, redox reactions prevailing at more oxidative conditions (O_2, NO_3^-/NO_2^- reduction, etc.) yield slightly more energy in the deep-sea low temperature-high pressure conditions, whereas those that dominate in more reducing conditions yield relatively less energy, compared to surface environments (25°C, 0.1 MPa) (Fang and Bazylinski 2008).

MOLECULAR, BIOMARKER, AND STABLE ISOTOPE TOOLS IN DEEP-SEA GEOMICROBIOLOGY

Molecular Phylogenetic Analysis

Molecular microbiology, lipid biomarkers, and stable carbon isotopes are indispensable tools in characterizing the diversity of piezophiles and gemicrobiological processes in the deep sea. Piezophilic bacterial diversity can be assessed through cultivation and cultivation-independent methods (lipid biomarkers and small subunit ribosomal RNA gene). Geomicrobiology has benefited tremendously from the development of the ever-maturing DNA sequencing capabilities and new methods to assay gene expression and protein function (molecular

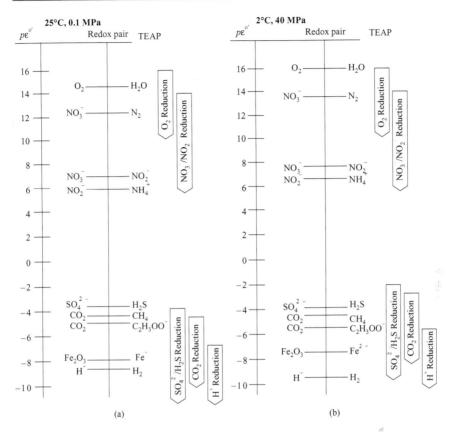

FIG. 2.4 Sequence of microbially-mediated reduction reactions based on values of electron activity at biological standard state ($p\varepsilon^{o'}$) at 25°C and 1 bar (a) and 2°C and 400 bar (b). The $p\varepsilon^{o'}$ values are calculated using SUPCRT92 (Amend and Teske, 2005).

geomicrobiology, Banfield et al. 2005). Typically, to determine microbial diversity of piezophilic bacteria in the deep ocean sediment, total DNA is extracted directly from a sediment sample. Molecular phylogenetic analysis using PCR amplified 16S rRNA gene sequences is then performed (Li et al. 1999). High-pressure microbial cultivation can be carried out using in-house built high-pressure bioreactors (e.g., Fang et al. 2006) or the advanced DEEPBATH system (deep-sea baro-piezo/thermophiles collecting and cultivating system, operated by JAMSTEC). It is noted that microbial community structure changes drastically under different pressure conditions of cultivation. Therefore, it is necessary to analyze microbial communities using molecular phylogenetic techniques in the original sample and in samples cultivated without decompression in order to determine the diversity of piezophilic bacteria in the deep-sea environment (Yanagibayashi et al. 1999). For example, in the original sediment obtained from the depth of 3,064 m of the Japan Sea, a number of bacterial peaks were observed in the terminal restriction fragment length polymorphism (t-RFLP). But after cultivation under high-pressure conditions, only a few peaks remained/or were amplified (Fig. 2.5, Arakawa et al. 2006). These peaks represent genera *Psychromonas*, *Moritella* and *Shewanella*, piezophiles typically found in the deep sea. A comparison of RFLP profiles in the amplified 16S rRNA genes with digestion of *Rsa*I and *Msp*I between piezophilic and non-piezophilic species in the five genera *Shewanella, Moritella, Colwellia, Psychromonas,* and *Photobacterium* is shown in Fig. 2.6. The RFLP profiles of all but *Shewanella* were the same, therefore it is necessary to combine the RFLP profiles with high-pressure cultivation and molecular phylogenetic analyses to distinguish between piezophiles and non-piezophiles.

The interlinked deep-sea physical, chemical, and microbiological processes by which bioactive elements are cycled in the ocean may be one

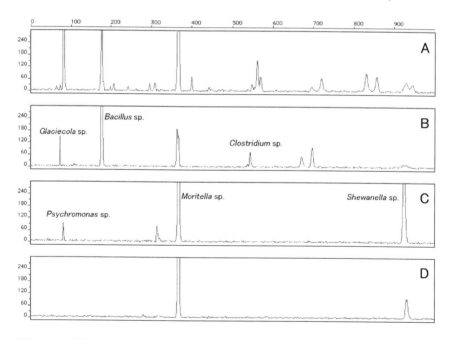

FIG. 2.5 T-RFLP (terminal-restriction fragment length polymorphism) analysis of bacterial community of the Japan Sea sediment sample under various pressure cultivations. A, microbial diversity of the sediment before cultivation; B, microbial community after cultivation at atmospheric pressure (0.1 MPa); C, microbial community after cultivation at 30 MPa; D, microbial community after cultivation at 50 MPa.

Piezophilic species, RFLP by *Rsa*I + *Msp*I

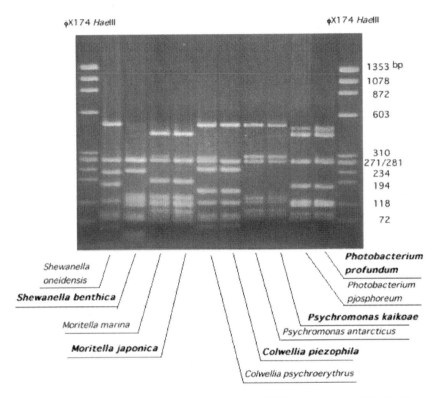

FIG. 2.6 Restriction fragment length polymorphism (RFLP) of the piezophilic (bold) and the related non-piezophilic bacterial 16S rRNA genes in five genera *Shewanella, Moritella, Colwellia, Psychromonas,* and *Photobacterium* on 3% agarose X gel electrophoresis. 16S rRNA genes were amplified by PCR using the primers, 27F-1492R (Kato et al., 1997c), then digested with *Rsa*I and *Msp*I.

of the most dynamic systems on Earth. Studying the deep-sea geomicrobiological processes requires an integrated approach combining the microbiological and geochemical characterization of the deep-sea environments. Our current understanding remains poor on the distribution and organization of piezophiles across spatial and temporal geochemical gradients in the deep sea (Fang and Bazylinski 2008). Comprehensive characterization of piezophilic bacteria diversity and metabolic potential in relation to geochemical gradients across different spatial and temporary scales is fundamentally important.

Lipid Profiles of Piezophilic Bacteria

Lipid biomarkers can provide important information on microbial community composition and metabolic status (e.g., Fang et al. 2006). In combining with stable carbon isotopes, lipids can shed light on biogeochemical and geomicrobiological processes taking place in an ecosystem.

The most abundant lipids detected in piezophilic bacteria are n-alkyl, acetogenic lipids (i.e., fatty acids). Fatty acids biosynthesized by piezophilic bacteria are also commonly found in surface bacteria (Fig. 2.2): C_{12-19} saturated, monounsaturated, terminal methyl-branched, β-hydroxyl, and cyclopropane fatty acids (DeLong and Yayanos, 1985, Kato et al. 1998, Fang et al. 2003, Fang and Bazylinski 2008). However, piezophilic bacteria are unique and distinctive in fatty acid composition in two ways: (1) that they synthesize abundant long-chain polyunsaturated fatty acids (PUFA), i.e., EPA ($20:5\omega3$) and DHA ($22:6\omega3$), and (2) that unsaturated fatty acids is abundant and can comprise up to 70% or more of the total fatty acids. Among the five genera of piezophilic bacteria, species of *Colwellia* and *Moritella* contain DHA, those of *Shewanella* and *Photobacterium* contain EPA, whereas piezophiles in the genus *Psychromonas* containing strain CNPT3 do not synthesize PUFA (Fang and Bazylinski 2008). Some newly isolated piezophilic bacterial species of the genera *Colwellia* and *Psychromonas* from the Japan Trench (e.g., *C. piezophila* and *P. kaikoae*) produce either no PUFA (Nogi et al., 2002) or low levels of both EPA and DHA (Nogi et al. 2004). Clearly, fatty acid composition has provided important information in characterizing piezophilic bacteria in phylogeny and taxonomy (e.g., Kato and Nogi 2001), piezoadaptation (Allen et al. 1999, Fang et al. 2004), and in biogeochemistry and geomicrobiology (Fang et al. 2000, 2002, 2006, Fang and Bazylinski 2008). Particularly, EPA and DHA can be used as biomarkers for detecting piezophilic bacteria in deep-sea sediment/water columns.

Stable Carbon Isotope Signature of Lipids

For biologically-mediated reactions, the isotopic signature can reflect the characteristics of specific enzymes in biochemical pathways (Hayes 1993). Analyzing the carbon isotopic composition of cell components and lipid biomarkers can provide information on modes of microbial metabolism and physiology. The carbon isotope signature is especially useful in deep-sea geomicrobiology for two reasons. First, environmental conditions (temperature and pressure) of the deep sea may exert a greater influence on carbon isotope fractionation in the biosynthesis of fatty acids and therefore the carbon isotopic ratios of individual compounds. Second, the deep-sea piezophilic bacteria possess two simultaneously operating biosynthetic pathways of fatty acids which dictate distinctive carbon isotope signature for the long-chain polyunsaturated fatty acids (Metz et al. 2001).

Carbon fractionation in biosynthesis of fatty acids of piezophilic bacteria is pressure dependent. Recently, Fang et al. (2006) examined carbon isotope fractionation during fatty acid biosynthesis in *Moritella japonica* strain DSK1. Fatty acids became progressively more depleted in ^{13}C with pressure. Carbon isotopic fractionation was in average –13.9, –14.5, and –18.3‰ at 10, 20, and 50 MPa, respectively. There was a strong linear correlation between carbon isotopic fractionation and hydrostatic pressure. Overall, piezophilic bacteria fractionate carbon isotopes significantly (14–18‰) more than surface heterotrophic bacteria. Thus, the recycling and resynthesis of fatty acids by piezophilic bacteria utilizing organic matter originating from primary production will greatly alter the carbon isotope signature of both short chain bacterial and long-chain planktonic fatty acids in oceanic environments and marine sediments. For example, if the carbon isotopic composition of phytoplankton-derived organic matter is –22‰, fatty acids synthesized by piezophilic bacteria would have δ^{13}C values of –36 to –40‰.

Further, PUFA had much more negative δ^{13}C values than other short-chain saturated and monounsaturated fatty acids. This was attributed to the operation of two different fatty acid biosynthetic systems in piezophilic bacteria: the FAS (fatty acid synthase)- and PKS (polyketide synthases)-based pathways. The FAS-based pathway is one common to surface bacteria which synthesizes typical bacterial fatty acids. The PKS-based pathway is a fundamentally different pathway which involves polyketide synthases (Metz et al., 2001) which catalyze the biosynthesis of long-chain polyunsaturated fatty acids. The PKS pathway apparently operate independently of FAS, elongase and desaturase activities to synthesize EPA and DHA without any reliance on fatty acyl intermediate such as 16:0-ACP (acyl carrier protein) (Wallis et al. 2002). The PKS pathway appear to be widely distributed in marine bacteria (Wallis et al. 2002) as genes with high homology to the *Shewanella* EPA gene cluster (*Shewanella* sp. SCRC-2738) (Yazawa 1996) have been found in *Photobacterium profundum* strain SS9 which synthesizes EPA (Allen et al. 1999) and in *Moritella marina* strain MP-1 which produces DHA (Tanaka et al. 1999). Thus, fatty acids (particularly EPA and DHA) and carbon isotope signatures become more informative in deep-sea geomicrobiology.

POTENTIAL BIOTECHNOLOGICAL APPLICATIONS

The deep-sea houses perhaps the most voluminous extreme environments on earth and represents a unique source of microorganisms. Research on piezobiology and their product or enzymes has a great potential of biotechnological applications and research significance (Aguilar, 1996, Pennisi, 1997, Cowan, 1998, Demirjian et al. 2001, Kato et al. 2006). The biotechnological potentials of piezophilic bacteria and piezophilic enzymes have been reviewed by Abe and Horikoshi (2001), Yano and Poulos (2003), Deming (1998), and Gomes and Steiner (2004). Recently, the National Research Council (2002) called for increasing more fundamental understanding of the biosynthetic capabilities of marine organisms for new drugs and agrichemical compounds, developing new paradigms for detecting marine natural products and biomaterials as potential pharmaceuticals, biopolymers, and biocatalysts, and developing new tools to solve environmental problems such as biofouling, pollution, ecosystem degradation, and hazards to human health.

HIGH PRESSURE ENZYMES

Piezophilic bacteria may possess a complex adaptation network for growth and survival in the deep sea (Campanaro et al. 2005). Expression studies on *P. profundum* SS9 performed at different pressure and temperature conditions reveal that piezoadaptation involved a great number of membrane transporters, metabolic processes, and amino acid biosynthesis and membrane modification enzymes (Campanaro et al. 2005). Therefore, studying piezophiles provides exciting insights into the fundamental concepts of molecule stability and cell survivability and offers the potential for exploring novel biomaterials in this immense deep-sea 'biofactory'.

Piezophilic enzymes, also called piezostable enzymes (Yano and Poulos 2003), have been isolated from psychrophilic and thermophilic piezophiles (Abe et al. 1999). Many enzymes exhibit higher catalytic activity and increased thermal stability at moderately high pressures. The increasing evidence that pressure can extend the stable temperature range for enzymes should expand the use of biocatalysts in the growing field of high-pressure bioorganic synthesis, among other applications (Gomes and Steiner 2004).

Proteins, from deep-sea piezophilic microorganisms, could be active under high-pressure conditions in general. Actually normal proteins can be inactive under higher-pressure conditions, ca. 500 MPa or higher. For bioprocessing under pressure, scientists are looking for pressure-tolerant enzymes, thus, 'piezophilic proteins' would be a focus in industrial applications. Comparative studies have been done of the same functional proteins and/or enzymes from *Escherichia coli* and deep-sea piezophiles (Ishii et al. 2002, Ishii et al. 2004, Kawano et al. 2004, Ohmae et al. 2004, De Poorter et al. 2004). Generally, piezophilic proteins were much more stable and/or active under higher-pressure conditions than proteins from *E. coli*. Our current studies on cell division protein, FtsZ, are described below.

Some rod-shaped bacteria, including *E. coli*, exhibit cell filamentation without septum formation under high hydrostatic pressure conditions, indicating that the cell division process is affected by hydrostatic pressure (Marquis 1976). We examined the effects of elevated pressure on FtsZ-ring formation in *E. coli* cells by indirect immunofluorescence microscopy. Elevated pressure completely repressed colony formation of *E. coli* cell at 40 MPa in our cultivation conditions and the cells exhibited obvious filamentous shapes. In the elongated cells, normal cell division processes appeared to be inhibited, because no FtsZ-rings were observed by indirect immnofluorescense staining. In addition, we also observed that hydrostatic pressure dissociated the *E. coli* FtsZ (ecFtsZ) polymers *in vitro* (Ishii et al. 2004) (Fig. 2.7). These results suggest that high-pressure directly affects cell survival and morphology through the dissociation of cytoskeletal frameworks.

In the experiment on piezophilic *S. violacea* DSS12, the growth occurred at high-pressure conditions (Kato et al. 1995a, Nogi et al. 1998b). To analyze the transcription upstream from the *ftsZ* gene, northern blot and primer extension analyses were performed and the results showed that gene expression was not pressure dependent. Western blot analysis also showed that the *S. violacea* FtsZ protein (svFtsZ) was equally expressed under several pressure conditions in the range of atmospheric (0.1 MPa) to high (50 MPa) pressures (Ishii et al. 2002). Using immunofluorescence microscopy, the svFtsZ ring was observed in the center of cells at pressure conditions of 0.1 to 50 MPa (Fig. 2.8). These results suggest that the svFtsZ protein function is not affected by elevated pressure in this piezophilic bacterium.

The cell morphology of *E. coli* under high-pressure is, however, quite different from *S. violacea*. The filamentous cells showed stoppage of cell division steps and it is expected that the hydrostatic pressure affect FtsZ protein function and causes inhibition of cell division. In fact, the C-termini of FtsZ in four bacteria compared were not conserved in each other and the region was essential for polymerization activity (Ishii et al. 2002). Therefore, it is expected that characterization of the biochemical features and polymerization activities of the FtsZ and any terrestrial

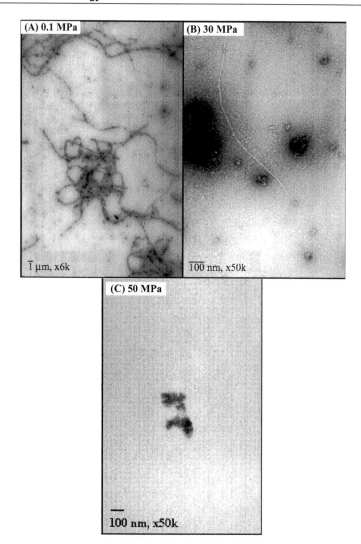

FIG. 2.7 Observation of ecFtsZ filaments assembled under each pressure condition with electron microscopy. The ecFtsZ filaments were fixed at A) 0.1 MPa, B) 30 MPa and C) 50 MPa. The scales and magnitudes were shown inside the pictures.

bacteria (eg. *E. coli*) *in vitro* help in understanding the effect of pressure on cell division steps *in vivo*.

POLYUNSATURATED FATTY ACIDS

Biosynthesis of polyunsaturated fatty acids (i.e., EPA and DHA) has been implicated as a physiological strategy in piezophilic adaptation to deep-

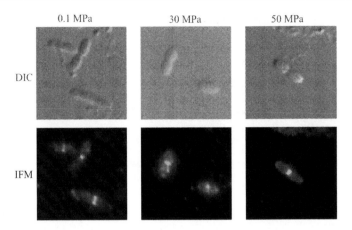

FIG. 2.8 Observation of FtsZ ring formation in *S. violacea*. Upper panel: Observation of *S. violacea* cells grown at 0.1, 30, and 50 MPa using differential interference contrast (DIC) microscopy. Lower panel: Observation of FtsZ ring formation using immunofluorescence microscopy (IFM).

sea low temperature and high pressure environments (DeLong and Yayanos, 1986, Fang et al. 2000). This is supported by laboratory studies (e.g., Fang et al. 2004). These methylene-interrupted polyunsaturated fatty acids had previously only been found in microeukaryotes. The detection of these PUFA in piezophiles suggests that these apparently unique fatty acids (to prokaryotes) were at least partially involved in facilitating growth of piezophilic bacteria at high pressures (DeLong and Yayanos 1986). Because of their low melting temperature and unique molecular geometry, PUFA may be needed for the maintenance of the correct phase of membrane lipids (homeophasic adaptation) because many bacterial phospholipids favor the formation of non-bilayer phases which would disrupt membrane packing (Russell and Nichols 1999). Therefore, these fatty acids play a dual role in bacterial adaptation to deep-sea high pressure environments by lowering the phase transition temperature (thereby keeping the membrane fluid) and by providing a higher degree of packing order (thereby preventing the formation of non-bilayer-phases) (Hazel 1995, Russell and Nichols 1999, Bartlett 2002). The highest levels of EPA and DHA in piezophilic isolates reported thus far are 37 and 20%, respectively, of total fatty acids (DeLong and Yayanos 1986).

The ω-3 PUFA are important to humans because of their biological activity, as essential cellular components, and as precursor for many hormone and hormone-like regulatory molecules (Albert et al. 1998,

Nichols et al. 2002, Valentine and Valentine 2004). Animals lack the ability to synthesize these compounds *de novo*, PUFA or their immediate C_{18} fatty acid precursors must be provided in the diet (Meyer et al. 1999). The current commercial sources of EPA and DHA are restricted to fish and algal-derived oil (Nichols et al. 2002). Bacterial production of PUFA represents an alternative source and provides the biotechnological opportunity to investigate the genes and enzymes responsible for PUFA production (Nichols et al. 2002). Research on gene expression of PUFA-producing strains in *Shewanella*, *Moritella*, and *Photobacterium* has identified genes responsible for PUFA biosynthesis (Takeyama et al. 1997, Nakasone et al. 1998, Tanaka et al. 1999, Allen and Bartlett 2000, Metz et al. 2001). The transfer of a gene cluster from *Shewanella putrefaciens* SCRC-2738 into *Escherichia coli* and the marine cyanobacterium *Synechococcus* sp. resulted in the successful expression of EPA in these organisms (Yazawa et al. 1996, Takeyama et al. 1997). Five *Shewanella* genes (ORFs 2, 5, 6, 7, and 8) were shown to be necessary for recombinant EPA synthesis in *Escherichia coli* and in *Synechococcus* sp. (Yazawa, 1996, Takeyama et al. 1997). These genes are related to microbial polyketide synthase (PKS) complexes and fatty acid synthase (FAS) enzymes (Metz et al. 2001), two independent fatty acid biosynthetic systems operating in piezophilic bacteria. The transgenic potential of bacteria PUFA genes can be significant in future biomaterial production.

OTHER APPLICATIONS

Some deep-sea microorganisms exhibit exceptional ability to degrade organic solvents (e.g., benzene, toluene, cyclohexane, etc.) (Kato et al. 1996c). These bacteria can tolerate and degrade toxic chemicals up to 20% (v/v) in water-solvent systems (Moriya and Horikoshi, 1993, Kato et al. 1996c). Strain DS-711, isolated from the Suruga Bay at 1,945 m, was able to degrade aromatic hydrocarbons and n-alkanes, 70-80% of the compounds were degraded in several days (Kato et al. 1996c). Unfortunately, no further studies have been done in exploring the efficacy of deep-sea bacterial degradation of organic contaminants.

FUTURE PERSPECTIVES AND CONCLUSIONS

Hydrostatic pressure is a significant environmental parameter of the deep ocean habitat. Deep-sea geomicrobiology has benefited greatly from advances in molecular biology and biogeochemistry and innovations in field and laboratory methods of microbial sample collection, isolation, and cultivation (e.g., Yayanos 1995, 2001). Since the first pressure-adapted bacterium (strain CNPT3) was successfully isolated from the deep-sea 25 years ago, there has been significant progress in understanding the biochemistry, physiology, and metabolism of piezophiles in the past several decades. However, some fundamental questions remain unanswered: Which microbes are there? What portion of the microbes is piezophilic? Are they actively reproducing and metabolizing in the deep sea? Answering these questions is fundamentally important to deep-sea geomicrobiology. Carbon isotopes are a useful tool for studying deep-sea geomicrobiological processes (e.g., Fang et al. 2006). More work is needed to investigate carbon isotope fractionations in lipid biosynthesis by piezophilic bacteria from the deep sea.

The study of piezophiles has added greatly to our understanding of the stability, structure, and function of key proteins involved in high pressure adaptation (Bartlett 2002). Piezostable enzymes have been isolated from many psychropiezophiles (Abe and Horikoshi, 2001) which have great economic potential in industrial processes (Yano and Poulos 2003). A prerequisite for large scale industrial applications is to first establish a committed effort in isolating piezophiles and pressure-adapted enzymes with commercial potential. In addition, piezophilic bacteria, together with psychrophiles may represent a major commercial source of EPA and DHA (Nichols et al. 2002). Further, the pressure-adapted microorganisms may be more tolerant to toxic chemicals and thus are better biodegrading agents for removing these chemicals from the environment. Additional research is needed to explore this potential.

REFERENCES

Abe, F. and K. Horikoshi. 2001. The biotechnological potential of piezophiles. Trends Biotechnol. 19: 102-108.

Abe, F., C. Kato, and K. Horikoshi. 1999. Pressure-regulated metabolism in microorganisms. Trends Microbiol. 7: 447-453.

Aguilar, A. 1996. Extremophile research in the European Union: From fundamental aspects to industrial expectations. FEMS Microbiol. Rev. 18: 89-92.

Albert, C.M., C.H. Hennekens, C.J. O'Donnell, U.A. Ajani, and V.J. Carey. 1998. Fish consumption and risk of sudden cardiac death. JAMA 279: 23-28.

Allen, E.E. and D.H. Bartlett. 2000. FabF is required for piezoregulation of *cis*-vaccenic acid levels and piezophilic growth of the deep-sea bacterium *Photobacterium profundum* strain SS9. J. Bacteriol. 182: 264–1271.

Allen, E.E., D. Facciotti, and D.H. Bartlett. 1999. Monounsaturated but not polyunsaturated fatty acids are required for growth of the deep-sea bacterium *Photobacterium profundum* SS9 at high pressure and low temperature. Appl. Environ. Microbiol. 65: 1710-1720.

Amend, J.P. and A. Teske. 2005. Expanding frontiers in deep subsurface microbiology. Paleogeogr., Peloclimatol., Paleoecol. 219: 131-155.

Arakawa, S., Y. Nogi, T. Sato, Y. Yoshida, R. Usami, and C. Kato. 2006. Diversity of piezophilic microorganisms in the closed ocean Japan Sea. Biosci. Biotechnol. Biochem. 70: 749-752.

Azam, F. and R.A. Long. 2001. Sea snow microorganisms. Nature 414: 496-498.

Azam, F., T. Fenchel, J.G. Field, J.S. Gray, L.A. Meyer-Reil, and F. Thingstad. 1983. The ecological role of water-column microbes in the sea. Marine Ecology Progress Series 10: 257-263.

Banfield, J.F. and K.H. Nealson. 1997. Geomicrobiology: Interactions between microbes and minerals. Reviews in Mineralogy, Vol. 35. Mineralogical Society of America, Washington, D.C.

Banfield, J.F., G.W. Tyson, E.E. Allen, and R.J. Whitaker. 2005. The search for a molecular-level understanding of the processes that underspin the Earth's biogeochemical cycles. *In*: J.F. Banfield, J. Cervini-Silvia, and K.M. Nealson [eds.]. Molecular Geomicrobiology. Reviews in Mineralogy and Geochemistry. Vol. 59, Mineralogical Society of America, Geochemical Society, Chantilly, Virginia, USA. pp. 1-7.

Bartlett, D.H. 2002. Pressure effects on *in vivo* microbial processes. Biochim. Biophys. Acta 1595: 367-381.

Bartlett, D.H., M. Wright, A.A. Yayanos, and M. Silverman. 1989. Isolation of a gene regulated by hydrostatic pressure in a deep-sea bacterium. Nature 342: 572-574.

Bartlett, D.H., E. Chi, and T.J. Welch. 1996. High pressure sensing and adaptation in the deep-sea bacterium *Photobacterium* species strain SS9. *In*: R. Hayashi and C. Balny [eds.]. High Pressure Bioscience and Biotechnology. Elsevier Science BV, The Netherlands, pp. 29-36.

Burton, S.K. and H.M. Lappin-Scott. 2005. Geomicrobiology, the hidden depths of the biosphere. Trends Microbiol. 13: 401.

Campanaro, S., A. Vezzi, N. Vitulo, F.M. Lauro, M. D'Angelo, F. Simonato, A. Cestaro, G. Malacrida, G. Bertoloni, G. Valle, and D.H. Bartlett. 2005. Laterally transferred elements and high pressure adaptation in *Photobacterium profundum* strains. BMC Genomics 6: 122-136.

Chi, E. and D.H. Bartlett. 1995. An rpoE-like locus controls outer membrane protein synthesis and growth at cold temperatures and high pressures in the deep-sea bacterium *Photobacterium* sp. strain SS9 Mol. Microbiol. 17: 713–726.

Cowan, D.A. 1998. Hot bugs, cold bugs and sushi. Trends Biotechnol. 16: 241-242.

De Baar, H.J.W., J.W. Farrington, and S.G. Wakeham. 1983. Vertical flux of fatty acids in the North Atlantic Ocean. J. Mar. Res. 41: 19-41.

DeLong, E.F. and A.A. Yayanos. 1985. Adaptation of membrane lipids of a deep-sea bacterium to changes in hydrostatic pressure. Science 228: 1101-1103.

DeLong, E.F. and A.A. Yayanos. 1986. Biochemical function and ecological significance of novel bacterial lipids in deep-sea prokaryotes. Appl. Environ. Microbiol. 51: 730-737.

DeLong, E.F., D.G. Franks, and A.A. Yayanos. 1997. Evolutionary relationship of cultivated psychrophilic and barophilic deep-sea bacteria. Appl. Environ. Microbiol. 63: 2105-2108.

Deming, J.W. 1998. Deep ocean environmental biotechnology. Curr. Opinion Biotechnol. 9: 283-287.

Deming, J.W. and J.A. Baross. 2000. Survival, dormancy, and nonculturable cells in extreme deep-sea environments. *In*: R.R. Colwell and D.J. Grimes [eds.]. Nonculturable Microorganisms in the Environment. ASM Press, Washington, D.C., USA. pp. 147-198.

Deming, J.W., H. Hada, R.R. Colwell, K.R. Luehrsen, and G.E. Fox. 1984. The ribonucleotide sequence of 5S rRNA from two strains of deep-sea barophilic bacteria. J. Gen. Microbiol. 130: 1911-1920.

Deming, J.W., L.K. Somers, W.L. Strauble, and M.T. McDonell. 1988. Isolation of an obligately barophilic bacterium and description of a new genus, *Colwell* gen. nov. Syst. Appl. Microbiol. 10: 152-155.

Demirjian, D.C., F. Moris-Varas, and C.S. Cassidy. 2001. Enzymes from extremophiles. Curr. Opinion Chem. Biol. 5: 144-151.

De Poorter, L.M.I., Y. Suzaki, T. Sato, H. Tamegai, and C. Kato. 2004. Effects of pressure on the structure and activity of isopropylmalate dehydrogenases from deep-sea *Shewanella* species. Mar. Biotechnol. 6: 190-194.

Edwards, K.J., W. Bach, and T.M. McCollom. 2005. Geomicrobiology in oceanography: Microbe–mineral interactions at and below the seafloor. Trends in Microbiology 13: 449-456.

Ehrlich, H.L. 1998. Geomicrobiology: Its significance for geology: Earth-Science Reviews 45: 45–60.

Fang, J. and D.A. Bazylinski. 2008. Deep-sea geomicrobiology. *In*: C. Michiels and D.H. Bartlett [eds.]. High-pressure Microbiology. American Society for Microbiology Press, Washington, D.C. pp. 237-264.

Fang, J., M.J. Barcelona, C. Kato, and Y. Nogi. 2000. Biochemical function and geochemical significance of novel phospholipids isolated from extremely barophilic bacteria from the Mariana Trench at a depth of 11,000 meters. Deep-Sea Res. 47: 1173-1182.

Fang, J., M.J. Barcelona, T.A. Abrajano, Jr, C. Kato, and Y. Nogi. 2002. Isotopic composition of fatty acids isolated from the extremely piezophilic bacteria from the Mariana Trench at 11,000 meters. Mar. Chem. 80: 1-9.

Fang, J., C. Kato, T. Sato, O. Chan, T. Peeples, and K. Niggemeyer. 2003. Phospholipid fatty acid profiles of deep-sea piezophilic bacteria from the deep sea. Lipids 38: 885-887.

Fang, J., C. Kato, T. Sato, O. Chan, and D.S. McKay. 2004. Polyunsaturated fatty acids in piezophilic bacteria: Biosynthesis or dietary uptake? Comp. Biochem. Physiol. B 137: 455-461.

Fang, J., M. Uhle, K. Billmark, D.H. Bartlett, and C. Kato. 2006. Fractionation of carbon isotopes in biosynthesis of fatty acids by a piezophilic bacterium *Moritella japonica* DSK1. Geochim. Cosmochim. Acta 70: 1753-1760.

Gomes, J. and W. Steiner. 2004. The biocatalytic potential of extremophiles and extremozymes. Food Technol. and Biotechnol. 42: 223-235.

Harvey, R.H. and S.A. Macko. 1997a. Catalysts or contributors? Tracking bacterial mediation of early diagenesis in the marine water column. Org. Geochem. 26: 531-544.

Harvey, R.H. and S.A. Macko. 1997b. Kinetics of phytoplankton decay during simulated sedimentation: Changes in lipids under oxic and anoxic conditions. Org. Geochem. 27: 129-140.

Hayes, J.M. 1993. Factors controlling ^{13}C contents of sedimentary compounds: Principles and evidence. Mar. Geol. 13: 111-125.

Hazel, J.R. 1995. Thermal adaption in biological membranes: Is homeoviscous adaption the explanation? Annu. Rev. Physiol. 57: 19-42.

Ikegami, A., K. Nakasone, M. Fujita, S. Fuji, C. Kato, R. Usami, and K. Horikoshi. 2000. Cloning and characterization of the gene encoding RNA polymerase sigma factor σ^{54} of deep-sea piezophilic *Shewanella violacea*. Biochim. Biophys. Acta 1491: 315–320.

Ishii, A., K. Nakasone, T. Sato, M. Wachi, M. Sugai, K. Nagai, and C. Kato. 2002. Isolation and characterization of the *dcw* cluster from the piezophilic deep-sea bacterium *Shewanella violacea*. J. Biochem. 132: 183-188.

Ishii, A., T. Sato, M. Wachi, K. Nagai, and C. Kato. 2004. Effects of high hydrostatic pressure on bacterial cytoskeleton FtsZ polymers *in vivo* and *in vitro*. Microbiology 150: 1965-1972.

Jannasch, H.W. and C.D. Taylor. 1984. Deep-sea microbiology. Ann. Rev. Microbiol. 38: 487-514.

Kato, C. and Y. Nogi. 2001. Correlation between phylogenetic structure and function: Examples from deep-sea *Shewanella*. FEMS Microbiol. Ecol. 35: 223-230.

Kato, C., T. Sato, and K. Horikoshi. 1995a. Isolation and properties of barophilic and barotolerant bacteria from deep-sea mud samples. Biodiv. Conserv. 4: 1-9.

Kato, C., S. Suzuki, S. Hata, T. Ito, and K. Horikoshi. 1995b. The properties of a protease activated by high pressure from *Sporosarcina* sp. strain DSK25 isolated from deep-sea sediment. JAMSTECR, 32: 7-13.

Kato, C., N. Masui, and K. Horikoshi. 1996a. Properties of obligately barophilic bacteria isolated from a sample of deep-sea sediment from the Izu-Bonin Trench. J. Mar. Biotechnol. 4: 96-99.

Kato, C., H.A. Tamegai, R. Ikegami, R. Usami, and K. Horikoshi. 1996b. Open reading frame 3 of the barotolerant bacterium strain DSS12 is complementary with cydD in *Escherichia coli*: cydD functions are required for cell stability at high pressure. J. Biochem. 120: 301-305.

Kato, C., A. Inoue, and K. Horikoshi. 1996c. Isolating and characterizing deep-sea marine micro-organisms. Trends Biotechnol. 14: 6-12.

Kato, C., A. Ikegami, M. Smorawinska, R. Usami, and K. Horikoshi. 1997a. Structure of genes in a pressure-regulated operon and adjacent regions from a barotolerant bacterium strain DSS12. J. Mar. Biotechnol. 5: 210-218.

Kato, C., L. Li, H. Tamegai, M. Smorawinska, and K. Horikoshi. 1997b. A pressure-regulated gene cluster in deep-sea adapted bacteria with reference to its distribution. Recent Res. Devel. in Agricultural & Biological Chem. 1: 25-32.

Kato, C., L. Li, J. Tamaoka, and K. Horikoshi. 1997c. Molecular analyses of the sediment of the 11000 m deep Mariana Trench. Extremophiles 1: 117-123.

Kato, C., L. Li, Y. Nogi, Y. Nakamura, J. Tamaoka, and K. Horikoshi. 1998. Extremely barophilic bacteria isolated from the Mariana Trench, Challenger Deep, at a depth of 11,000 meters. Appl. Environ. Microbiol. 64: 1510-1513.

Kato, C., L. Li, Y. Nogi, N. Nakasone, and D.H. Bartlett. 2000. Marine microbiology: Deep sea adaptations. *In*: Y. Taniguchi, H.E. Stanley and H. Ludwig [eds.]. Biological Systems under Extreme Conditions. Springer, Berlin, Germany. pp. 205-218.

Kato, C., T. Sato, A. Ishii, H. Kawano, F. Abe, E. Ohmae, and K. Nakasone. 2006. Proteins, under high-pressure environments — Discoveries of deep-sea piezophiles, and their pressure adapted enzymes. The Proceedings of International Symposium on Extremophiles and Their Applications, in press.

Kawano, H., K. Nakasone, M. Matsumoto, R. Usami, C. Kato, and F. Abe. 2004. Differential pressure resistance in the activity of RNA polymerase isolated from *Shewanella violacea* and *Escherichia coli*. Extremophiles 8: 367-375.

Laureillard, J., L. Pinturier, J. Fillaux, and A. Saliot. 1997. Organic geochemistry of marine sediments of the Sub-Antarctic Indian Ocean sector: Lipid classes — sources and fate. Deep-Sea Research 44: 1085-1108.

Li, L. and C. Kato. 1999. Microbial diversity in the sediments collected from cold-seep areas and from different depths of the deep sea. *In*: K. Horikoshi and K. Tsujii [eds.]. Extremophiles in Deep-Sea Environments, Springer-Verlag, Tokyo, Japan. pp. 55-58.

Li, L., C. Kato, Y. Nogi, and K. Horikoshi. 1998. Distribution of the pressure-regulated operons in deep-sea bacteria. FEMS Microbiology Letters 159: 159-166.

Li, L., C. Kato, Y. Nogi, and K. Horikoshi. 1999. Microbial diversity in sediments collected from the deepest cold-seep area, the Japan Trench. Mar. Biotechnol. 1: 391-400.

Liesack, W.H., W. Weyland, and E. Stackebrandt. 1991. Potential risks of gene amplification by PCR as determined by 16S rDNA analysis of a mixed-culture of strict barophilic bacteria. Microb. Ecol. 21: 191-198.

Macalady, J. and J.F. Banfield. 2003. Molecular geomicrobiology: Genes and geochemical cycling. Earth Planet. Sci. Lett. 209: 1-17.

MacDonell, M.T. and R.R. Colwell. 1985. Phylogeny of the Vibrionaceae, and recommendation for two new genera, *Listonella* and *Shewanella*. Syst. Appl. Microbiol. 6: 171-182.

Marquis, R.E. 1976. High-pressure microbial physiology. Advances in Microbiology and Physiology 14: 159-241.

Matsueda, H. and N. Handa. 1986. Source of organic matter in the sinking particles collected from the Pacific Sector of the Antarctic Ocean by sediment trap experiment. Memoirs of National Institute of Polar Research, Special Issue 40: 364-379.

Metz J.G., P. Roessler, D. Facciotti, C. Leverine, F. Dittrich, M. Lassner, R. Valentine, K. Lardizabal, F. Domergue, A. Yamada, K. Yazawa, V. Knauf, and J. Browse. 2001. Production of polyunsaturated fatty acids by polyketide synthases in both prokaryotes and eukaryotes. Science 293: 290-293.

Meyer, B.J., E. Tsivis, P.R.C. Howe, L. Tapsell, and G.D. Calvert. 1999. Polyunsaturated fatty acid content of foods: Differentiating between long and short chain omega-3 fatty acids. Food Australia 51: 82-95.

Moriya, K. and K. Horikoshi. 1993. A benzene-tolerant bacterium utilizing sulfur compounds isolated from deep sea. J. Ferment. Bioeng. 76: 397-399.

Nakasone, K., A. Ikegami, C. Kato, R. Usami, and K. Horikoshi. 1998. Mechanisms of gene expression controlled by pressure in deep-sea microorganisms. Extremophiles 2: 149-154.

Nakasone, K., A. Ikegami, C. Kato, R. Usami, and K. Horikoshi. 1999. Analysis of *cis*-elements upstream of the pressure-regulated operon in the deep-sea barophilic bacterium *Shewanella violacea* strain DSS12. FEMS Microbiology Letters 176: 351–356.

Nakasone, K., A. Ikegami, H. Kawano, R. Usami, C. Kato, and K. Horikoshi. 2002. Transcriptional regulation under pressure conditions by the RNA polymerase σ^{54} factor with a two-component regulatory system in *Shewanella violacea*. Extremophiles 6: 89-95.

National Research Council. 2002. Marine Biotechnology in the Twenty-First Century. National Academies Press, Washington, D.C.

Newman, D.K. and J.F. Banfield. 2002. Geomicrobiology: How molecular-scale interactions underpin biogeochemical systems. Science 296: 1071-1077.

Nichols, D.S., J.L. Brown, P.D. Nichols, and T.A. McMeekin. 1997. Production of eicosapentaenoic and arachidonic acids by an Antarctic bacterium: response to growth temperature. FEMS Microbiological Letters 152: 349-354.

Nichols, D.S., K. Sanderson, A.D. Buia, J.L. van de Kamp, P.E. Holloway, J.P. Bowman, M. Smith, C.A. Mancuso-Nichols, P.D. Nichols, and T.A. McMeekin. 2002. Bioprospecting and biotechnology in Antarctica, Conference Proceedings—The Antarctic: Past, Present and Future, University of Tasmania, Hobart, Tasmania. pp. 85-105.

Nogi, Y. and C. Kato. 1999. Taxonomic studies of extremely barophilic bacteria isolated from the Mariana Trench, and *Moritella yayanosii* sp. nov., a new barophilic bacterial species. Extremophiles 3: 71-77.

Nogi, Y., N. Masui, and C. Kato. 1998a. *Photobacterium profundum* sp. nov., a new, moderately barophilic bacterial species isolated from a deep-sea sediment. Extremophiles 2: 1-7.

Nogi, Y., C. Kato, and K. Horikoshi. 1998b. Taxonomic studies of deep-sea barophilic *Shewanella* species, and *Shewanella violacea* sp. nov., a new barophilic bacterial species. Arch. Microbiol. 170: 331-338.

Nogi, Y., C. Kato, and K. Horikoshi. 1998c. *Moritella japonica* sp. nov., a novel barophilic bacterium isolated from a Japan Trench sediment. J. Gen. Appl. Microbiol. 44: 289-295.

Nogi, Y., C. Kato, and K. Horikoshi. 2002. *Psychromonas kaikoi* sp. nov., isolation of novel piezophilic bacteria from the deepest cold-seep sediments in the Japan Trench. Int. J. Syst. Evol. Microbiol. 52: 1527-1532.

Nogi, Y., S. Hosoya, C. Kato, and K. Horikoshi. 2004. *Colwellia piezophila* sp. nov., a novel piezophilic *Colwellia* species from deep-sea sediments of the Japan Trench. Int. J. Syst. Evol. Microbiol. 54: 1627-1631.

Ohmae, E., K. Kubota, K. Nakasone, C. Kato, and K. Gekko. 2004. Pressure-dependent activity of dihydrofolate reductase from a deep-sea bacterium *Shewanella violacea* strain DSS12. Chem. Lett. 33: 798-799.

Patching, J.W. and D. Eardly. 1997. Bacterial biomass and activity in the deep waters of the eastern Atlantic—evidence of a barophilic community. Deep-Sea Research 44: 1655-1670.

Pennisi, E. 1997. How a marine bacterium adapts to multiple environments. Science 311: 1697.

Perry, G.J., J.K. Volkman, and R.B. Johns. 1979. Fatty acids of bacterial origin in contemporary marine sediments. Geochimica et Cosmochimica Acta 43: 1715-1725.

Pinturier-Geiss, L., J. Laureillard, C. Riaux-Gobin, J. Fillaux, and A. Saliot. 2001. Lipids and pigments in deep-sea surface sediments and interfacial particles from the Western Crozet Basin. Marine Chemistry 75: 249-328.

Qureshi, M.H., C. Kato, and K. Horikoshi. 1998a. Purification of a novel *ccb*-type quinol oxidase specifically induced in a deep-sea barophilic bacterium, *Shewanella* sp. strain DB-172F. Extremophiles. 2: 93-99.

Qureshi, M.H., C. Kato, and K. Horikoshi. 1998b. Purification and two pressure-regulated *c*-type cytochromes from a deep-sea barophilic bacterium, *Shewanella* sp. strain DB-172F. FEMS Microbiol. Lett. 161: 301-309.

Russell, N.J. and D.S. Nichols. 1999. Polyunsaturated fatty acids in marine bacteria—a dogma rewritten. Microbiology 145: 767-779.

Saitou, N. and M. Nei. 1987. The neighbor-joining method: A new method for reconstructing phylogenetic trees. Mol. Biol. Evol. 4: 406-425.

Smith, G.A., P.D. Nichols, and D.C. White. 1986. Fatty acid composition and microbial activity of benthic marine sediment from McMurdo Sound, Antarctica. FEMS Microbiology Ecology 38: 219-231.

Smith, G.A., P.D. Nichols, and D.C. White. 1989. Triacylglycerol fatty acid and sterol composition of sediment microorganisms from McMurdo Sound, Antarctica. Polar Biology 9: 273-279.

Takeyama, H., D. Takeda, K. Yazawa, A. Yamada, and T. Matsunaga. 1997. Expression of the eicosapentaenoic acid synthesis gene cluster from *Shewanella* sp. in a transgenic marine cyanobacterium, *Synechococcus* sp. Microbiology 143: 2725-2731

Tamegai, H., C. Kato, and K. Hirokoshi. 1998. Pressure-regulated respiratory system in barotolerant bacterium, *Shewanella* sp. strain DSS12. J. Biochem. Mol. Biol. Biophys. 1: 213-220.

Tamegai, H., H. Kawano, A. Ishii, S. Chikuma, K. Nakasone, and C. Kato. 2005. Pressure-regulated biosynthesis of cytochrome *bd* in piezo- and psychrophilic deep-sea bacterium *Shewanella violacea* DSS12. Extremophiles 9: 247-253.

Tanaka, M., A. Ueno, K. Kawasaki, I. Yumoto, S. Ohgiya, T. Hoshino, K. Ishizaki, H. Okuyama, and N. Morita. 1999. Isolation of clustered genes that are notably homologous to the eicosapentaenoic acid biosynthesis gene cluster from the docosahexaenoic acid-producing bacterium *Vibrio marinus* strain MP-1. Biotechnol. Lett. 21: 939-945.

Valentine, R.C. and D.L. Valentine. 2004. *Omega*-3 fatty acids in cellular membranes: A unified concept. Progress in Lipid Research 43: 383-402.

Venkateswaran, K., D.P. Moser, M.E. Dollhopf, D.P. Lies, D.A. Saffarini, B.J. MacGregor, D.B. Ringelberg, D.C. White, M. Nishijima, H. Sano, J. Burghardt, E. Stackebrandt, and K.H. Nealson. 1999. Polyphasic taxonomy of the genus *Shewanella* and description of *Shewanella oneidensis* sp. Nov. Int. J. Syst. Bacteriol. 49: 705-724.

Vezzi, A., S. Campanaro, M. D'Angelo, F. Simonato, N. Vitulo, F.M. Lauro, A. Cestaro, G. Malacrida, B. Simionati, N. Cannata, C. Romualdi, D.H. Bartlett, and G. Valle. 2005. Life at Depth: *Photobacterium profundum* genome sequence and expression analysis. Science 307: 1459-1461.

Wakeham, S.G. and E.A. Canuel. 1988. Organic geochemistry of particulate matter in the eastern tropical North Pacific Ocean: Implications for particle dynamics. J. Mar. Res. 46: 183-213.

Wakeham, S.G. and C. Lee. 1993. Production, transport, and alteration of particulate organic matter in the marine water column. *In*: M.H. Engel and S.A. Macko [eds.]. Organic Geochemistry, Principles and Applications. Plenum Press, New York, USA. pp. 145-169.

Wallis J.G., J.L. Watts, and J. Browse. 2002. Polyunsaturated fatty acid synthesis: What will they think of next? Trends Biochem. Sci. 27: 467-473.

Welch, T.J. and D.H. Bartlett. 1996. Isolation and characterization of the structural gene for OmpL, a pressure-regulated porin-like protein from the deep-sea bacterium *Photobacterium* species strain SS9. Journal of Bacteriology 178: 5027-5031.

Welch, T.J. and D.H. Bartlett. 1998. Identification of a regulatory protein required for pressure-responsive gene expression in the deep-sea bacterium *Photobacterium* species strain SS9. Mol. Microbiol. 27: 977-985.

Xu, Y., Y. Nogi, C. Kato, Z. Liang, H.-J. Rüger, D.D. Kegel, and N. Glansdorff. 2003a. *Moritella profunda* sp. nov. and *Moritella abyssi* sp. nov., two psychropiezophilic organisms isolated from deep Atlantic sediments. Int. J. Syst. Bacteriol. 53: 533-538.

Xu, Y., Y. Nogi, C. Kato, Z. Liang, H.-J. Rüger, D.D. Kegel, and N. Glansdorff. 2003b. *Psychromonas profunda* sp. nov., a psychropiezophilic bacterium from deep Atlantic sediments. Int. J. Syst. Bacteriol. 53: 527-532.

Yamada, M., K. Nakasone, H. Tamegai, C. Kato, R. Usami, and K. Horikoshi. 2000. Pressure regulation of soluble cytochromes c in a deep-sea piezophilic bacterium, *Shewanella violacea*. J. Bacteriol. 182: 2945-2952.

Yanagibayashi, M., Y. Nogi, L. Li, and C. Kato. 1999. Changes in the microbial community in Japan Trench sediment from a depth of 6292 m during cultivation without decompression. FEMS Microbiol. Lett. 170: 271-279.

Yano, J.K. and T.L. Poulos. 2003. New understandings of thermostable and piezostable enzymes. Curr. Opinion Biotechnol. 14: 360-365.

Yayanos A.A. 1986. Evolutional and ecological implications of the propeties of deep-sea barophilic bacteria. Proc. Natl. Acad. U.S.A. 83: 9542-9546.

Yayanos, A.A. 1995. Microbiology to 10500 meters in the deep sea. Ann. Rev. Microbiol. 49: 777-805.

Yayanos, A.A. 2001. Deep-sea piezophilic bacteria. *In*: J.H. Paul [ed.]. Methods in Microbiology. Vol. 30. Academic Press, San Diego, USA. pp. 615-635.

Yayanos, A.A. and A.S. Dietz. 1982. Death of a hadal deep-sea bacterium after decompression. Science 220: 497-498.

Yayanos, A.A., A.S. Diets, and R. Van Boxtel. 1979. Isolation of a deep-sea barophilic bacterium and some of its growth characteristics. Science 205: 808-810.

Yayanos, A.A., A.S. Diets, and R. Van Boxtel. 1981. Obligately barophilic bacterium from the Mariana Trench. Proc. Natl. Acad. Sci. U.S.A. 78: 5212-5215.

Yazawa, K. 1996. Production of eicosapentaenoic acid from marine bacteria. Lipids 31: S297-S300.

CHAPTER 3

Chalcopyrite Bioleaching: The Changing Face of Copper Treatment

Todd J. Harvey

INTRODUCTION

The mining industry has traditionally been relatively slow in adopting new technologies; most companies choosing to be fast-followers instead of first-adopters. Given the inherent risks associated with most mining projects this strategy is easily justified. However, a few forward-thinking companies have paved the way for the commercialization of *'enabling technologies'* and their widespread adoption by others. Companies like Phelps Dodge, with their development of pressure leaching technology (Carter 2003), and BHP Billiton and Codelco, with their Alliance Copper JV biohydrometallurgical BioCop™ process (Dreisinger 2006), are leading the field in development and application of new technologies for the treatment of chalcopyrite concentrates. Others too have recognized the limitations of existing pyrometallurgical techniques and have engaged in process development of alternative treatment processes (Dreisinger 2006). However, if the value chain of copper production is examined (Fig. 3.1) from mine to metal it is evident that the concentrator adds a significant cost both as capital and operating charges. Perhaps, the mining industry needs to look further back the chain for the true value

Vice President, Base Metal Technologies, GeoBiotics, LLC, 12345 W. Alameda Pkwy #310, Lakewood, CO, USA 80228, E-mail: tharvey@geobiotics.com

capture and think about processes that remove the need for concentration. This obviously reduces the number of available process routes to those capable of treating whole ore economically, and likely reduces the selection to a single choice, heap leaching. This further implies that thermophilic bioleaching must be employed, as chalcopyrite does not react sufficiently with other chemical or biological regimes (Hackl et al. 1995, Tshilombo and Dixon 2003).

The following discussion outlines the future potential of thermophilic bioleaching and the process routes available for its use with respect to becoming an alternative to conventional smelting techniques for copper concentrates. In order to determine which technology will rise to the status of *'enabling'* their merits must be evaluated both from a technical and economic standpoint.

The development of bioleaching or biooxidation processes for the treatment of sulfide minerals has followed a path typical of countless mineral processing technologies: although developed centuries ago, commercial application has been achieved only recently. The treatment of refractory sulfide gold ores via biooxidation processes is now widely accepted as a viable commercial process and the heap bioleaching of secondary copper sulfide is also widespread throughout the world but does biohydrometallurgy offer a solution to conventional pyrometallurgical techniques for primary copper sulfides? The recent construction and operation of the Alliance Copper BioCop™ pilot plant would indicate that it does, as would the previous BacTech pilot plant at Penoles (Business Wire 2001). Furthermore, there are several other companies who believe that biological treatment of primary copper sulfide minerals is a cause worth significant R&D investment including GeoBiotics, LLC who are currently busy commercializing their whole ore and concentrate processes.

There are two primary methods of evaluating new technologies: (1) technical achievement and (2) economic merit. This chapter considers these variables while examining the current state of the industry, what hurdles have yet to be crossed and what the future potential may be.

Industry Status

There are currently 10-15 processes being marketed as replacements for traditional smelting technology (Peacey et al. 2003, Dreisinger 2006). The vast majority of these are based on sulfate hydrometallurgy with the balance usually relying on mixed chloride/sulfate systems. Table 3.1 shows a summary of several of the available processes.

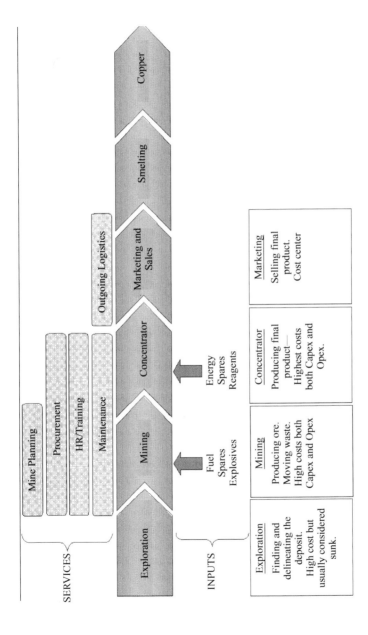

FIG. 3.1 Typical Mine to Metal Value Chain

TABLE 3.1 Copper Mineral Treatment Options

Process	Owner	Status	Temp (°C)	Pressure	Particle Size (μm)	Comments
Activox	WMT	P	90-110	10-12	5-10	Fine grinding combined with oxygen overpressure
Albion	APT	P	85	1	5-10	Atmospheric ferric leaching of fine concentrate
AAC–UBC	UBC/Anglo	P	150	10-12	10-15	Fine grinding combined with oxygen overpressure and surfactants
BacOx	Bactech/Mintech	P	40-55	1	5-10	Bioleach with fine grinding at moderate regime
BioCop	BHPB	C	65-80	1	20-40	Bioleach with moderate grinding at thermophile regime
CESL	CESL	D	140-150	10-12	20-40	Chloride leach producing basic copper sulfate ppt.
Dynatec Process	Dynatec	P	150	10-12	20-40	Low grade coal as additive to pressure leach
Mt. Gordon	Birla	C	90	8	100	Whole ore leach with ferric and oxygen
Platsol	Polymet	P	220-230	30-40	15	Pressure leach using NaCl, also leached precious metals
Sepon	Oxiana	C	80-230	30-40	50-100	Pressure ferric leaching of moderately fine concentrate
Total POX	Many	C	200-230	30-40	20-40	High temperature and pressure oxidation

Table 3.1 Contd.

Table 3.1 Contd.

Intec	Intec	P	60	20-40	Halide leaching with chloride and bromide
HydroCopper	OK	P	90	20-40	Chloride leaching with chloride recycle and H2 ppt
Sumitomo	SMM	P	65-80	20-40	Chlorine leach that also extracts precious metals
GEOCOAT	GeoBiotics	P	55-70	35-100	Thermophilic heap bioleach for concentrates
GEOLEACH	GeoBiotics	P	55-70	35-100	Thermophilic heap bioleach for whole ore

Note: P = Pilot Scale, C = Commercial, D = Development of Commercial Plant

Although the number of available options is large, few of these processes offer low cost, effective alternatives to smelting. Many processes do overcome other smelting limitations such as: lower capital costs than a greenfield smelter, better payment terms for by-product contained metals, ability to handle impurities, reduced SO_2 emissions, no TC/RC (treatment charge/refining charge) risk and provide control over metal marketing.

The adoption of several of these processes has been fueled by site-specific needs. Many have been commercialized by mines that had a unique set of circumstances that required a tailored process route. These processes are generally only suitable for a small number of deposits, for example, the Mt. Gordon Process (Arnold et al. 2003). It is likely that many of the processes in Table 3.1 will never be commercialized for a variety of reasons mainly associated with economics but also due to the risk adverse nature of mine management and the complicated nature of the process (Kuhn 1998). Furthermore, effective marketing of process technologies to mining companies is often a challenge.

Despite these issues, bioleaching has gained a commercial foothold and there are of the order of 12 commercial BIOX® plants operating or being commissioned today for the treatment of refractory gold sulfides. There are a similar number of heap bioleach facilities treating secondary copper sulfide minerals. Recently, BHPB and Codelco assembled a joint venture company to commercialize bioleaching for primary copper sulfide concentrates. The Alliance Copper Company was founded based on BHPB's BioCop™ technology. The goal of Alliance was to commercialize the BioCop™ process through the construction of a demonstration plant with the initial intent of deploying this technology at Codelco's Mansa Mina mine (now called Alejandro Hales). Mansa Mina contains significant concentrations of arsenic in the mineral form enargite and a processing route that could tolerate these elevated levels was required.

BIOCOP™—COMMERCIAL THERMOPHILIC BIOLEACHING

The BioCop™ plant was commissioned in 2003 and is located in Calama Chile. The plant was designed to treat 77,200 tonnes per annum of copper concentrates in a stirred tank system producing 20,000 tpa of copper cathode. The plant operated effectively for in excess of two years and has yielded exceptional results. Table 3.2 shows the plant statistics.

TABLE 3.2 BioCop™ Key Statistics

Statistic	Units
Capacity (tpa)	
Conc	77, 200
Copper	20, 000
Grade	
Cu	33-36%
S(total)	32-35%
Recovery	
Cu	95-97%
S =	97%

The plant was a technical success with many engineering barriers overcome during the course of engineering and operation. BioCop™ has led the way in many areas but most notably; this is the only commercial plant utilizing thermophilic organisms to leach any type of sulfide material. This plant provided some unique challenges to the engineering team including: the need to find a method to keep the dissolved oxygen high in the tanks, dealing with a hot corrosive slurry and feed grade fluctuations. Many of the engineering solutions were borrowed from other extreme leaching environments; the BioCop™ tankage employs a similar structure to that of most pressure leaching vessels, layers of acid proof brick on top of a fiberglass coating. However, the oxygen demand of the system resulted in the need for a cryogenic oxygen plant. This additional service greatly detracts from the processes potential simplicity and adds significantly to the capital and operating cost. Furthermore, the use of oxygen presented some new challenges in that the microorganisms have a low tolerance for elevated oxygen levels. Alliance overcame this with a patented oxygen control strategy. Figures 3.2, 3.3, and 3.4 show the plant.

The flowsheet for the BioCop™ plant consists of a preleach where pregnant leach solution is recycled from the bioleach discharge back to the head of the circuit where it is contacted with fresh concentrate to remove carbonates, easily leached copper and arsenic minerals. The slurry then undergoes solid liquid separation with the solution being directed to solvent extraction and the solid slurry to the main thermophilic bioleach reactors. Figure 3.5 shows the flowsheet schematic.

The BioCop™ process, despite its technical achievement, appears to be an orphan plant. Codelco announced initially that they were proceeding with the application for environmental permitting and then recently announced that it was canceling plans to build the

FIG. 3.2 BioCop™ Pilot Plant

FIG. 3.3 BioCop™ Agitator Drive (Dreisinger 2006)

FIG. 3.4 Specialized Agitator, Air Distribution and Cooling Coils (Dreisinger 2006)

US $320 million plant at Mansa Mina, citing poor process economics (Business News Americas 2006).

The pilot plant was built for US $52 million (excluding SX/EW) which places the cost per annual tonne of copper at approximately US $2,600, near the cost of Phelps Dodge's Bagdad pressure leach plant and edging close to traditional smelting costs (obviously very heavily dependant on the grade of the concentrate). However, it was probably not the capital cost that caused the Mansa Mina project to be placed on hold, more likely it was the operating costs. Codelco needs a process to treat arsenic minerals and their closest available smelter just does not provide that in an environmentally friendly manner. However, they can choose to invest in modern smelting technology as one alternative. BioCop™'s operating costs, although not officially published, have been rumored to be in the range of US $0.80 per pound. These costs are driven in large part by the high energy demand of the agitators and the oxygen plant plus significant acid and lime usage. The BioCop™ design team made a conscious decision not to recycle raffinate from the SX circuit due to the contained organics' toxicity on the microorganism, that and the high acid concentration of it. Commercial operations would have to find a solution to this potential problem.

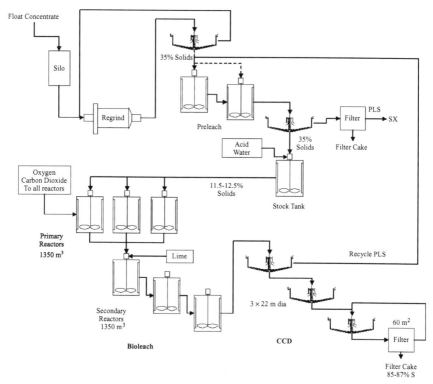

FIG. 3.5 BioCop™ Pilot Plant Flowsheet

The BioCop™ pilot plant is certainly a technical success, unfortunately, Codelco does not view the economics to be completive. It is likely that the next generation of BioCop™ plants would be able incorporate many design improvements that could reduce costs of both capital and operations.

Thermophilic bioleaching of chalcopyrite concentrates, using stirred tank systems, provides an excellent technical alternative to conventional smelting but the economics of the process remain in debate.

THE WAY FORWARD—HEAP BIOLEACHING

The application of heap leaching for oxide copper treatment and now for secondary sulfide copper minerals greatly changed the landscape of the copper industry. Heap leaching provides an economic processing route for low grade ores. There has been a succession of process developments over the last 100 years that have enabled the economic exploitation of the available copper resources. As technologies improved, lower grades became economic and growth of the copper output was sustained (see Fig. 3.6).

Solvent extraction (SX) has likely had the greatest impact as its development facilitated the propagation of heap leaching. Bioleaching has successfully built on these established technologies and expanded the application to the leaching of secondary sulfide ores (chalcocite, covelite) (Jergensen 1999). Copper bioleaching consists of three facets: (1) heap bioleaching of secondary copper sulfide minerals, (2) heap bioleaching of primary copper sulfide minerals and (3) stirred tank bioleaching of primary copper sulfide minerals and concentrates. Only tier 1 has been commercialized with tiers 2 and 3 at various demonstration plant scales. The ability to heap bioleach primary sulfides will be the next major breakthrough in the copper industry.

Pressure oxidation of copper has not been considered as an enabling technology. In the author's opinion, pressure oxidation of primary sulfides is not new; it is primarily a borrowed technology from the refractory gold and zinc industry. More importantly, it does not facilitate the economic processing of some previously untreatable ore type; it provides only an alternative processing route to smelting (it does have advantages but not of the same order of magnitude). This does not detract in any way the technical achievments of those who have commercialized this technology. However, pressure oxidation does not

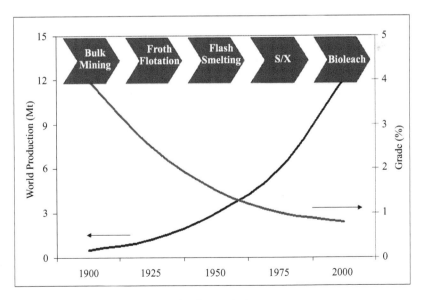

FIG. 3.6 Enabling Technologies in the Copper Industry

impact the grade tonnage relationship to the same extent as those technologies shown in Fig. 3.6 Why is heap bioleaching the next big enabler? The economics of heap leaching are just so much better than any other processing alternative.

It is not always easy to quantify the advantages of one process over another; there are often a myriad of trade-offs that have to be analyzed. However, when comparing similar project scopes it is best to use quantifiable facts as much as possible and in particular the economics of the process. Figure 3.7 shows a small sampling of South American heap leach and concentrator projects comparing their capital cost per annual tonne of ore capacity (compiled from publicly available data).

As shown in Figure 3.7, heap-leaching projects generally require a considerably lower capital investment on a tonnage treated basis than traditional concentrators. The tonne weighted average capital cost (normalizing the project costs based on tonnage treated) for the concentrators examined was US \$38.00 per annual tonne versus US \$17.66 per tonne for the heap leach operations. Several additional factors need to be applied to this data to make a fair comparison; namely, concentrators and heap leach plants do not achieve the same degree of metal recovery and a heap leach produces a final copper metal product.

If consideration is given to the fact that a heap leach process generally would have to treat a larger tonnage of ore to achieve the same annual

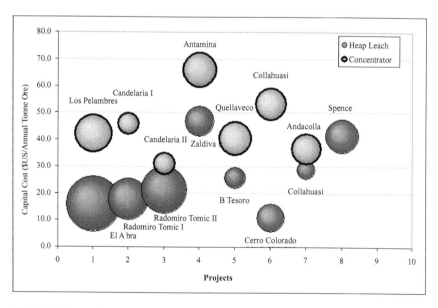

FIG. 3.7 Copper Processing Plant Capital Costs (Area represents annual ore capacity)

metal production as a concentrator, the average cost of the heap leach increases to US $20.49/t (as a result of metal recovery differences). The average heap recovery for this data was 76.4% and the average concentrator recovery was 90.4% with a 98% payment from the smelter.

A more significant difference between these two process options is the form of the final product. Since a heap leach produces a saleable metal product, a comparison cannot be made of the two processes unless the operating costs of both processes are compared based on a similar final product. This requires that the operating cost differences be converted to a capital cost. The most significant cost difference is the treatment charge and refining charge (TC/RC) for producing metal from the concentrate. If the TC/RC operating cost of the smelter is capitalized and added to the concentrator capital expenditure, the capital cost per annual tonne of ore treated increases dramatically. The following example illustrates this impact.

Hypothetical Concentrator Case

Copper concentrate produced: 350.6 kt/yr

Contained copper: 298.9 kt/yr

TC/RC: US $80/t of concentrate plus $0.08/lb contained copper = $0.18/lb of copper produced

Mine Life: 20 years

Discount Rate: 10%

Present value of TC/RC = $1,009,580,853

Additional capital cost per annual tonne of ore = US $34.25

For the projects examined, the weighted average cost per annual tonne due to TCRC (80/8) capitalization was US $37.68/t. This implies that for a given tonnage treated to produce similar quantities of copper metal the capital cost per tonne of ore treated is US $20.81/t. for heap leaching versus US $75.68/t for concentrators.

Heap leaching not only has capital cost benefits but also tends to have operating cost benefits as well. Figure 3.8 shows the producer cash cost curves without credits for 2004. The dotted areas indicate heap leach copper production. Obviously these are very general statements and each process route has to be examined with site specifics in mind. Heap leaching is not currently the best choice if the ore contains significant credit elements such as molybdenum or precious metals.

Some may argue that the comparison of a heap leach to a concentrator project is not valid as the two processes are not treating similar feedstock. Today it is commonly accepted that the heap bioleaching of low grade copper secondary sulfides tends to be the most

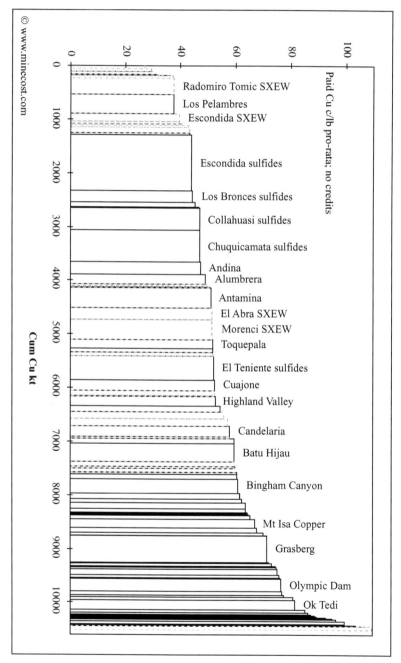

FIG. 3.8 Producer Cost Curves—Without Credits (heap leach in dots)

economic option. However, a conventional acid heap leach is not capable of treating primary sulfide ores effectively. The argument could be made that several companies, Phelps Dodge, BHPB and GeoBiotics, believe that this is not the case. There is no reason to believe that processes once reserved for low grade 'waste' cannot be capable of processing high grade 'ore'. The state of the art has evolved and despite the conventional wisdom of the average process engineer, recovery is not always the variable that matters most—ultimately it is profit which matters and that analysis needs to include not only recovery but capital and operating costs.

Given the huge cost advantage (both capital and operating) that heap bioleaching has over conventional mill, float, and smelt, it seems logical that many companies should be pursuing this avenue with vigor. GeoBiotics is currently commercializing both the GEOCOAT® (for concentrates) and GEOLEACH™ (for whole ore) technologies for copper. Figure 3.9 shows a typical GEOLEACH™ flowsheet.

Most process engineers could easily identify this flowsheet as that typical of conventional heap leaching and one which is commonly employed today. The enabling aspects of the GEOLEACH™ technology are not visible within the flowsheet unit operations but exist in the operation and management of the system as a whole combined with some novel instrumentation.

Heap bioleaching for primary copper sulfides still has many engineering hurdles to overcome (Harvey, 2005). The main issue with heap bioleaching is: can the heat balance be controlled/manipulated to a significant degree to allow the heaps to reach thermophilic temperatures? This is a two-fold problem, first, the ore must contain enough sulfide energy for the temperature rise to be thermodynamically possible and second, the heat generation rate (kinetics) must be fast enough and the heat losses controlled. There will be low grade ores that simply do not have enough fuel to be treated via heap bioleaching. Controlling the heat losses and reaction rates is a fairly simple task on the surface but faced with hundreds of hectares of stacked ore in the high altitudes of the Chilean Andes, this task becomes a significant challenge.

THE CHALLENGES

The application of heap bioleaching to primary copper sulfides still has several operational and engineering hurdles to overcome (Harvey 2005, Domic 2007). The following is a synopsis of the areas that represent the most significant challenges.

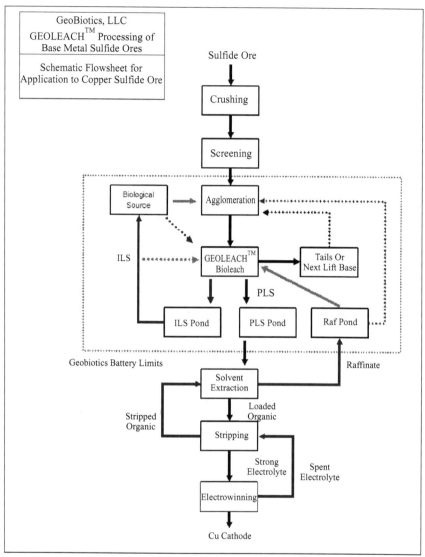

FIG. 3.9 GEOLEACH™ Bioheap Flowsheet for Primary Copper Concentrates

LEACHING RATE AND RECOVERY

The ultimate achievement for bioheap leaching would be to establish a process that can rapidly leach low grade run-of-mine ore primary copper sulfides in a dump leach environment. This is the fastest growing application for bioleaching, unfortunately, the results are typically not very spectacular. Consider the Escondida Sulphide Project being designed to treat up to 120 million tonnes per annum of 0.52% copper whole ore and recover only 36% of the copper (BHPB 2004). Compounding the poor recovery, the leaching rates of these types of heaps are incredibly slow, taking up to a year or more to reach this level of extraction. Despite all of these hurdles, and a capital cost of US $870 million, the Escondida Sulphide Project looks highly economic (for reasons to do with the allocation of waste mining and hauling costs and of course the recent rise in copper prices).

The typical low recovery and long leach times of the whole ore systems is the net result of a large number of variables including: the particle size, the solution and gas distribution, aeration, heap temperatures and ambient conditions. However, the areas that can be controlled need to be better managed. Work needs to be conducted into the heat management of these large heaps. Primary copper sulfide does not leach well except at elevated temperatures (Hackl et al. 1995, Peterson and Dixon 2002) and with the low grades of these heaps, there is very little energy to waste.

Managing the heat produced from sulfide mineral oxidation is a complex process with multiple interconnected processes. Sulfides will not oxidize at any appreciable rate without microbial assistance, so microbes must be present. The pH of the environment must be suitable for microbial growth (pH of 1.0-1.5) (Mousavi et al. 2005). The microbes require oxygen for growth and mineral oxidation. The reaction products need to be removed effectively to prevent product inhibition. The optimization of these parameters is governed by the delivery of solution, air and both stream will remove heat from the system. Thus a balancing act exists between the optimization of the microbial conditions to enable energy production and the removal of that energy through the addition of liquid and gas.

ACID BALANCE

The biooxidation of sulfide ores may or may not result in net acid production, ore mineralogy plays a key role. The bioleaching of chalcopyrite is a net acid consumer while pyrite is a net acid producer. Also consider that in a whole ore system the sulfide may account for one percent or less of the total mineral composition. Thus, it is the gangue minerals that typically cause the high acid demand. Many of the minerals associated with copper ore bodies consume acid, even those not directly containing carbonate may dissolve to significant extents, such as feldspars (Jansen and Taylor 2003).

The acid demand of the heap leads to further complications related to acid delivery. In a typical copper heap leaching system, the raffinate from the solvent extraction circuit returns to the heap. This high acid solution (<1 pH) is delivered to the top of the heap via the irrigation system but in the process of acid adjusting the heap it also works effectively to sterilize the biological population or at the very least to place them in an acid environment beyond their optimum. The ability to effectively acid stabilize the heap is a real difficulty. The complete heap must be at a pH level that is conducive to biological activity, large pH fronts move through the heap as a result of the acid being applied only from the heap surface. For this reason, concentrated acid is added to the agglomeration process (typically 50-75% of the total acid demand) but this too is a very effective method of sterilizing the biological population present in the ore or recycled solutions.

Methods need to be developed to manage acid consumption and to understand the relationships between it and particle size, leaching rates, biological impacts, etc.

OXYGEN UTILIZATION

Oxygen utilization has always been an issue with stirred tank processes. The difficulty arises due to the need to stir the slurry with minimal shear so as not to harm the bacteria but with enough force to distribute air effectively. Most recently, higher temperature stirred tank leaching processes (>60°C) for copper have found that the oxygen utilization when using air has decreased to such an extent that pure oxygen is now required. The problem of diffusion has been compounded by lower oxygen solubility at higher temperature. The engineering focus is on developing better agitation to increase oxygen utilization and to potentially recycle oxygen off-gas to increase the overall utilization.

Oxygen utilization is not a major issue in heap systems because generally low pressure air is employed as the oxygen source. This is a cheap source of oxygen and delivered at very low pressures. The problem with heaps is the method of air delivery and the quantities required when high altitude implications are factored in. Higher aeration for a given oxygen delivery means potentially more cooling of the system.

HEAP DESIGN

There are several aspects of heap design that need careful examination. These can be broken down into two categories: physical characteristics and mechanical systems. The physical characteristics include things such as crush size, size distribution and heap height. The mechanical systems include aeration, irrigation, drainage, agglomeration and stacking.

The main goal of a bioleach design is provide a suitable environment for the microbes to thrive. This infers that the heap must be at a uniform pH and have adequate oxygen available. Heaps with pore permeability simply will not be able to provide this. To achieve and maintain high heap permeability requires that the agglomerants remain stable throughout the leach cycle.

The delivery of oxygen to the heap is a function of the permeability and the aeration design. The delivery method that is typically employed, is to use a perforated drain pipe connected to a fan. In the past, very little effort went into understanding these delivery systems and they subsequently did not work well. The focus has shifted to careful air system design to ensure that air can be delivered in measured amounts to the areas of the heap that require it. There are still potential problems associated with the precipitation of soluble salts and other materials at or near to the air pipes potentially plugging the outlets. The air system, although relatively simple in its delivery system (low pressure fans and perforated pipes), must be fairly complex in its capabilities to be effective. It must be able to deliver oxygen to areas of the heap that require it and it must be part of the heat control system.

The irrigation system also requires careful design. In most heap leach systems emitters are now the preferred option. Their main advantage is that they have lower ambient evaporation losses than conventional sprinklers, they may also reduce the destruction of agglomerants due to low impact forces. Their disadvantage is that they apply solution to a single point and many closely spaced points of application are required. The spacing of the irrigation distribution points needs to be closely examined; there is a trade-off between cost and efficiency. The use of conventional rotating sprinklers should be reexamined. These provide a uniform solution distribution pattern, large droplets to reduce evaporation losses and a visual check of their performance (a plugged emitter is not easily located). Uniform irrigation is necessary for a uniform heap environment.

Other areas of heap design that require extensive investigation are drainage and lining systems. Early heap leaching systems did not incorporate any drainage systems and the results could be spectacular 'blowouts' (the complete failure of the side of a heap due a perched water table within the heap). Under heap drainage piping helps reduce hydraulic head, reduce solution holdup (inventory), and allows control of the PLS grades through the segregation of streams. Careful design of these systems is necessary to ensure adequate drainage rates and failure due to crushing and scale (Kampfa et al. 2002).

Significant work has been conducted on heap leach liner systems and base preparation (Thiel and Smith 2004). A variety of liner systems are available and each has its merits. The goal of a good design is to ensure no leakage over the life of the operation.

ENERGY CONSUMPTION

The last engineering hurdle to overcome is the power consumption of stirred tank processes. These processes have high power requirements, potentially as much as pressure leaching, due to the large low shear agitation required, the potential need for oxygen, long retention times and cooling requirements. Power consumption issues are being addressed through the design of more efficient agitators and higher temperature leaching (less cooling).

Power consumption in metal extraction portion of the heap bioleach is typically small as low pressure air is the primary oxidant. However, the recovery of metal using solvent extraction and electrowinning is going to result in significantly higher power usage. The electrowinning costs are a function of energy prices, process efficiency and electrochemistry. Unfortunately, there is little that can be done in a sulfate system to manipulate these variables beyond their current levels. This is one major reason why some processes are examining the production of monovalent copper, which requires roughly half the energy to produce copper metal as typical divalent sulfates. The production of monovalent copper (Cu^+) is usually achieved with the addition of chlorides to the leach system. Chlorides present a whole new level of challenges is beyond the scope of this chapter.

FUTURE PERSPECTIVES

There is little doubt that bioleaching of primary copper ores will become a commercial technology, the economic driving force is too great. The development of heap leaching as a modern processing alternative has taken 30 years. Similar development periods were required for stirred tank biooxidation. Both stirred tank and heap-based systems show significant promise for future application but more work is required. The commercialization of these processes is going to require further innovation from the technology developers in combination with progressive mining companies. Despite many mining companies claiming value technology, their adoption rates are the lowest among all industries and the number of companies actively developing their own technology is decreasing every year. Furthermore, the number of graduates entering this field has been under a steady decline. Is it any wonder that it took the mining industry 30 years to adopt new technologies?

Copper bioleaching, as a replacement or an alternative to smelting, provides many advantages including the long overlooked environmental impact of smelting, cost reductions and maximized resource utilization. The world may be viewed as flat in today's technologically advanced environment but it still takes three weeks to deliver concentrate to China from Chile and it still consumes energy doing it. Delivering concentrate to antiquated smelters with little or no pollution abatement still pollutes the globe we live on. Reducing processing capital and operating costs reduces the cut-off grade and thus increases the reserve base and thus maximizes the world's resource base.

Some companies see the advantages that bioleaching can bring and have either embraced the technology or are working to commercialize the next phase. Companies such as BHPB, Tech-Cominco, GeoBiotics, and Codelco among others, are working to bring these technologies to market. Both GeoBiotics and BHPB recognize that heap bioleaching for the treatment of primary copper ores is the equivalent of the 'holy grail' and both have significant developments underway on this initiative. BHPB with their Escondida Sulphide Project and GeoBiotics with their heap based GEOCOAT® and GEOLEACH™ processes both represent major milestones on this path. Stirred tank biooxidation has passed the technical hurdles on the way to commercial development but has yet to show unbiased economic proof of its viability. It is likely to be a successful technology given the right application.

REFERENCES

Arnold, S.N., J.R. Glen and G. Richmond. 2003. Improving the Flexibility of the Mount Gordon Ferric Leaching Process, Eight Mill Operators Conference, AusIMM.

BHPB Website New Release. 2004. http://www.bhpbilliton.com/bb/investors Media/news/2004/bhp BillitonApprovesEscondidaSulphideLeachCopper Project.jsp

Business News Americas, May 5. 2006. Codelco drops bio-leach project with BHP Billiton – Chile. http://www.bnamericas.com/news/mining/Codelco_drops_bio-leach_project_with_BHP_Billiton

Business Wire, Dec 6. 2001. BacTech Produces LME "A" Grade Commercial-Sized Copper Cathodes Using Proprietary Bioleaching Technology. http://www.allbusiness.com/company-activities-management/company-structures-ownership/6192958-1.html

Carter, R., Sept. 2003. Pressure Leach Plant Shows Potential. Engineering and Mining Journal. http://findarticles.com/p/articles/mi_qa5382/is_200309/ai_n21342740/

Canchoa, L., M.L. Blázquez, A. Ballestera, F. Gonzáleza and J.A. Muñoz. 2007. Bioleaching of a chalcopyrite concentrate with moderate thermophilic microorganisms in a continuous reactor system. Hydrometallurgy, Vol. 87, 3-4.

Domic, E. 2007. A review of the developments and current status of copper bioleaching operations. *In:* D.E. Rawlings and D.B Johnson [eds.]. Chile: 25 Years of Successful Commercial Implications, Biomining. Springer, New York p. 81.

Dreisinger, D. 2006. New Developments in Cu and Ni Hydrometallurgy. Presentation to JOGMEC, Japan.

Hackl, R.P., D. Dreisinger, E. Peters and J.A. King. 1995. Passivation of chalcopyrite during oxidative leaching in sulfate media. Hydrometallurgy 39(1): 25-48(24).

Harvey, J.T., March 2005. It's a Bug's Life, World Mining Equipment. http://technology.infomine.com/biometmine/biopapers/biomet_issues.pdf

Jansen, M. and A. Taylor. 2003. Overview of gangue mineralogy issues in oxide copper heap leaching, Conference Proceedings, ALTA Conference.

Jergensen II, G.V. 1999. Copper Leaching, Solvent Extraction, and Electrowinning. SME, Littleton, CO.

Kampfa, S.K., M. Salazarb and S.W. Tyler. 2002. Preliminary investigations of effluent drainage from mining heap leach facilities. Vadose Zone Journal 1: 186-196.

Kuhn, M.C. 1998. Managing innovation in the minerals industry. Society for Mining, Metallurgy and Exploration Inc, Littleton, CO.

Mousavia, S.M., S. Yaghmaeia, M. Vossoughia, A. Jafarib and S.A. Hoseinia. 2005. Comparison of bioleaching ability of two native mesophilic and thermophilic bacteria on copper recovery from chalcopyrite concentrate in an airlift bioreactor. Hydrometallurgy 80(1-2): 139-144.

Peacey, J., X.J. Guo, and E. Robles. 2003. Copper Hydrometallurgy—Current Status, Preliminary Economics, Future Direction and Positioning versus Smelting, Copper VI: Hydrometallurgy of Copper, CIMM. http://www.amazon.com/Copper-VI-Hydrometallurgy-Modeling-Extraction/dp/1894475410#

Peterson, J. and D.G. Dixon. 2002. Thermophilic heap leaching of a chalcopyrite concentrate. Minerals Engineering 15(11): 777-785.

Thiel, R. and M.E. Smith. 2004. State of the Practice Review of Heap Leach Pad Design Issues. Geotextiles and Geomembranes 22(6): 555-568.

Tshilombo, A.F. and D.G. Dixon. 2003. Kinetic study of chalcopyrite passivation during electrochemical and chemical leaching. Sixth International Symposium on Electrochemistry in Mineral and Metal Processing, Paris.

CHAPTER 4

Mineral Biotechnology of Sulphides

Ata Akcil[1*] and Haci Deveci[2]

INTRODUCTION

Microorganisms have been involved in the natural transformation of the Earth's surface, such as the biological oxidation of sulphide ore deposits when sulphide minerals are in contact with air and water (Ehrlich 1996, 2004, Johnson 2006). In the late 1940s the recognition of the role of microorganisms in the formation of acid mine waters by the degradation of sulphide minerals (Colmer and Hinkle 1947) eventually led to the introduction of biotechnological principles into hydrometallurgy and to the development of mineral biotechnology known as biohydrometallurgy. It is, therefore, based on the exploitation of the economic potential of the interactions between microorganisms and metals/minerals in aqueous environments. In this context, mineral biotechnology embraces a diverse range of disciplines, mainly hydrometallurgy, geomicrobiology, microbial ecology and microbial biochemistry (Cupp 1985, Rossi 1990).

Bioprocesses using microorganisms have become a well-established technology over the years and have found applications for the extraction

[1]BIOMIN Group, Mineral Processing Division, Department of Mining Engineering, Suleyman Demirel University, Isparta TR 32260, Turkey, E-mail: ata@mmf.sdu.edu.tr
[2]Mineral Processing Division, Department of Mining Engineering, Karadeniz Technical University, Trabzon TR 61080, Turkey, E-mail: hdeveci@ktu.edu.tr
*Corresponding author

of copper and uranium from low grade ores or old tailings at industrial scale in heap, dump and in-situ operations (Campell et al. 1985, Tuovinen and Bhatti 1999, Brierley and Brierley 2001, Olson et al. 2003, Akcil 2004, Watling 2006). Likewise, biooxidative pre-treatment of refractory gold ores and concentrates is now commercially practised as an alternative to the traditional processes such as oxidative roasting (Miller 1997, van Aswegen and Marais 1999, Olson et al. 2003, Rawlings et al. 2003). In recent years, bioleaching technology has been extended to extract other metals including zinc, nickel, cobalt and manganese from sulphide ores and concentrates (Torma and Bosecker 1982, Torma 1987, Brigss and Millard 1997, Miller et al. 1997, Steemson et al. 1997). Nonetheless, bioprocesses are proven to utilize for the desulphurization of coal prior to combustion by removing pyrite and ash contents (Bos and Kuenen 1990). Incorporation of biotechnology into these fields is on the basis of its potential to offer a propitious solution to the challenges facing the metallurgical industries in that it enables the treatment of low grade and refractory ores with reduced environmental impact and lower fixed capital costs.

Although a variety of microorganisms are utilized within the aforementioned bioprocesses, the most prominent types, of significance to mineral biotechnology of sulphides, are acidophilic bacteria and archaea capable of chemolithotrophic growth on inorganic substrates, e.g. ferrous iron and reduced sulphur compounds to obtain their energy (Johnson 1998, Robbins 2000, Rawlings 2002). In general, these species are acidophilic, aerobic autotrophs that require oxygen and cabon dioxide in an acidic environment (pH 0.5-5.0) to sustain their growth. One of the most significant characteristics of these microorganisms, and the main reason for their exploitation in the mining and minerals industry, is their ability to adapt to tolerate high concentrations of metal ions.

In biohydrometallurgical processes, the function of bacteria involves either the direct dissolution of metal values from ores or concentrates, known as bioleaching or the oxidation of mineral constituents to release the encapsulated valuable metal from mineral matrix, known as biooxidation. The former can be exemplified by the bioleaching of a chalcopyrite flotation concentrate while the latter by the biooxidation of refractory gold-bearing mineral sulphide concentrates.

ACIDOPHILIC MICROORGANISMS USED IN MINERAL BIOTECHNOLOGY OF SULPHIDES

Chemolithotrophic acidophiles with the capability of oxidizing iron and reduced sulphur compounds as well as surviving in an environment of high metal concentrations and acidity (Brierley 1978, Brierley 1982, Johnson 2001, Akcil and Ciftci 2006) play a vital role in the bioleaching of sulphide minerals. These microorganisms can grow autotrophically and/or mixotropically by obtaining their cellular carbon from atmospheric carbon dioxide and/or at least a proportion of it from organic compounds. Although autotrophs are the most desirable microorganisms for bioleaching processes, a number of mixotrophic and/or heterotrophic species such as fungi, yeast, algae and protozoa prevail in association with sulphide environments to a less extent (Lundgren and Silver 1980, Ehrlich 1991, Johnson 1998).

The existence of a certain type of bacteria in an environment is intimately associated with physical and chemical characteristics of that environment. Accordingly, classification and description of the bacteria utilized in the biohydrometallurgical processes are based upon the characteristics of the leaching environment, e.g. temperature, acidity and presence or absence of air. Classical description of particular microorganisms shows very specific differences between microorganisms, e.g. shape and size, substrate requirements. In practice, the acidophilic chemolithotrops of interest to mineral biotechnology are categorized into mesophiles, moderate thermophiles and extreme thermophiles according to the range of temperatures in which their optimum growth is sustained (Table 4.1).

Mesophiles

Although it was only recently realized that a wide variety of microorganisms partake in the degradation of sulphide minerals (Norris 1997, Rawlings 1997, Johnson 2001), the bacteria belonging to the genus *Acidithiobacillus* (formerly *Thiobacillus* (Kelly and Wood 2000)) and *Leptospirillum* are well known and the most extensively investigated microorganisms of commercial interest in this group of bacteria. *Acidithiobacillus ferrooxidans* is a chemolithotrophic, gram-negative, rod-shaped microorganism with a cell size in range of 0.3 to 0.8 μm in diameter and of 0.9 to 2 μm in length (Brierley 1982, Norris 1990, Rossi 1990, Barrett et al. 1993). The optimum growth temperature of these microorganisms lies in the range of 20 to 40°C depending, to some extent, upon the particular strain and other growth conditions such as acidity of environment (Harrison 1982, Norris 1990). The bacteria can grow in acidic environments ranging from pH 1 to 5 with the optimum growth at pH 1.8-2.5 (Torma 1987, Rossi 1990, Leduc and Ferroni 1994). *Acidithiobacillus ferrooxidans* derives its energy required for growth and other metabolic functions from the oxidation of ferrous iron and a variety of inorganic sulphur compounds including elemental sulphur,

TABLE 4.1 Acidophilic bacteria and archaea of considerable interest for mineral biotechnology of sulphides

Group	Culture	Characteristics
Mesophiles (20-40°C)	*Acidithiobacillus ferrooxidans*	autotroph, iron/sulphur oxidizer
	Leptospirillum ferrooxidans	autotroph, iron oxidizer
	Acidithiobacillus thiooxidans	autotroph, sulphur oxidizer
	Ferroplasma acidiphilum	autotroph, iron oxidizer (archaea)
Moderate Thermophiles (40-55°C)	*Sulfobacillus acidophilus*	autotroph/mixotroph, iron/sulphur oxidizer
	S. termosulfidooxidans	autotroph/mixotroph, iron/sulphur oxidizer
	Acidithiobacillus caldus	autotroph/mixotroph, sulphur oxidizer
	Acidimicrobium ferrooxidans	autotroph/mixotroph, iron oxidizer
Extreme Thermophiles (Archaea) (55-85°C)	*Sulfolobus*-like archaea	
	Sulfolobus metallicus	autotroph, iron/sulphur oxidizer
	Acidianus brierleyi	autotroph/mixotroph, iron/sulphur oxidizer

thiosulphate, sulphite and polythionates (Brierley 1982, Rossi 1990). It secures all of its cellular carbon by fixing atmospheric carbon dioxide. Although *At. ferrooxidans* is generally considered to be strictly aerobic, it has been demonstrated that some strains, at least, of the species are capable of anaerobic growth on elemental sulphur in extremely acidic environments causing dissimilatory reduction of ferric iron by using it as a terminal electron acceptor (Pronk and Johnson 1992).

At. thiooxidans exhibits almost the same morphological characteristics as *At. ferrooxidans* and similar growth conditions. *At. thiooxidans* can grow in environments of extreme acidity as low as pH 1. The principal difference between these two species lies in that *At. thiooxidans* exclusively thrives on elemental sulphur and some soluble sulphur compounds but is unable to oxidize ferrous iron. In pure culture form, these microorganisms are reported to be ineffective for degradation of pyrite and chalcopyrite (Norris 1990, Akcil and Ciftci 2003a,b, Akcil et al. 2007).

Another important mesophile is *Leptospirillum ferrooxidans* which occurs as curved (vibroidal) rods (Harrison 1984, Rawlings et al. 2003). The main feature of *L. ferrooxidans* is its inability to utilize sulphur compounds as energy source. *L. ferrooxidans* grows solely on ferrous iron albeit more slowly than *At. ferrooxidans* (Norris 1990). *Leptospirillum*-like microorganisms were however shown to degrade pyrite more extensively even at pH levels inhibitory to *At. ferrooxidans* due to its low sensitivity to end product inhibition by ferric iron and increasing acidity (Norris 1983, Norris et al. 1988). *At. ferrooxidans, At. thiooxidans* and *L. ferrooxidans* as well as acidophilic archaea (e.g. *Ferroplasma* spp.) and heterotrophs (e.g. *Acidiphilium* spp.) can coexist in a leaching environment (Johnson and Roberto 1997, Golyshina et al. 2000, Hallberg and Johnson 2003) and mixed cultures of these microorganisms can degrade sulphide minerals such as pyrite and chalcopyrite more effectively than single culture (Fig. 4.1) (Norris 1983, Akcil and Ciftci 2003a,b, Qui et al. 2005, Akcil et al. 2007). However, recent investigations into the characteristics of the mixed populations used in the commercial operations demonstrated that *Leptospirillum*-type bacteria dominated the microbial population with *At. thiooxidans* to lesser extent whereas, if detectable, *At. ferrooxidans* was present in small numbers (Sand et al. 1993, Dew et al. 1997, Battaglia-Brunett et al. 1998). This was attributed to the operating conditions of temperature (~40°C), high acidity (~pH 1.6) and ferric iron concentration in the bioreactors (Rawlings et al. 1999a, b).

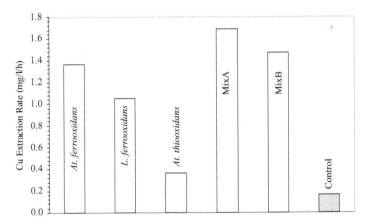

FIG. 4.1 Extraction rate of copper from a pyritic copper concentrate using pure and mixed cultures of mesophilic bacteria (MixA: *At. ferrooxidans, At. thiooxidans* and *L. ferrooxidans*; MixB: *At. thiooxidans* and *L. ferrooxidans* (Akcil et al. 2007)).

Moderate Thermophiles

The moderately thermophilic bacteria have been initially isolated from a variety of sites; the acidic thermal springs and from the 'hot spots' of copper leach dumps/heaps, sulphide ore deposits, acid mine drainage (AMD) and self heating coal spoils where the temperature often exceeds ~45°C (Le Roux et al. 1977, Brierley 1978, Marsh and Norris 1983, Ghauri and Johnson 1991, Bond et al. 2000, Robertson et al. 2002, Hallberg and Johnson 2003, Johnson et al. 2003).

These microorganisms are heterogeneous in genotype and some strains of moderate thermophiles have yet to be taxonomically identified and hence are known as the code names i.e. strain YTF1. The genera *Sulfobacillus* and *Acidimicrobium* have been characterized including *Sulfobacillus thermosulfidooxidans* (strains TH1 and BC1), *Sulfobacillus acidophilus* (strains ALV and THWX) and *Acidimicrobium ferrooxidans* (strains TH3 and ICP) (Norris et al. 1986, Norris et al. 1996, Clark and Norris 1996). Moderately thermophilic species exhibit a wide diversity of morphology and physiology ranging from obligately chemolithotrophic *Acidithiobacillus*-like strains to facultatively heterotrophic strains. These microorganisms can only flourish autotrophically with supplementary organic compounds such as yeast extract (i.e. mixotrophic growth) and/or high concentrations of carbon dioxide (Norris 1997). Many of the isolates were shown capable of heterotrophic growth on yeast extract and exhibit varying degrees of mixotrophic growth on ferrous iron (Marsh and Norris 1983, Ghauri and Johnson 1991). The strains of *S. acidophilus* is less 'heterotrophic' in character than the strains of *Am. ferrooxidans* (Ghauri and Johnson 1991, Clark and Norris 1996).

Autotrophic growth for some strains on ferrous iron can be maintained if provided with a source of reduced sulphur compounds, i.e. tetrathionate (Norris 1990). Norris et al. (1996) and Clark and Norris (1996) provided the visual evidence that the strains of *Sulfobacillus* and *Acidimicrobium* displayed morphological variations in response to growth substrate/conditions. For instance, iron oxidizing cells of *S. thermosulfidooxidans* was observed to increase significantly in size in the presence of yeast extract and when growing heterotrophically on the yeast extract as a sole substrate. Similarly, these authors also noted that the cells of *Sulfobacillus*-like species appeared to be elongated or to form chains during the autotrophic growth on sulphide minerals as the acidity increased.

Moderate thermophiles are, in general, rod shaped with a gram positive character. The cell rods vary in size depending on a particular strain (Brierley 1978, Norris et al. 1996). Their optimum growth occurs in a temperature range of 40-55°C and in a pH range of 1.3-3 dependent upon the individual strain (Barrett et al. 1993, Norris 1997, Johnson et al. 2001, Deveci et al. 2003a). Ewart (1990) reported the optimum conditions of 40-45°C and pH 1.6 using a mixed culture for the dissolution of iron from a gold bearing pyrite concentrate. The bioleaching activity of moderate thermophiles was also noted to be adversely affected at <pH 1.6 (Deveci et al. 2004a), but, these microorganisms can be readily adapted to operate at low pHs (Deveci et al. 2004b). The BacTech® process using moderate thermophiles was designed for the oxidation of pyrite/arsenopyrite concentrates to operate at a temperature of 45°C and pH 1.3-1.5 (Miller 1997). Moderate thermophiles generally exhibit higher tolerance to acidity generated by more extensive dissolution of pyrite compared with mesophiles (Norris et al. 1986). However, Norris et al. (1986) noted that strain ALV was more sensitive to acidity with a significant decrease in the oxidizing ability of the strain below pH 1.7. Apart from pyrite ores/concentrates, moderate thermophiles have been reported to oxidize a variety of sulphide ores/concentrates containing minerals such as chalcopyrite and sphalerite (Norris et al. 1986, Norris and Owen 1993, Gomez et al. 1999a, Deveci et al. 2004a,b, Dopson and Lindstrom 2004). It was also pointed out that moderate thermophiles can oxidize ferrous iron, elemental sulphur and mineral sulphides more efficiently in the presence of yeast extract (0.01-0.02% w/v) or of carbon dioxide enriched air (~1% v/v) (Norris 1989a, Ghauri and Johnson 1991, Clark and Norris 1996, Witne and Phillips 2001).

The recently described moderately thermophilic bacterium, *Acidithiobacillus caldus* which grows optimally at 45°C, was found to play an important role in many industrial bioleaching operations (Rawlings 1997). *At. caldus* is in many respects similar to the mesophilic *At. thiooxidans* since it can only oxidize inorganic sulphur compounds including elemental sulphur, tetrathionate, sulphite and thiosulphate (Hallberg et al. 1996). Several isolates of *At. caldus* appear to require supplementary yeast extract for growth in pure culture form and can tolerate high acidity as low as pH 1 (Hallberg et al. 1996). *At. caldus* has been reported to be the most common sulphur oxidizer encountered in continuous biooxidation reactors operating at temperatures of 40-50°C (Rawlings et al. 1999a,b). Dopson and Lindström (1999) revealed that mixed cultures of *At. caldus* and *S. thermosulfidooxidans* degraded arsenopyrite more effectively than the pure cultures in the presence or absence of yeast extract.

The kinetics of leaching reactions are generally enhanced with increased temperature and high temperature environments develop in most dump/heap/tank leaching operations due to the exothermic nature of oxidation of sulphide minerals. The indications are that commercial application of the moderately thermophilic bacteria for bioleaching processes has a great potential for improving rates of metal extraction from sulphide minerals. Accordingly, in recent years, considerable interest has been shown to develop bioleaching processes using moderate thermophiles for the extraction of base metals such as copper, nickel and cobalt from ores/concentrates both in the laboratory and by successful pilot scale demonstrations (van Staden 1998, Miller et al. 1999).

Extreme Thermophiles

Extreme thermophiles with their ability to operate even at higher temperatures and acidity than the moderate thermophiles are potentially applicable for bioleaching processes (Norris et al. 2000, Rawlings et al. 2003). This group of microorganisms is capable of thriving at temperatures exceeding 55°C (optimum at ~70°C) and at pH values in the range 1-5.9 (optimum at ~pH 2) (Brierley 1978, Kelly 1988). Some extreme thermophiles have been reported to inhabit in hot springs even at temperatures approaching the boiling point of water (Norris et al. 2000).

The genus, *Sulfolobus* with the pertinent species of *Sulfolobus acidocaldarius, Sulfolobus solfatarious, Metaosphaera sedula* and *Sulfolobus brierleyi* (which was subsequently categorized as *Acidanus brierleyi* (Segerer et al. 1986)) can be considered as important in biohydrometallurgy (Norris et al. 2000). However, Norris (1997) cautioned that several of the species of *Sulfolobus*-like microorganisms do not oxidize sulphur and these are not closely related to this genus for classification. The ability of *Sulfolobus* BC to oxidize sulphide minerals has been extensively demonstrated (Brierley 1978, Barr et al. 1992, Norris and Owen 1993, Gericke and Pinches 1999, Witne and Phillips 2001, Deveci et al. 2004a). This strain is reported to be an isolate of *Sulfolobus metallicus,* which inhabits hot springs and coal spoil heaps (Norris 1997).

Extreme thermophiles exhibit a very distinctive morphological characteristic with a different cell wall structure compared with *At. ferrooxidans.* The absence of peptidoglycan in the cell wall places them in the group called Archaeabacteria (Brierley 1982). These microorganisms are gram-negative (albeit not firmly established), immotile and possess no flagella (Barrett et al. 1993). The cells are spherical but irregular (coccoid) in shape with a diameter of 1-1.5 mm (Brierley 1978, Segerer et al. 1986). *Sulfolobus* species have been observed to possess clusters of pili on the cell surface which aid the attachment of the cell to mineral surfaces (Brierley 1978).

These species can grow in autotrophic, mixotrophic or heterotrophic environments deriving energy for growth and other metabolic functions from the oxidation of elemental sulphur, ferrous iron or sulphide minerals and under autotrophic conditions carbon dioxide serves as a source of cellular carbon (Norris 1997). The growth rate of these microorganisms may be enhanced under mixotrophic conditions in the presence of 0.01-0.02% yeast extract or other organic substances (Brierley 1978, Barrett et al. 1993).

Extreme thermophiles are able to readily solubilize minerals known to be resistant to acid attack such as chalcopyrite and molybdenite owing to their ability to operate at high temperatures (65-85°C) (Norris et al. 2000). Konishi et al. (1999) observed that the extent and kinetics of copper dissolution from a chalcopyrite concentrate using *Acidianus brierleyi* were superior to those using *At. ferrooxidans*. Similar findings have been reported by other authors (Le Roux and Wakerley, 1988, Norris and Owen 1993, Dew et al. 1999, Witne and Phillips 2001). Based on the technical and economic analysis, Lawrence and Marchant (1988) concluded that extreme thermophiles are a potentially superior alternative to the mesophiles for the biooxidation of refractory gold ores/ concentrates. Reduced cooling requirements and kinetic improvements reflect significant savings for capital and operating costs. However, some findings (Gericke and Pinches 1999, Nemati et al. 2000) suggest that these microorganisms are very sensitive to solids concentration (probably due to mechanical damage to the cells) and only superior at low pulp densities. Presumably, the lack of a rigid cell wall and other unfavourable growth conditions such as limited availability of oxygen and carbon dioxide at high temperatures adversely affect its performance. In recent years though, several commercial processes using extreme thermophiles have been developed for the extraction of base metals (Cu, Ni, Co, Zn) and successfully operated at pilot scale with the reports of imminent commercial realization (Miller et al. 1999, Gilbertson 2000, Clark et al. 2006).

MICROBIOLOGICAL ASPECTS OF METAL/MINERAL OXIDATION

Microorganisms require energy to sustain their growth. The energy required for their metabolic functions is obtained from either the metabolism of organic compounds or the oxidation of inorganic compounds and/or elements depending on the type of microorganisms. In this respect, ferrous iron, reduced sulphur compounds or mineral sulphides can serve as substrate (source of energy) for acidophilic bacteria.

The processes involved in the transfer of electrons from substrate into bacterial cells and the assimilation of energy within the cell are well documented in the literature (Ingledew 1982, Norris 1989b, Rossi 1990, Ehrlich 1996). The oxidation of any substrate occurs through a series of reactions (biochemical pathways) and electrons liberated during the oxidation are transferred to a terminal electron acceptor via the electron transport system of the microorganism (Rossi 1990). The energy generated from the oxidation of a substrate through the reactions of biochemical pathways is utilized for assembling the organic compounds (i.e. ATP) and other physiological functions (Ehrlich 1996).

The transmission of electrons to the terminal electron acceptor occurs via a complex set of redox reactions between donor and acceptor compounds within the electron transport chain (Jordan 1993). A potential difference across the plasma membrane is necessary to drive the series of redox reactions. This potential difference is provided by the pH gradient present between the two different environments inside and outside the cell. The internal pH of a cell approaches neutrality (~ pH 6.5) whilst the external environment is more acidic (< pH 2.5 in most cases). The internal pH is maintained by the removal of protons on reduction together with oxygen.

OXIDATION OF FERROUS IRON

The process of oxidation of ferrous iron by *At. ferrooxidans* is well documented in the literature (Ingledew 1982, 1986, Norris 1989b, 1990, Ehrlich 1996, 2004). *At. ferrooxidans* can utilize ferrous iron as an electron donor coupling the energy from the oxidation process to the synthesis of ATP for build-up of cellular material. The free energy yield of ferrous iron oxidation is estimated to be approximately 6.5 kcal/mol, which is scarcely sufficient to form 1 mole of ATP, which requires ~ 7 kcal/mol (Leduc and Ferroni 1994). Assuming that the assimilation of 1 mole carbon dioxide entails 120 kcal (at 100% efficiency) large quantities of iron have to be oxidized by bacteria to fulfil the energy requirement for overall metabolic functions (Ingledew 1982). The efficiency of the utilization of available energy (by bacteria) from oxidation of ferrous iron may range from 3.2 to 30% (Leduc and Ferroni 1994).

A number of models for the oxidation of ferrous iron by *At. ferrooxidans* have been postulated (Ingledew 1986, Blake and McGinness 1993) and a critical review of these models was presented by Ehrlich (1996). The most widely accepted model (Fig. 4.2) assumes that oxidation of ferrous iron takes place at the surface of the outer membrane of the cell with the transfer of electrons from ferrous iron through a polynuclear Fe^{3+} layer which is structurally bound in the outer membrane (Ingledew 1982). The electrons from ferrous iron in the outer membrane are transferred to periplasmic cytochrome c through a small copper protein rusticyanin mediated by a catalytic component (X). Rusticyanin being stable at low pH values is deemed as the principal component of respiratory electron transport system of *At. ferrooxidans* (Blake and McGinness 1993, Leduc and Ferroni 1994). The reduced cytochrome c at the outer surface of the membrane conveys the electrons across the membrane to cytochrome oxidase (cytochrome a_1). The reduced cytochrome oxidase subsequently reacts with molecular oxygen and protons resulting in the formation of water. Leduc and Ferroni (1994) asserted that the participation of the unknown enzymatic component (X) in the mediation of transfer of electrons to rusticyanin is a requisite in that the rate of transition of electrons to periplasmic cytochrome c by rusticyanin would otherwise be too slow to elucidate the observed rate of ferrous iron oxidation by bacteria.

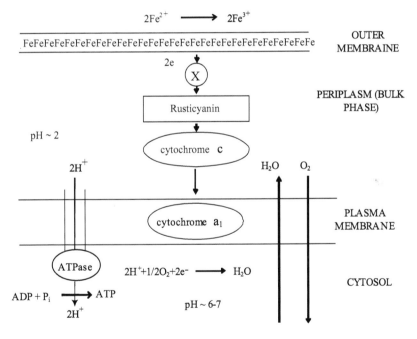

FIG. 4.2 A model for ferrous iron oxidation in *At. ferrooxidans* (Ehrlich 1996)

OXIDATION OF SULPHUR

Sulphur is an important element for the cell where it performs the function of stabilizing the protein structure and transferring hydrogen by enzymes in redox metabolism (Ehrlich 1996). Moreover, reduced forms of inorganic sulphur compounds are utilized by chemolithotrophic bacteria as sources of energy.

Several different enzymes can be involved in the oxidation of inorganic sulphur compounds by *At. ferrooxidans* possibly through different biochemical pathways (Kelly 1985, Suzuki 2001). Inorganic sulphur compounds are metabolized by bacteria through a series of intermediate products (although they have not been completely determined) many of which are considered to be highly unstable under oxidative conditions (Torma 1977). Suzuki et al. (1993) demonstrated that the oxidation of sulphide to sulphate by *At. thiooxidans* occurs through the formation of elemental sulphur and sulphite as the intermediates. These reactions are catalyzed by the sulphur and sulphite oxidizing enzyme systems. During this process, the formation of thiosulphate was observed (Suzuki et al. 1993). Sand et al. (1995) suggested that thiosulphate is the first intermediate in the oxidation (by ferric iron) of pyrite followed by the appearance of a variety of other sulphur compounds such as tetrathionate.

Oxidation of thiosulphate is assumed to be primarily based on the presence of two enzymes, namely rhodonese and thiosulphate oxidation enzymes (Silver 1978). Rhodonese catalyses the cleavage of thiosulphate leading to the formation of membrane associated sulphur (S^0) and sulphite (SO_3^{2-}) prior to the oxidation to sulphate. The thiosulphate oxidizing enzyme mediates the formation of tetrathionate ($S_4O_6^{2-}$) using two molecules of thiosulphate (Ehrlich 1996). The resulting tetrathionate is, by a series of hydrolytic and oxidative steps, transformed into sulphate with transient accumulations of intermediary sulphur from polythionates (Silver 1978). The details of these reactions are documented elsewhere (Rossi 1990, Ehrlich 1996). In light of the foregoing discussion, the following scheme may well be presented for the oxidation of inorganic sulphur compounds by *At. ferrooxidans* (Torma 1977, Kelly 1985).

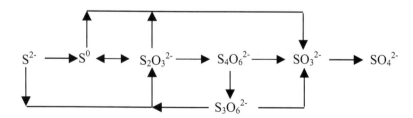

It has also been shown that *At. ferrooxidans* can oxidize elemental sulphur possibly as well as the other sulphur compounds under anaerobic conditions with ferric iron as terminal electron acceptor (Pronk and Johnson 1992).

MECHANISM OF OXIDATION OF SULPHIDE MINERALS BY ACIDOPHILIC BACTERIA

Direct *vs* Indirect Mechanism

The inherent characteristic of acidophilic bacteria is their ability to derive the energy required for growth from the oxidation of ferrous iron and reduced (inorganic) sulphur compounds in acidic environments. Sulphide minerals are potential sources of these energy-yielding substrates for bacteria.

Oxidation of sulphide minerals by bacteria takes place in multistages and involves the breakdown of the mineral crystal lattice, the penetration of oxidizing agent into the degraded crystal structure and further the oxidation of mineral sulphide whereby electrons are transferred from iron or sulphur components of the mineral to oxygen (Karavaiko et al. 1977). The actual role of bacteria in the oxidation process has not been completely resolved and until recently the direct and indirect mechanisms have been extensively discussed in the literature (Sand et al. 1995, 2001, Ehrlich 1996, Fowler et al. 1999, Tributsch 2001, Suzuki 2001, Crundwell 2003). In direct mechanism it is assumed that sulphide mineral is oxidized directly by the enzymatic action of bacteria. The sulphur moiety of the mineral is microbiologically converted to sulphate with no detectable intermediates and metal is released into solution. Attachment of bacteria to sulphide minerals is often interpreted as the indication of the direct mode of bacterial leaching (eq 1).

$$MS + 2O_2 \xrightarrow{\text{Bacteria}} MSO_4 \qquad (1)$$

$$MS + 2Fe^{3+} \longrightarrow M^{2+} + S^0 + 2Fe^{2+} \qquad (2)$$

$$2FeSO_4 + 1/2O_2 + H_2SO_4 \xrightarrow{\text{Bacteria}} Fe_2(SO_4)_3 + H_2O \qquad (3)$$

$$S^0 + 3/2O_2 + H_2O \xrightarrow{\text{Bacteria}} H_2SO_4 \qquad (4)$$

In contrast, the indirect mechanism assumes the oxidation of sulphides by ferric iron which produces ferrous iron and elemental sulphur (eq 2). The function of bacteria in this process is the regeneration of ferric iron from the oxidation of ferrous iron released (eq 3) and the conversion of elemental sulphur formed into sulphuric acid (eq 4).

Recently significant advances in the knowledge of sulphur chemistry, solid state physics and electrochemistry provided further elucidation of the mechanism whereby bioleaching of sulphide minerals occurs. Sand et al. (2001) proposed that ferric iron (Fe^{3+}) and/or protons (H^+) are the only (chemical) agents to attack the sulphide minerals leading to the dissolution in bacterial leaching systems. This hypothesis rules out the involvement of the direct mechanism in the dissolution process and suggests that the bacteria perform the function of the regeneration of ferric iron and/or protons, which are concentrated at the mineral/water and/or the mineral/bacteria interface. This results in the enhanced dissolution of sulphides in the presence of bacteria. There is a growing agreement for this proposal.

Based on the key intermediates involved Sand et al. (2001) propounded two indirect leaching models; namely the thiosulphate and the polysulphide mechanism depending upon the sulphide mineral involved, as illustrated in Fig. 4.3. The thiosulphate mechanism involves the reaction of metal sulphides with ferric iron leading to the formation of ferrous iron and thiosulphate as the primary sulphur intermediate. The thiosulphate formed is further converted via other intermediate sulphur compounds such as tetrathionate, sulphane monosulphonic acid and trithionate into sulphate and small amounts of elemental sulphur as well as pentathionate. The ferrous iron is recycled to ferric iron by the iron oxidizers e.g. *At. ferrooxidans* and *L. ferrooxidans* (eq 3). In the case of pyrite, the main reactions in thiosulphate mechanism can be summarized as given below (eqs 5-6) (Sand et al. 2001):

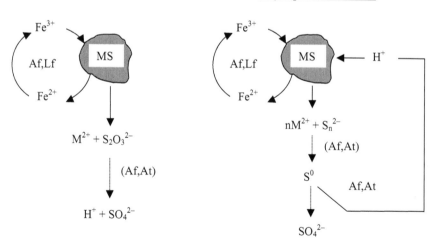

FIG. 4.3 Thiosulphate and polysulphide mechanisms for bacterial leaching of sulphide minerals (after Sand et al. 2001). MS: Metal sulphide; $S_2O_3^{2-}$: Thiosulphate; S_n^{2-}: Polysulphide with a chain length (n); S^0: Elemental sulphur; Af, At, Lf: Oxidation by *At. ferrooxidans*, *At. thiooxidans* and *L. ferrooxidans* and (Af, At): Possibility of bacterial oxidation.

$$FeS_2 + 6Fe^{3+} + 3H_2O \rightarrow S_2O_3^{2-} + 7Fe^{2+} + 6H^+ \tag{5}$$

$$S_2O_3^{2-} + 8Fe^{3+} + 5H_2O \rightarrow 2SO_4^{2-} + 8Fe^{2+} + 10H^+ \tag{6}$$

The polysulphide mechanism assumes that polysulphides are the primary intermediate sulphur compounds to appear as a consequence of ferric iron and acid leaching of metal sulphides. The further chemical and/or microbiological oxidation of the polysulphides results in the production of mainly elemental sulphur and sulphate to a lesser extent. The polysulphide mechanism may be represented by the reactions below (Sand et al. 2001, Rohwerder et al. 2003):

$$MS + Fe^{3+} + H^+ \rightarrow M^{2+} + 0.5H_2S_n + Fe^{2+} \quad (n \geq 2) \tag{7}$$

$$0.5H_2S_n + Fe^{3+} \rightarrow S^0 + H^+ + Fe^{2+} \quad \text{(for } n = 2\text{)} \tag{8}$$

$$S^0 + 1.5O_2 + H_2O \xrightarrow{\text{Bacteria}} SO_4^{2-} + 2H^+ \tag{9}$$

The thiosulphate mechanism was defined to apply to the bioleaching of sulphide minerals such as pyrite (FeS_2), molybdenite (MoS_2) and tungstenite (WS_2), whereas the polysulphide mechanism applies for the sulphides including sphalerite (ZnS), chalcopyrite ($CuFeS_2$) and galena (PbS) (Sand et al. 2001). Further evidence supporting the models proposed above has appeared in technical literature. Hackl et al. (1995) observed the formation of polysulphides on the surface of chalcopyrite during leaching in acidic sulphate media. Similarly, Osseo-Asare (1992) reported that oxidation of metal sulphides leads to the formation of different sulphur compounds depending on the mineral type.

Sand et al. (2001) investigated the formation of sulphur compounds as a result of the ferric iron leaching of a variety of pure sulphides at pH 1.9. The findings indicated that the majority of sulphide sulphur was converted to sulphate in case of pyrite and molybdenite i.e. 82% SO_4^{2-} compared to 16% S^0 for pyrite whereas ferric leaching produced over 90% elemental sulphur (S^0) for the minerals such as sphalerite (ZnS), chalcopyrite ($CuFeS_2$) and galena (PbS). Deveci et al. (2006) provided further supporting evidence to these postulations through SEM studies that elemental sulphur is the main product of the oxidation of sphalerite in bioleaching processes; yet the oxidation of pyrite does not produce elemental sulphur in significant (detectable) amounts. The dissolution behaviour of sulphide minerals, i.e. via thiosulphate and polysulphide in response to the chemical (ferric and/or acid) attack appears to be related to their crystal structure and hence electronic properties (Hackl et al. 1995, Sand et al. 2001, Tributsch 2001, Crundwell 2003).

The recent findings (Boon and Heijnen 1998b) on the biooxidation of pyrite also conform to the model proposed above. These authors argued that the oxidation process proceeds through the indirect mechanism where the main role of bacteria is the maintenance of high Fe^{3+}/Fe^{2+} ratio. Fowler et al. (1999) carried out experiments where redox conditions were artificially kept constant. They observed that the rate of dissolution of pyrite was consistently higher in the presence of bacteria where all other conditions were identical, i.e. Fe^{3+}/Fe^{2+} ratio. From the analysis of the rate data, the investigators inferred that the dissolution of pyrite was dependant on the pH and the higher rate of leaching with bacteria was due to the function of bacteria increasing the pH at the surface of the pyrite.

ROLE OF EXTRACELLULAR POLYMERIC SUBSTANCES (EPS) IN BIOLEACHING

Extracellular polymeric substances (EPS) excreted by the leaching bacteria may be involved in the bacterial leaching processes. Crundwell (1996) reported the formation of biofilms on the mineral surface composed of a layer of extracellular polysaccharides. The author also observed a ferric hydroxide layer formed underneath the biofilm and deduced that bacterial attachment was not required to oxidize pyrite on the grounds that the iron was cycled between the bacteria and the pyrite within the biofilm. On examination of the morphological properties of the bioleaching patterns, Rodriguez-Levia and Tributsh (1988) postulated that the corrosion occurs in a thin layer formed by bacterial secretion of organic substances (i.e. EPS) between the cell and sulphide interface.

Sand et al. (2001) suggested that iron species may form complexes within exopolymeric substances presumably by glucuronic acids as a complexing agent and the concentration of iron in this layer can be as high as 53 g/l. This may well suggest that the exopolymeric layer with complexed ferric iron provides a reaction zone where the dissolution of the sulphides occurs. Sand et al. (2001) also reported that living cells with EPS and ferric iron produced the highest increase in the surface potential on the pyrite crystal while living cells stripped of EPS resulted in a minimal increase in the absence of ferric iron. The investigators also noted that the cells devoid of EPS could restore high oxidation activity in the presence of ferric iron presumably producing new EPS. Pogliani and Donati (1999) demonstrated that the loss of extracellular polymeric substances adversely affected the dissolution of copper from covellite and the leaching ability of *At. ferrooxidans* was revivified within the course of the dissolution process.

CONCEPTUAL MODELS FOR BIOLEACHING OF SULPHIDE MINERALS

The models proposed by Tributsh (2001) may well be adopted to elucidate the role of bacteria in the dissolution process. Accordingly, the oxidation process may proceed via:

- indirect leaching, i.e. attack by ferric iron and acid generated by bacteria,
- contact leaching through extracellular polymeric layer,
- cooperative leaching, i.e. combination of both indirect and contact leaching.

These models are illustrated in Fig. 4.4. From all the indications in technical literature it is most likely that cooperative leaching assumes the most of bioleaching processes.

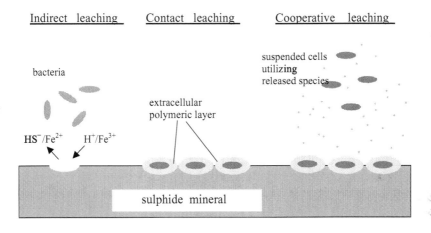

FIG. 4.4 The models depicting bacterial leaching of sulphide minerals proposed by Tributsh (2001)

IMPORTANCE AND BACTERIAL PRODUCTION OF FERRIC IRON

Ferric iron is a well known oxidant for sulphide minerals and uranium oxides in acid solution. The ferric-ferrous couple represents a redox component in leach solutions with the standard potential of ~ 0.77 V for the half reaction ($Fe^{3+} + e^- \rightarrow Fe^{2+}$) according to the Nernst equation:

$$E = E_o + (RT/nF) \ln ([Fe^{3+}]/[Fe^{2+}]) \tag{10}$$

Oxidation potential of this couple even for a Fe^{3+}/Fe^{2+} ratio of 10^{-6} is approximately 0.4 volt (eq 10) at 25°C (Dutrizac and MacDonald 1974). The value reflects the oxidizing power of ferric iron even at such low concentrations. Many sulphide minerals may well be oxidized at this potential. The Nernst expression implies that the reduction of ferric to ferrous iron is independent of the pH. However, the pH controls the solubility and speciation of ferric iron which presents as various soluble sulphato-complexes (pH < 2.5) in sulphate systems (Barrett et al. 1993). The hydrolysis of ferric iron (eq 11) becomes important at pH values above ~ 2.5.

$$Fe_2(SO_4)_3 + 6H_2O \rightarrow 2Fe(OH)_{3(s)} + 3H_2SO_4 \qquad (11)$$

In acid solutions the oxidation of metal sulphides by ferric iron (eq 2) produces metal sulphate, ferrous iron and varying quantities of elemental sulphur and sulphate depending on the mineral. The order of reaction with respect to ferric iron concentration is around one-half for most sulphide minerals (Crundwell 2003). If formed, the oxidation of elemental sulphur does not appear to be kinetically feasible possibly due to its hydrophobic peculiarity (Dutrizac and MacDonald 1974). In this regard, the elemental sulphur generated as the reaction product on the mineral surface may interfere with the progress of the dissolution process. This may be attributed to its stability under a wide range of conditions, e.g. at temperatures even approaching the melting point of sulphur. In the presence of sulphur oxidizing bacteria, however, elemental sulphur formed can be continuously removed by its conversion into sulphuric acid (eq 4).

The amount of ferric iron required to completely oxidize the sulphide minerals present may exceed even its solubility and also the maintenance of strong oxidation conditions, i.e. high Fe^{3+}/Fe^{2+} ratio, is of prime importance in order to sustain the kinetics of the dissolution process at optimum. In this respect, ferrous iron as the resultant product of ferric leaching of sulphides needs to be oxidized for the regeneration of ferric iron. This can be performed by a variety of chemical oxidants such as MnO_2 and H_2O_2 or by bacteria (Dutrizac and MacDonald 1974). In the absence of bacteria or chemical oxidants, the rate of ferrous iron oxidation is reported to be extremely slow, even stable (Tuovinen 1990). Conversely, it is accelerated by an estimated factor of 10^6 in the presence of bacteria (Singer and Strumm 1970).

Kinetics of bacterial oxidation of ferrous iron has been the scope of many investigations (Norris et al. 1988, Harwey and Crundwell 1997, Boon et al. 1999b, Nemati and Harrison 2000). The oxidation reaction necessitates the provision of aerobic, autotrophic and acidic environment to proceed at optimum. During the oxidation, the growth of bacteria occurs and follows Monod-type (Monod 1949) growth pattern. The growth rate of acidophilic bacteria is reported to be in the range of 0.03-1.8 h^{-1} (Nemati et al. 1998). The kinetics of bacterial ferrous iron oxidation have been observed to depend on the pH, temperature, the strain and type of bacteria, the metal ions present, the inhibition by ferric iron as a reaction product and the population of bacteria (cell numbers) (Norris et al. 1988, Jensen and Webb 1995, Nemati et al. 1998, Ojumu et al. 2006).

ELECTROCHEMICAL ASPECTS OF BACTERIAL LEACHING OF SULPHIDE MINERALS

Bacterial leaching of sulphides involves a series of oxidative dissolution reactions with the breakage of chemical bonds in the solid and the transfer of charged species across the phase boundary. Therefore, the dissolution process is deemed as electrochemical in nature (Crundwell 2003). The electronic structure of sulphide minerals accords semiconductor or conductor characteristics to these minerals leading to their behaviour as electrodes in aqueous solutions (Osseo-Asare 1992).

When a sulphide mineral is immersed in a solution containing its ionic species in two different oxidation states, an equilibrium potential will develop on open circuit (Natarajan and Iwasaki 1974). This potential is termed as the rest potential/electrode potential/redox potential. The rest potential of mineral sulphides (Table 4.2) assumes paramount importance in the dissolution process. A mineral sulphide undergoes oxidation when the oxidation potential of the system is higher than the electrode potential of the mineral. Therefore, the rest potential values of mineral electrodes may indicate the electrochemical activity of the mineral in a given system. Consequently, various sulphide minerals may

Table 4.2 Rest potential of various sulphide minerals at ambient temperature (Hiskey and Wadsford 1975)

Mineral	Structure	Solution	Rest potential (V vs SHE)	Relative activity
Pyrite	Cubic	1.0 M H_2SO_4	0.63	Noble
Chalcopyrite	Tetragonal	1.0 M H_2SO_4	0.52	
Chalcocite	Orthorhombic	1.0 M H_2SO_4	0.44	
Covellite	Hexagonal	1.0 M $HClO_4$	0.42	
Galena	Cubic	1.0 M H_2SO_4	0.28	
Sphalerite	Cubic	1.0 M H_2SO_4	−0.24	Active

be arranged in galvanic series with respect to their relative electrochemical activity as shown in Table 4.2 (Noble → Active). It can be deduced that pyrite and chalcopyrite are relatively noble minerals and difficult to oxidize in aqueous solutions whilst sphalerite and pyrrhotite are relatively electro-active minerals and can be readily oxidized. Accordingly the selective dissolution of base metal sulphides can be achieved by the exploitation of their electrochemical properties. Figure 4.5 illustrates the visual evidence for the preferential oxidation of sphalerite over pyrite during moderately thermophilic bioleaching of a complex ore.

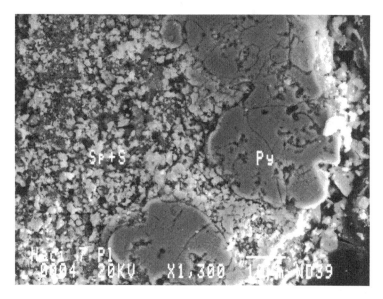

FIG. 4.5 A SEM photograph depicting more extensive (selective) of oxidation of sphalerite (Sp) than pyrite (Py) after three days of bioleaching using a mixed moderately thermophilic culture (Deveci et al. 2006)

Galvanic interactions can play a significant role in the bioleaching of sulphide ores/concentrates. When different sulphide minerals are in contact with each other in an aqueous solution, a number of electrochemical cells are presumably established. The minerals behave as electrodes in the system. The mineral with a lower rest potential exhibits anodic behaviour and undergoes anodic corrosion whilst that with a higher rest potential acts as the cathode and is galvanically protected. Possible anodic and cathodic reactions can be presented as below:

Cathodic reaction:	$0.5O_2 + 2H^+ + 2e^- \rightarrow H_2O$	(12)
Anodic reaction:	$MeS \rightarrow Me^{2+} + S^0 + 2e^-$	(13)
The overall reaction:	$MeS + 0.5O_2 + 2H^+ \rightarrow Me^{2+} + S^0 + H_2O$	(14)

The rationale or driving force behind the galvanic dissolution process is the difference in the rest potentials between the minerals in contact. The higher the difference in the rest potentials between the sulphides, the higher the rate of anodic corrosion (dissolution) of the mineral with a lower rest potential. Several factors such as difference in rest potentials of the minerals in contact, relative surface areas of anode and cathode, medium characteristics such as pH and dissolved salts, conductivity of the mineral and electrolyte and the presence or absence of bacteria also control the galvanic interactions and the rate of anodic dissolution of more active mineral (Natarajan 1990).

The presence of bacteria in the system has been reported to accelerate galvanic interactions several fold (Natarajan 1988). This enhancement in galvanic interactions induced by bacteria may be ascribed to the bacterial oxidation of ferrous iron and elemental sulphur formed in anodic reactions possibly eliminating the passivating sulphur layer. Jyothi et al. (1989) pointed out that under the same conditions, when contacted with chalcopyrite, the dissolution rate of sphalerite was lower than that when contacted with pyrite. The difference most likely stems from its relative electropotential to the mineral in combination. Ahonen and Tuovinen (1993) also noted that at high redox potentials (600-650 mV vs SCE), pyrite leaching was promoted while the oxidation of chalcopyrite was slower. On maintaining a lower redox potential (500-550 mV vs SCE) the suppression of the oxidation of pyrite and the preferential/accelerated oxidation of chalcopyrite were observed by these investigators.

PRACTICAL SIGNIFICANCE OF MINERALOGY TO BACTERIAL LEACHING PROCESSES

Ore bodies exhibit unique and indigenous characteristics of mineralogical composition and concentrations of metals. Despite the fact that any sulphide ore/concentrate can be theoretically oxidized by bacteria, the mineralogical characteristics should be assessed in detail prior to the development of any bioleaching process.

Iron content of sulphide ores/concentrates could be of practical importance for bacterial leaching processes (Deveci et al. 2004a) because the oxidation of pyrite or pyrrhotite provides soluble iron which is converted to ferric iron by bacteria. Ferric iron is an important oxidant for the dissolution of sulphide minerals and the oxidation of some uranium minerals. Pyrite (FeS_2) is generally present in association with other common sulphides including arsenopyrite (FeAsS), sphalerite (ZnS), chalcopyrite ($CuFeS_2$) and galena (PbS) possibly as binary, ternary or quaternary combinations. In such systems, the galvanic interactions, selective or preferential oxidation of the mineral sulphides such as sphalerite and arsenopyrite prior to pyrite are presumed to occur as discussed above. Biooxidation of pyrite (eq 15) produces acid leading to a gradual increase in the acidity of the leaching medium. Similarly an excessive decrease in the pH may result in the inhibition of the activity of bacteria.

$$4FeS_2 + 15O_2 + 2H_2O \xrightarrow{\text{Bacteria}} 4Fe^{3+} + 8SO_4^{2-} + 4H^+ \qquad (15)$$

The presence of bacterially toxic elements such as As, Ag and Hg and the increasing concentrations of metal ions in the leach solutions can inhibit the activity of bacteria and the achievable rate and extent of extraction (Norris 1990). However, the bacteria utilized in the leaching process are in general versatile and can develop resistance (by adaptation) to toxic metal ions (Rossi 1990).

Natural occurrence, i.e. degree of crystallinity, variations in the metal and sulphur content, presence of impurities in the sulphide matrix will determine the varying susceptibility of a sulphide mineral to bioleaching. Boon et al. (1999a) reported that *At. ferrooxidans* was able to oxidize framboidal pyrite whilst the microorganism failed to degrade euhedral (highly crystalline) pyrite. Similarly, Andrews and Merkle (1999) reported that the oxidation rate of arsenopyrite was determined by its arsenic and sulphur content i.e. As-rich or S-rich and the presence of cobalt (0.01%) in the mineral matrix enhanced the dissolution rate of arsenic.

Complex sulphide ores are often characterized as low grade with metal values intergrown and finely disseminated in association with carbonaceous and/or siliceous gangue minerals. These gangue minerals contribute to acid consumption and, if abundant, bioleaching of such ores could be impractical due to the high acid consumption. In effect, the presence of carbonate minerals in certain quantity could be beneficial in that carbon dioxide released from the reaction of these minerals with acid would promote the growth of autotrophic bacteria. The weathering of silicate minerals in bioleaching systems release the cations such as K^+ and Na^+ which are then involved in the formation of the detrimental jarosites of these cations (eq 16) (Tuovinen and Bhatti 1999, Deveci et al. 2004a). Similarly, the presence of clay minerals, which have a significant cation-exchange capacity, can lead to the loss of dissolved metal values by cation exchange (Hiskey 1994). This problem and the formation of ferric precipitates may be partially alleviated by a proper control of pH given the pH-dependent nature of ion-exchange and precipitation processes.

$$Na^+/K^+ + 3Fe^{3+} + 2SO_4^{2-} + 6H_2O \rightarrow Na/KFe_3(SO_4)_2(OH)_6 + 6H^+ \quad (16)$$

PARAMETERS OF PRACTICAL IMPORTANCE FOR BIOLEACHING OF SULPHIDE MINERALS

Bioleaching of sulphide minerals is inherently a complex process with the involvement of acidophilic bacteria and archaea to mediate the oxidative dissolution of sulphides. These microorganisms themselves establish optimum conditions under which they optimally thrive. Although the performance of bioleaching processes is closely controlled by the activity/growth of bacteria in the leaching environment, the growth conditions may be manipulated within a certain range to maximize the rate and extent of metal dissolution from sulphide ores/concentrates. The factors deemed to be of fundamental significance to bioleaching of sulphides are temperature, acidity, growth media, particle size and pulp density, oxygen and carbon dioxide availability and toxicity of metals present in the system (Bailey and Hansford 1993, Bosecker 1997, Dew et al. 1997, Boon and Heijnen 1998a, Rawlings 2002, Deveci et al. 2003a).

Optimum activity of each type of bacteria takes place in a well-defined range of temperatures (Table 4.1) at which these microorganisms perform most efficiently (e.g. Figure 4.6). This indicates the temperature dependent character of bioleaching processes. Although the rate and extent of dissolution of sulphides tends to increase with temperature, beyond an optimum temperature range the oxidative activity of bacteria with the resultant dissolution of sulphides decreases due to the likely denaturation of proteins and thermal death of microorganisms. Biooxidation of gold concentrates is commercially practised at 40°C (Dew et al. 1997).

Acidity of the bioleaching environment controls the activity of bacteria and the solubility of ferric iron (Deveci et al. 2004a). The optimum pH for the growth of acidophilic bacteria and archaea varies in a pH range of 1.5-2.5 (Torma 1977, Bosecker 1997). However, in practice the operating pH is often lower than the optimum values for growth, e.g. pH 1.2-1.8 for the BIOX® process (Dew et al. 1997) and pH 1.3-1.5 for the BacTech® process (Miller 1997) to curb the formation of undesirable ferric precipitates, jarosites in particular.

The bacteria and archaea of practical importance to the biodegradation of sulphide minerals are, in general, autotrophic aerobes and hence the oxidizing activity of these microorganisms depends largely on the availability of oxygen and carbon dioxide. Oxygen

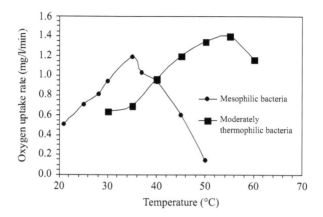

FIG. 4.6 Effect of temperature on the rate of oxygen uptake during the oxidation of ferrous iron by the mixed mesophilic and moderately thermophilic cultures (Deveci et al. 2003a)

functions as the terminal electron acceptor while carbon derived from the fixation of the carbon dioxide is utilized in the synthesis of biomass (Rossi 1990). Oxygen and carbon dioxide transfer is one of the most important factors in the bioleaching processes. A minimum level of dissolved oxygen of >1-2 mg/l is to be maintained for an optimum operation in a given bioleaching system (Deveci et al. 2003a). Similarly, adequate supply of carbon dioxide is a prerequisite for cell growth. Norris (1989a) and Witne and Phillips (2001) reported significant improvements in the thermophilic bioleaching of sulphides with the introduction of CO_2-enriched air. Boon and Heijnen (1998a) concluded that the observed decrease in the biooxidation rates at high solids densities is caused largely by insufficient transfer of carbon dioxide to bioleaching media.

A culture medium, essentially a nutrient salt solution, is required to provide microorganisms with all the elements required for cell mass production and sufficient energy for biosynthesis and maintenance. As shown in Table 4.3, many different formulations of growth media mainly as the modifications of '9K' medium (Silverman and Lundgren 1959) are used in laboratory studies and in practical applications. However, the requirement of growth media, i.e. the concentrations of salts to be added appears to be determined by the quantity of the substrate available (i.e. head grade and/or pulp density) for bacterial oxidation (Gomez et al. 1999b, Deveci et al. 2003b).

TABLE 4.3 Growth media used in bioleaching/biooxidation of sulphides

Growth media	$(NH_4)_2SO_4$ (g/l)	$MgSO_4 \cdot 7H_2O$ (g/l)	KH_2PO_4 (g/l)	KCl (g/l)	$Ca(NO_3)_2 \cdot H_2O$ (g/l)
9K[*]	3	0.5	0.5	0.1	0.01
T and K[§]	0.4	0.4	0.4	–	–
ES[γ]	0.2	0.4	0.1	0.1	–
Leathen[£]	0.15	0.5	0.01	0.05	0.05
Norris[δ]	0.2	0.2	0.2	–	–

[*]Silverman and Lundgren (1959); [§]Tuovinen and Kelly (1973); [γ]Norris and Barr (1985); [£]Leathen et al. (1956); [δ]Gomez et al. (1999b)

Bacterial oxidation of sulphide minerals occurs through surface chemical reactions and the increased surface area, though particle size reduction would lead to a higher rate and extent of extraction. In practice biooxidation of refractory gold concentrates is carried out at a particle size of –75 μm (Dew et al. 1997). However, the optimum particle size is determined by the size reduction costs versus the improved kinetics and recoveries. It is also of particular importance to operate bioleaching processes at high pulp densities due to process economics. Operating pulp density is often limited to a threshold level of 20% w/w in industrial stirred tank biooxidation practice (Dew et al. 1997) due to a number of factors including the decrease in bacteria-to-solid ratio, mechanical damage to bacterial cells by solid particles, the accumulation of toxic metal ions in solution and the limited availability (i.e. transfer) of oxygen and carbon dioxide (Komnitsas and Pooley 1991, Bailey and Hansford 1993, Boon and Heijnen 1998a, Deveci 2002a, 2004, Deveci et al. 2003b).

Metal cations and anions (e.g. Cl⁻ in solution at certain levels may exert a toxic effect on acidophilic bacteria and arhaea albeit different strains of bacteria exhibit varying sensitivity to toxicants (Sampson and Phillips 2001, Deveci 2002b). *At. ferrooxidans* was reported to grow on ferrous iron in the presence of (10 g/l) of Zn, Ni, Cu, Co, Mn and Al while Ag and anions of Te, As and Se were proved to inhibit the oxidizing activity of the culture at concentrations of only 50-100 mg/l (Tuovinen et al. 1971). Bacterial cultures can be adapted to a particular environment to mitigate for the inhibitory effects of toxic metal ions or of increasing concentrations of metals (Das et al. 1997). The development of specially adapted strains with tolerance for metals; 50 g/l Cu, 72 g/l Ni and 124 g/l Zn was reported (Natarajan 1990, Deveci et al. 2004b).

CURRENT STATUS AND FUTURE TRENDS IN MINERAL BIOTECHNOLOGY OF SULPHIDES

Based upon the success of the early operations, bioleaching processes have gradually gained acceptance in the extractive metal industry. Approximately 18-25% of total copper is produced through bacterially assisted heap or dump leaching of low grade copper sulphide ores (Brierley and Brierley 2001, Watling 2006). Similarly, it was estimated in 1993 that 12-13% of total uranium production resulted from the bioleaching processes (Mwaba 1993). Biooxidation of refractory gold concentrates has proven to be a feasible alternative to roasting and pressure oxidation on a commercial scale (Rawlings et al. 2003, Clark et al. 2006). Nonetheless, bioleaching of base metal sulphides has been extensively investigated at laboratory and pilot scales and have proven feasible processes (Steemson et al. 1997, Gericke and Pinches 1999, van Staden 1998, Gilbertson 2000, Clark et al. 2006).

The commercial application of bioleaching processes has been extended to extract base metals including cobalt and nickel from sulphide concentrates. A bioleaching process within the Kasese Project in Uganda was developed to treat pyritic cobalt concentrates for the recovery of cobalt (Brigss and Millard 1997, Brierley and Brierley 2001). The BioNIC® process developed by Billiton was demonstrated at pilot scale (Miller et al. 1997) as an alternative process to conventional smelting for the recovery of nickel from sulphide concentrates. More recent developments have been the announcements of the BioCOP® and BioZINC® processes by Billiton for the extraction of copper and zinc respectively from concentrates (Gilbertson 2000, Batty and Rorke 2006, Clark et al. 2006). The GEOCOAT process, developed by Geobiotics offers the extension of bioheap concept in a unique way to treat gold, copper and zinc sulphide concentrates (Harvey et al. 2002a, b).

Although bioleaching processes have already proved to be a most promising option for the extraction of metals from marginal and/or submarginal resources over the years there are still areas where extensive research is demanded to maintain the development of bioprocesses as well as to expand the range of applications in the future. In this regard, the application of thermophilic bacteria (50-80°C) has a potential for improving kinetics of metal extraction and minimizing cooling requirements within the bioleaching systems. Improvements in dump or heap design are required to permit better control of bioleaching process which may in turn enhance extent and rate of the dissolution process. The severe detractions to bioleaching processes are the slow rates of dissolution coupled with low solid density which makes the overall process less attractive to the industry since both capital and operating costs will be aggravated. In this regard, engineering aspects of reactor design need to be established to optimize the design technology in an attempt to attain better leaching kinetics at higher solid concentrations.

As a consequence, since the recognition of bacterial involvement in the generation of AMD in the early 1950s, considerable progress has been made in biological extraction of metals from sulphide ores/concentrates and old tailings. The evolution of biohydrometallurgy appears to continue in the integration of biological processes with metal extraction, waste/water treatment, degradation of free cyanide, desulphurization of coals and particularly in handling complex sulphide ores and concentrates. These may well be considered as future areas of development in addition to the current industrial practice of gold/silver, copper and uranium recovery from sulphide ores/concentrates.

REFERENCES

Ahonen, L. and O.H. Tuovinen. 1993. Redox potential controlled bacterial leaching of chalcopyrite ores. *In:* A.E. Torma, J.E. Wey, and V.I. Lakshmanan [eds.]. Biohydrometallurgical Technologies, Vol. 1: Proc. of Int. Biohydrometallurgy Symp. The Minerals, Metals & Materials Society, Warrendale, Pennsylvania USA. pp. 571-578.

Akcil, A. 2004. Potential bioleaching developments towards commercial reality: Turkish metal mining's future. Minerals Engineering 17: 477-480.

Akcil, A. and H. Ciftci. 2003a. Bacterial leaching of Kure copper ore. The Journal of The Chamber of Mining Engineers of Turkey 42: 15-25 (Turkish).

Akcil, A. and H. Ciftci. 2003b. Effect of sulphur and iron-oxidizing bacteria on metal recovery in leaching of Kure pyritic copper ore. The Bulletin of Earth Sciences Application and Research Centre of Hacettepe University 29: 181-192 (Turkish).

Akcil, A. and H. Ciftci. 2006. Mechanisms of bacterial leaching in metal recovery. The Journal of the Chamber of Mining Engineers of Turkey 45: 19-27.

Akcil, A., H. Ciftci, and H. Deveci. 2007. Role and contribution of pure and mixed cultures of mesophiles in bioleaching of a pyritic chalcopyrite concentrate. Minerals Engineering 20: 310-318.

Andrews, L. and R.K.W. Merkle. 1999. Mineralogical factors affecting arsenopyrite oxidation rate during acid ferric sulphate and bacterial leaching of refractory gold ores. *In:* R. Amils and A. Ballester [eds.]. Biohydrometallurgy and the Environment toward the Mining of the 21st Century, Part A, IBS'99, Elsevier, Amsterdam, The Netherlands. pp. 109-117.

Bailey, A.D. and G.S. Hansford. 1993. Factors affecting the biooxidation of sulphide minerals at high concentrations of solids: A review. Biotechnology and Bioengineering 12: 1164-1174.

Barr, D.W., M.A. Jordan, P.R. Norris, and C.V. Phillips. 1992. An investigation into bacterial cell, ferrous iron, pH and Eh interactions during thermophilic leaching of copper concentrates. Minerals Engineering 5: 557-567.

Barrett, J., M.N. Hughes, G.I. Karavaiko, and P.A. Spencer. 1993. Metal Extraction by Bacterial Oxidation of Minerals. Ellis Horwood Ltd., London, UK.

Battaglia-Brunett, F., P. d'Hugues, T. Cabral, P. Cezac, J.L. Garcia, and D. Morin. 1998. The mutual effect of mixed *Thiobacilli* and *Leptosprilli* populations on pyrite bioleaching. Minerals Engineering 11: 195-205.

Batty, J.D. and G.V. Rorke. 2006. Development and commercial demonstration of the BioCOP^{TM} thermophile process. Hydrometallurgy 83: 83-89.

Blake, R.C. and S. McGinness. 1993. Fundamental aspects of the solubilisation of minerals by bacteria. *In:* Proc. of the 4th Int. Symp. on Hydrometallurgy. SME. Littleton, USA. pp. 727-741.

Bond, P.L., G.K. Druschel, and J.F. Banfield. 2000. Comparison of acid mine drainage microbial communities in physically and geochemically distinct ecosystems. App. and Env. Microbiology 66: 4962-4971.

Boon, M. and J.J. Heijnen. 1998a. Gas–liquid mass transfer phenomena in biooxidation experiments of sulphide minerals: A review of literature data. Hydrometallurgy 48: 187-204.

Boon, M. and J.J. Heijnen. 1998b. Chemical oxidation kinetics of pyrite in bioleaching processes. Hydrometallurgy 48: 27-41.

Boon, M., H.J. Brasser, G.S. Hansford, and J.J. Heijnen. 1999a. Comparison of the oxidation kinetics of different pyrites in the presence of *T. ferrooxidans* or *L. ferrooxidans*. Hydrometallurgy 53: 57-72.

Boon, M., T.A. Meeder, C. Thone, C. Ras, and J.J. Heijnen. 1999b. The ferrous iron oxidation kinetics of *Thiobacillus ferrooxidans* in batch cultures. App. Microbiol. Biotechnology 51: 813-819.

Bos, P. and J.B. Kuenen. 1990. Microbial treatment of coal. *In:* H.L. Ehrlich and C.L. Brierley [eds.]. Microbial Mineral Recovery. McGraw-Hill, New York, USA. pp. 343-377.

Bosecker, K. 1997. Bioleaching: metal solubilization by microorganisms. FEMS Microbiology Reviews 20: 591-604.

Brierley, C.L. 1978. Bacterial leaching. CRC Critical Reviews in Microbiology 6: 207-262.

Brierley, C.L. 1982. Microbial mining. Scientific American 247: 42-53.

Brierley, J.A. and C.L. Brierley. 2001. Present and future commercial applications of biohydrometallurgy. Hydrometallurgy 59: 233-239.

Brigss, A.P. and M. Millard. 1997. Cobalt recovery using bacterial leaching at the Kasese Project in Uganda. *In:* Biomine 97, Int. Biohydrometallurgy Symposium, IBS97. Australian Mineral Foundation. Glenside, Australia. pp. 1-12.

Campell, M.C., H.W. Parson, A. Jongejan, V. Sanmugasunderam, and M. Silver. 1985. Biotechnology for the mineral industry. Can. Met. Quarterly 24: 115-120.

Clark, D.A. and P.R. Norris. 1996. *Acidimicrobium ferrooxidans* gen. nov., sp. nov.: mixed culture ferrous iron oxidation with *Sulfobacillus* species. Microbiology 142: 785-790.

Clark, M.E. J.D. Batty, C.B. van Buuren, D.W. Dew, and M.A. Eamon. 2006. Biotechnology in minerals processing: Technological breakthroughs creating value. Hydrometallurgy 83: 3-9.

Colmer, A.R. and M.F. Hinkle. 1947. The role of micro-organisms in acid mine drainage. Science 106: 253-256.

Crundwell, F.K. 1996. The formation of biofilms of iron-oxidising bacteria on pyrite. Minerals Engineering 9: 1081-1089.

Crundwell, F.K. 2003. How do bacteria interact with minerals. Hydrometallurgy 71: 75-81.

Cupp, C.R. 1985. After *Thiobacillus ferrooxidans* what? Can. Met. Quarterly 24: 109-113.

Das, A., J.M. Modak, and K.A. Natarajan. 1997. Studies on multi-metal ion tolerance of *Thiobacillus ferrooxidans*. Minerals Engineering 10: 743-749.

Deveci, H. 2002a. Effect of solids on viability of acidophilic bacteria. Minerals Engineering 15: 1181-1189.

Deveci, H. 2002b. Effect of salinity on the oxidative activity of acidophilic bacteria during bioleaching of a complex Zn/Pb sulphide ore. The European Journal of Mineral Processing and Environmental Protection 2: 141-150.

Deveci, H. 2004. Effect of particle size and shape of solids on the viability of acidophilic bacteria during mixing in stirred tank reactors. Hydrometallurgy 71: 385-396.

Deveci, H., A. Akcil, and I. Alp. 2003a. Parameters for control and optimisation of bioleaching of sulphide minerals. *In:* F. Kongoli, B. Thomas, and K. Sawamiphakdi [eds.]. Materials Science & Technology 2003 Symposium: Process Control and Optimization in Ferrous and Non-ferrous Industry. TMS. Warrendale, P.A., USA. pp. 77-90.

Deveci, H., I. Alp, and T. Uslu. 2003b. Effect of surface area, growth media and inert solids on bioleaching of complex zinc/lead sulphides. *In:* G. Ozbayoglu [ed.]. Proc. of the 18th Int. Mining Congress and Exhibition of Turkey, IMCET 2003. The Chambers of Mining Engineers of Turkey. Antalya, Turkey. pp. 415-423.

Deveci, H., A. Akcil, and I. Alp. 2004a. Bioleaching of complex zinc sulphides using mesophilic and thermophilic bacteria: Comparative importance of pH and iron. Hydrometallurgy 73: 293-303.

Deveci, H., I. Alp, and E.Y. Yazici. 2004b. Bench-scale bioleaching of a complex zinc sulphide ore in stirred tank reactors. *In:* A. Akar, U. Ipekoglu, I. Cocen, and M. Polat [eds.]. Proc. of the Xth International Mineral Processing Symposium (IMPS). The Chambers of Mining Engineers of Turkey. Izmir, Turkey. pp. 523-529.

Deveci, H., T. Ball, I. Alp, T. Uslu, and E.Y. Yazici. 2006. A SEM study on the oxidation patterns of sphalerite, pyrite and galena during chemical and bacterial leaching of a sulphide ore. *In:* G. Onal, N. Acarkan, M.S. Çelik,

F. Arslan, G. Ateşok, A. Güney, A.A. Sirkeci, A.E. Yüce, and K.T. Perek [eds.]. Proc. of The XXIII. Int. Mineral Processing Congress, IMPC 2006, Vol. 2, Promed Advertising Agency. Istanbul, Turkey. pp. 1494-1499.

Dew, D.W., E.N. Lawson, and J.L. Broadhurst. 1977. The BIOX® process for biooxidation of gold bearing ores or concentrates. *In*: D.E. Rawlings [ed.]. Biomining: Theory, Microbes and Industrial Processes. Springer-Verlag, Berlin, Germany. pp. 45-79.

Dew, D.W., C. van Buuren, K. McEwan, and C. Bowker. 1999. Bioleaching of base metal sulphide concentrates: A comparison of mesophilic and thermophilic bacterial cultures. *In:* R. Amils and A. Ballester [eds.]. Biohydrometallurgy and the Environment toward the Mining of the 21st Century, Part A, IBS'99 Elsevier, Amsterdam, The Netherlands. pp. 229-238.

Dopson, M. and E.B. Lindström. 1999. Potential role of *Thiobacillus caldus* in arsenopyrite bioleaching. App. and Env. Microbiology 65: 36-40.

Dopson, M. and E.B. Lindström. 2004. Analysis of community composition during moderately thermophilic bioleaching of pyrite, arsenical pyrite and chalcopyrite. Microbial Ecology 48: 19-28.

Dutrizac, J.E. and R.J.C. MacDonald. 1974. Ferric ion as a leaching medium. Minerals Sci. Eng. 6: 59-95.

Ehrlich, H.L. 1991. Microbes for biohydrometallurgy. *In:* R.W. Smith and M. Misra [eds.]. Mineral Bioprocessing: Proceedings of the Conference, TMS. Warrendale, P.A., USA. pp. 27-41.

Ehrlich, H.L. 1996. Geomicrobiology. 3rd ed. Marcel-Dekker Inc., New York, USA.

Ehrlich, H.L. 2004. Beginnings of rational bioleaching and highlights in the development of biohydrometallurgy: A brief history. The European Journal of Mineral Processing and Environmental Protection 4: 102-112.

Ewart, D.K. 1990. Studies on a moderately thermophilic mixed culture of bacteria and its application to the biooxidation of gold bearing minerals. PhD Thesis, Kings College, University of London, UK.

Fowler, T.A., P.R. Holmes, and F.K. Crundwell. 1999. Mechanism of pyrite dissolution in the presence of *T. ferrooxidans*. App. and Env. Microbiology 65: 2987-2993.

Gericke, M. and A. Pinches. 1999. Bioleaching of a copper sulphide concentrate using extreme thermophilic bacteria. Minerals Engineering 12: 893-904.

Ghauri, M.A. and B.D. Johnson. 1991. Physiological diversity amongst some moderately thermophilic iron oxidising bacteria. FEMS Microbiology Ecology 85: 327-334.

Gilbertson, B. 2000. Creating value through innovation: Biotechnology in mining. IMM Trans. C 109: 61-67.

Golyshina, O.V., T.A. Pivovarova, G.I. Karavaiko, T.F. Kondrateva, E.R.B. Moore, W. Abraham, H. Lunsdorf, K.N. Timmis, M.M. Yakimov, and P.N. Golyshin. 2000. Ferroplasma *acidiphilum* gen. nov., sp. nov., an acidophilic, autotrophic, ferrous-iron-oxidizing, cell-wall-lacking, mesophilic member of the Ferroplasmaceae fam. nov., comprising a distinct lineage of the *Archaea*. Int. J. of Syst. and Evol. Microbiology 50: 997-1006.

Gomez, E., A. Ballester, M. Blazquez, and F. Gonzales. 1999a. Silver catalysed bioleaching of a chalcopyrite concentrate with mixed cultures of moderately thermophilic microorganisms. Hydrometallurgy 51: 37-46.

Gomez, C., M.L. Blazquez, and A. Ballester. 1999b. Bioleaching of a Spanish complex sulphide ore—bulk concentrate. Minerals Engineering 12: 93-106.

Hackl, R.P., D.B. Dreisinger, E. Peters, and J.A. King. 1995. Passivation of chalcopyrite during oxidative leaching in sulphate media. Hydrometallurgy 39: 25-28.

Hallberg, K.B. and D.B. Johnson. 2003. Novel acidophiles isolated from moderately acidic mine drainage waters. Hydrometallurgy 71: 139-148.

Hallberg, K.B., M. Dopson, and E.B. Lindstrom. 1996. Reduced sulphur compound oxidation by *Thiobacillus caldus*. J. Bacteriology 178: 6-11.

Harrison, A.P. 1982. Genomic and physiological diversity amongst strains of *Thiobacillus ferrooxidans* and genomic comparison with *Thiobacillus thiooxidans*. Arch. Microbiology 131: 68-76.

Harrison, A.P. 1984. The acidophilic *Thiobacilli* and other acidophilic bacteria that share their habitat. Annu. Rev. Microbiol. 38: 265-292.

Harvey, T.J., N. Holder, and T. Stanek. 2002a. Thermophilic bioheap leaching of chalcopyrite concentrates. The European Journal of Mineral Processing and Environmental Protection 2: 253-263.

Harvey, T.J., W. Van Der Merwe, and K. Afewu. 2002b. The application of the GeoBiotics GEOCOAT® biooxidation technology for the treatment of sphalerite at Kumba resources' Rosh Pinah mine. Minerals Engineering 15: 823-829.

Harwey, P.I. and F.K. Crundwell 1997. Growth of *Thiobacillus ferrooxidans*: A novel experimental design for batch growth and bacterial leaching studies. App. and Env. Microbiology 63: 2586-2592.

Hiskey, J.B. 1994. In-situ leaching recovery of copper—what's next? *In:* Biomine'94, Int. Conference and Workshop Applications of Biotechnology to the Minerals Industry. Australian Mineral Foundation, Adelaide, Australia.

Hiskey, J.B. and M.E. Wadsford. 1975. Galvanic conversion of chalcopyrite. IMM Met. Trans. B 6B: 183-190.

Ingledew, W.J. 1982. *Thiobacillus ferrooxidans*: The bioenergetics of an acidophilic chemolithotroph. Biochimica et Biophysica Acta 683: 89-117.

Ingledew, W.J. 1986. Ferrous iron oxidation by *Thiobacillus ferrooxidans*. In: H.L. Ehrlich and D.S. Holmes [eds.]. Workshop on Biotechnology for the Mining Metal Refining and Fossil Fuel Processing Industries, Biotechnology and Bioengineering Symp., No. 16. John Wiley & Sons, New York, USA. pp. 22-33.

Jensen, A.B. and C. Webb. 1995. Ferrous sulphate oxidation using *Thiobacillus ferrooxidans*: A review. Process Biochemistry 30: 225-236.

Johnson, D.B. 1998. Biodiversity and ecology of acidophilic microorganisms. FEMS Microbiology Ecology 27: 307-317.

Johnson, D.B. 2001. Importance of microbial ecology in the development of new mineral technologies. Hydrometallurgy 59: 147-157.

Johnson, D.B. 2006. Biohydrometallurgy and the environment–Intimate and important interplay. Hydrometallurgy 83: 153-166.

Johnson, D.B. and F.F. Roberto. 1997. Heterotrophic acidophiles and their roles in the bioleaching of sulphide minerals. *In:* D.E. Rawlings [ed.]. Biomining: Theory, Microbes and Industrial Processes. Springer-Verlag, Berlin, Germany. pp. 259-279.

Johnson, D.B, D.A. Body, T.A.M. Bridge, D.F. Bruhn, and F.F. Roberto. 2001. Biodiversity of acidophilic moderate thermophiles isolated from two sites in Yellowstone National Park and their roles in the dissimilatory oxido-reduction of iron. *In:* A.L. Resenbach and A. Voytek [eds.]. Biodiversity, Ecology and Evolution of Thermophiles in Yellowstone National Park. Plenum Press, New York, USA. pp. 23-29.

Johnson, D.B., N. Okibe, and F.F. Roberto. 2003. Novel thermo-acidophilic bacteria isolated from geothermal sites in Yellowstone National Park: physiological and phylogenetic characteristics. Arch. Microbiology 180: 60-68.

Jordan, M.A. 1993. The oxidation of base metal sulphides and mechanisms and preferential release of ferrous iron. PhD Thesis, Camborne School of Mines, University of Exeter, UK.

Jyothi, N., K.N. Sudha, G.P. Brahmaprakash, and G.R. Rao. 1989. Electrochemical aspects of bioleaching of mixed sulphides. *In:* B.J. Scheiner, F.M. Doyle, and S.K. Kawatra [eds.]. Biotechnology in Mineral and Metal Processing Society of Mining Engineers, AIME, Colorado, USA. pp. 9-16.

Karavaiko, G.I., S.I. Kuznetsov, and A.I. Golonizik. 1977. The Bacterial Leaching of Metals from Ores. Translated by Burns W., Technicopy Ltd., London. UK.

Kelly, D.P. 1985. Physiology of the *Thiobacilli:* Elucidating the sulphur oxidation pathways. Microbiological Sciences 2: 105-109.

Kelly, D.P. 1988. Evaluation of the understanding of the microbiology and biochemistry of the mineral leaching habitat. *In:* P.R. Norris and D.P. Kelly. [eds.]. Biohydrometallurgy 1987: Proc. of the Int. Symp., Science and Technology Letters, Kew, Surrey, UK. pp. 3-13.

Kelly, D.P. and A.P. Wood. 2000. Reclassification of some species of *Thiobacillus* to the newly designated genera *Acidithiobacillus* gen. nov., *Halothiobacillus* gen. nov. and *Thermithiobacillus* gen. nov. Int. J. of Syst. and Evol. Microbiology 50: 511-516.

Komnitsas, C. and F.D. Pooley. 1991. Optimisation of the bacterial oxidation of an arsenical gold sulphide concentrate from Olympias, Greece. Minerals Engineering 4: 1297-1303.

Konishi, Y., M. Tokushige, and A. Asai. 1999. Bioleaching of chalcopyrite concentrate by acidophilic thermophile *Acidianus brierleyi. In:* R. Amils and A. Ballester [eds.]. Biohydrometallurgy and the Environment toward the Mining of the 21st Century, Part A, IBS'99. Elsevier, Amsterdam, The Netherlands. pp. 777-786.

Lawrence, R.W. and P.B. Marchant. 1988. Comparison of mesophilic and thermophilic oxidation systems for the treatment of refractory gold ores and concentrates. *In:* P.R. Norris and D.P. Kelly [eds.]. Biohydrometallurgy 1987: Proc. of the Int. Symp., Science and Technology Letters, Kew, Surrey, UK. pp. 359-374.

Leathen, W., N.A. Kinsel, and I.A. Braley. 1956. *Ferrobacillus ferrooxidans*: A chemosynthetic autotrophic bacterium. J. Bacteriology 72: 700-704.

Le Roux, N.W. and D.S. Wakerley. 1988. Leaching of chalcopyrite ($CuFeS_2$) at 70°C, using *Sulfolobus*. *In:* P.R. Norris and D.P. Kelly [eds.]. Biohydrometallurgy 1987: Proc. of the Int. Symp., Science and Technology Letters, Kew, Surrey, UK. pp. 305-317.

Leduc, L.G. and G.D. Ferroni. 1994. The chemolithotrophic bacterium *Thiobacillus ferrooxidans*. FEMS Microbiology Reviews 14: 103-120.

Le Roux, N.W., D.S. Wakerley, and S.D. Hunt. 1977. Thermophilic *Thiobacillus*-type bacteria from Icelandic thermal areas. J. Gen. Microbiology 100: 197-201.

Lundgren, D.G. and M. Silver. 1980. Ore leaching by bacteria. Annu. Rev. Microbiol. 34: 263-283.

Marsh, R.M. and P.R. Norris. 1983. The isolation of some thermophilic, autotrophic iron- and sulphur-oxidising bacteria. FEMS Microbilogy Letters 17: 311-315.

Miller, D.M., D.W. Dew, A.E. Norton, P.M. Cole and G. Benetis. 1997. The BioNIC process: Descriptions of the process and presentation of pilot plant results. *In:* W.C. Cooper and I. Mihaylov [eds.]. Nickel–Cobalt 97: Hydrometallurgy and Refining of Nickel and Cobalt CIM. Montreal, Canada. pp. 97-110.

Miller, P.C. 1997. The design and operating practice of bacterial oxidation plant using moderate thermophiles. *In*: D.E. Rawlings [ed.]. Biomining: Theory, Microbes and Industrial Processes. Springer-Verlag, Berlin, Germany. pp. 81-100.

Miller, P.C., M.K. Rhodes, R. Winby, A. Pinches, and P.J. Van Staden. 1999. Commercialisation of bioleaching for metal extraction. Minerals and Metallurgical Processing 16: 42-50.

Monod, J. 1949. Growth of bacterial cultures. Annu. Rev. Microbiol. 3: 371-394.

Mwaba, C.C. 1993. Application of biotechnology in the mineral, metal refining and fossil-fuel processing industries. *In:* XVIII Int. Mineral Processing Congress, 23-28 May, Sydney, Australia. pp. 1101-1109.

Natarajan, K.A. 1988. Electrochemical aspects of bioleaching multisulphide minerals. Minerals and Metallurgical Processing 5: 61-65.

Natarajan, K.A. 1990. Electrochemical aspects of bioleaching of base metal sulphides. *In*: H.L. Ehrlich and C.L. Brierley [eds.]. Microbial Mineral Recovery. McGraw-Hill, New York, USA. pp. 79-106.

Natarajan, K.A. and I. Iwasaki. 1974. Eh measurements in hydrometallurgical systems. Minerals Sci. Eng. 6: 35-44.

Nemati, M. and S.T.L. Harrison. 2000. A comparative study on thermophilic and mesophilic biooxidation of ferrous iron. Minerals Engineering 13: 19-24.

Nemati, M., S.T.L. Harrison, G.S. Hansford, and C. Webb. 1998. Biological oxidation of ferrous sulphate by *Thiobacillus ferrooxidans*: a review on the kinetic aspects. Biochemical Engineering Journal 1: 171-190.

Nemati, M., J. Lowenadler, and S.T.L. Harrison. 2000. Particle size effects in bioleaching of pyrite by acidophilic thermophile *Sulfolobus metallicus* (BC). App. Microbial Biotechnology 53: 173-179.

Norris, P.R. 1983. Iron and mineral oxidation with *Leptospirillum*-like bacteria. *In:* G. Rossi and A.E. Torma [eds.]. Recent Progress in Biohydrometallurgy. Associazione Mineraria Sarda. Iglesias, Italy. pp. 83-96.

Norris, P.R. 1989a. Factors affecting bacterial mineral oxidation: The example of carbon dioxide in the context of bacterial diversity. *In:* J. Salley, R.G.L. McCready and P.L. Wichlacz [eds.]. Biohydrometallurgy, SP89-10. CANMET Ottawa, Canada. pp. 3-14.

Norris, P.R. 1989b. Mineral oxidising bacteria: Metal-organism interactions. *In:* R.K. Poole and G.M. Gadd [eds.]. Metal–Microbe Interactions. Society for General Microbiology, IRL Press, Oxford, UK. pp. 99-119.

Norris, P.R. 1990. Acidophilic bacteria and their activity in mineral sulphide oxidation. *In:* H.L. Ehrlich and C.L. Brierley [eds.]. Microbial Mineral Recovery. McGraw-Hill, New York, USA. pp. 3-27.

Norris, P.R. 1997. Thermophiles and bioleaching. *In:* D.E. Rawlings [ed.]. Biomining: Theory, Microbes and Industrial Processes. Springer-Verlag, Berlin, Germany. pp. 247-258.

Norris, P.R. and D.W. Barr. 1985. Growth and iron oxidation by acidophilic thermophiles. FEMS Microbiology Letters 28: 221-224.

Norris, P.R. and J.P. Owen. 1993. Mineral sulphide oxidation by enrichment cultures of novel thermoacidophilic bacteria. FEMS Microbilogy Reviews 11: 51-56.

Norris, P.R., L. Parrot, and R.M. Marsh. 1986. Moderately thermophilic mineral oxidising bacteria. *In:* H.L. Ehrlich and D.S. Holmes [eds.]. Workshop on Biotechnology for the Mining Metal Refining and Fossil Fuel Processing Industries, Biotechnology and Bioengineering Symp., No. 16. John Wiley & Sons, New York, USA. pp. 253-262.

Norris, P.R., D.W. Barr, and D. Hindson. 1988. Iron and mineral oxidation by bacteria: Affinities for iron and attachment to pyrite. *In:* P.R. Norris and D.P. Kelly [eds.]. Biohydrometallurgy 1987: Proc. of the Int. Symp., Science and Technology Letters, Kew, Surrey, UK. pp. 43-59.

Norris, P.R., D.A. Clark, J.P. Owen, and S. Waterhouse. 1996. Characteristics of *Sulfobacillus acidophilus* sp. nov. and other moderately thermophilic mineral-sulphide-oxidising bacteria. Microbiology 142: 775-783.

Norris, P.R., N.P. Burton, and N.A.M. Foulis. 2000. Acidophiles in bioreactor mineral processing. Extremophiles 4: 71-76.

Ojumu, T.V., J. Petersen, G.E. Searby, and G.S. Hansford. 2006. A review of rate equations proposed for microbial ferrous-iron oxidation with a view to application to heap bioleaching. Hydrometallurgy 83: 21-28.

Olson, G.J., J.A. Brierley, and C.L. Brierley. 2003. Bioleaching review, Part B: Progress in bioleaching: applications of microbial processes by the minerals industries. Appl. Microbiol. Biotechnol. 63: 249-257.

Osseo-Asare, K. 1992. Semiconductor electrochemistry and hydrometallurgical dissolution process. Hydrometallurgy 29: 61-90.

Pogliani, C. and E. Donati. 1999. The role of exopolymers in the bioleaching of a non-ferrous metal sulphide. J. Ind. Microbiology & Biotechnology 22: 88-92.

Pronk, J.T. and D.B. Johnson. 1992. Oxidation and reduction of iron by acidophilic bacteria. Geomicrobiology Journal 10: 153-171.

Qui, M., S. Xiong, W. Zhang, and G. Wang. 2005. A comparison of bioleaching of chalcopyrite using pure culture or a mixed culture. Minerals Engineering 18: 987-990.

Rawlings, D.E. 1997. Mesophilic, autotrophic, bioleaching bacteria: Description, physiology and role. *In*: D.E. Rawlings [ed.]. Biomining: Theory, Microbes and Industrial Processes. Springer-Verlag, Berlin, Germany. pp. 229-245.

Rawlings, D.E. 2002. Heavy metal mining using microbes. Annu. Rev. Microbiol. 56: 65-91.

Rawlings, D.E., H. Tributsch, and G.S. Hansford. 1999a. Reasons why *Leptospirillum*-like species rather than *Thiobacillus ferrooxidans* are the dominant iron oxidising bacteria in many commercial processes for the biooxidation of pyrite and related ores. Microbiology 145: 5-13.

Rawlings, D.E., N.J. Coran, M.N. Gardner, and S.M. Deane. 1999b. *Thiobacillus caldus* and *Leptospirillum ferrooxidans* are widely distributed in continuous flow biooxidation tanks used to treat a variety of metal containing ores/ concentrates. *In:* R. Amils and A. Ballester [eds.]. Biohydrometallurgy and the Environment toward the Mining of the 21st Century, Part A, IBS'99. Elsevier, Amsterdam, The Netherlands. pp. 777-786.

Rawlings, D.E., D. Dew, and C. du Plessis. 2003. Biomineralization of metal containing ores and concentrates. Trends in Biotechnology 21: 38-44.

Robbins E.I. 2000. Bacteria and archaea in acidic environments and a key to morphological identification. Hydrobiologia 433: 61-89.

Robertson, W.J., P.H.M. Kinnunen, J.J. Plumb, P.D. Franzmann, J.A. Puhakka, J.A.E. Gibson, and P.D. Nichols. 2002. Moderately thermophilic iron oxidising bacteria isolated from a pyritic coal deposit showing spontaneous combustion. Minerals Engineering 15: 815-822.

Rodriguez-Leiva, M. and H. Tributsch. 1988. Morphology of bacterial leaching patterns by *Thiobacillus ferrooxidans* on synthetic pyrite. Arch. Microbiology 149: 401-405.

Rossi, G. 1990. Biohydrometallurgy. McGraw-Hill, Hamburg, Germany.

Rohwerder, T., T. Gehrke, K. Kinzler, and W. Sand. 2003. Bioleaching review, Part A: Progress in bioleaching—fundamentals and mechanism of bacterial metal sulphide oxidation. Appl. Microbiol. Biotechnol. 63: 239-248.

Sampson, M.I. and C.V. Phillips. 2001. Influence of base metals on the oxidising ability of acidophilic bacteria during the oxidation of ferrous sulfate and mineral sulfide concentrates using mesophiles and moderate thermophiles. Minerals Engineering 14: 317-340.

Sand, W., T. Gehrke, and R. Hallman. 1993. In-situ bioleaching of metal sulphides: Importance of *Leptospirillum ferrooxidans*. *In:* A.E. Torma, J.E. Wey, and V.I. Lakshmanan [eds.]. Biohydrometallurgical Technologies, Vol. 1: Proc. of Int. Biohydrometallurgy Symp., The Minerals, Metals and Materials Society. Jackson Hole, Wyoming, USA. pp. 15-27.

Sand, W., T. Gehrke, R. Hallman, and A. Schippers. 1995. Sulphur chemistry, biofilm and the (in) direct attack mechanism—a critical evaluation of bacterial leaching. Appl. Microbiol. Biotechnology 43: 961-966.

Sand, W., T. Gehrke, P.G. Jozsa, and A. Schippers. 2001. (Bio)chemistry of bacterial leaching—direct vs. indirect bioleaching. Hydrometallurgy 59: 159-175.

Segerer, A., A. Neuner, J.K. Kristjanssen, and K.O. Stetter. 1986. *Acidianus infernus* gen. nov., sp. nov., and *Acidianus brierleyi* comb. nov.: Facultatively aerobic extremely acidophilic thermophilic sulphur metabolising archaebacteria. Int. J. Sys. Bacteriology 36: 559-564.

Silver, M. 1978. Metabolic mechanisms of iron oxidising *Thiobacilli*. *In:* E.L. Murr A.E. Torma, and J.A. Brierley [eds.]. Metallurgical Applications of Bacterial Leaching and Related Microbiological Phenomena. Academic Press, New York, USA. pp. 3-17.

Silverman, M.P. and D.G. Lundgren. 1959. Studies on the chemolithotrophic iron bacterium *Ferrobacillus ferrooxidans*: I. An improved medium and harvesting procedure for securing high cell yields. J. Bacteriology 77: 642-677.

Singer, P.C. and W. Strumm. 1970. Acidic mine drainage: The rate determining step. Science 167: 1121-1123.

Steemson, M.L,. F.S. Wong, and B. Goebel, 1997. The integration of zinc bioleaching with solvent extraction for the production of zinc metal from zinc concentrates, M1.4. *In:* Biomine 97, Int. Biohydrometallurgy Symposium, IBS97. Australian Mineral Foundation, Glenside, Australia. pp. 1-10.

Suzuki, I. 2001. Microbial leaching of metals from sulphide minerals. Biotechnology Advances 19: 119-132.

Suzuki, I., C.W. Chan, R. Vilar, and T.L. Takeuchi. 1993. Sulphur and sulphide oxidation by *Thiobacillus thiooxidans*. *In:* A.E. Torma, J.E. Wey, and V.I. Lakshmanan [eds.]. Biohydrometallurgical Technologies, Vol. 1: Proc. of Int. Biohydrometallurgy Symp., The Minerals, Metals and Materials Society. Jackson Hole, Wyoming, USA. pp. 109-116.

Torma, A.E. 1977. The role of *Thiobacillus ferrooxidans* in hydrometallurgical processes. *In:* T.K. Ghose, A. Fiechter, and N. Blakebrough [eds.]. Advances in Biochemical Engineering. Springer-Verlag, Berlin, Germany. pp. 1-37.

Torma, A.E. 1987. Impact of biotechnology on mineral extractions. Mineral Processing and Extractive Metallurgy Review 2: 230-289.

Torma, A.E. and K. Bosecker. 1982. Bacterial leaching. Progress in Ind. Microbiol. 6: 77-118.

Tributsch, H. 2001. Direct versus indirect bioleaching. Hydrometallurgy 59: 177-185.

Tuovinen, O.H. 1990. Biological fundamentals of mineral leaching processes. . *In*: H.L. Ehrlich and C.L. Brierley [eds.]. Microbial Mineral Recovery. McGraw-Hill, New York, USA. pp. 55-57.

Tuovinen, O.H. and D.P. Kelly. 1973. Studies on the growth of *Thiobacillus ferrooxidans*: I. Use of membrane filters and ferrous iron agar to determine viable numbers and comparison with CO_2 fixation and iron oxidation as measure of growth. Arch. Microbiology 88: 285-298.

Tuovinen, O.H. and T.M. Bhatti. 1999. Microbiological leaching of uranium ores. Minerals and Metallurgical Processing 16: 51-60.

Tuovinen, O.H., S.I. Niemela, and H.G. Gyuenberg. 1971. Tolerance of *Thiobacillus ferrooxidans* to some metals. Antonie van Leeuwenhoek 37: 489-496.

van Aswegen, P.C. and P.C. Marais. 1999. Advances in the application of the BIOX® process for refractory gold ores. Minerals and Metallurgical Processing 16: 61-68.

van Staden, P.J. 1998. The Mintek Bactech copper bioleach process. *In:* ALTA 98: Copper Sulphides Symposium. ALTA Metallurgical Services, Brisbane, Australia.

Watling, H.R. 2006. The bioleaching of sulphide minerals with emphasis on copper sulphides—A review. Hydrometallurgy 84: 81-108.

Witne, J.Y. and C.V. Phillips. 2001. Bioleaching of Ok Tedi copper concentrate in oxygen- and carbon dioxide-enriched air. Minerals Engineering 14: 25-48.

CHAPTER 5

Petroleum Geomicrobiology

Elijah Ohimain

INTRODUCTION

Petroleum, which consists of both crude oil and natural gas, is the world's most important energy source accounting for over 50% of world energy consumption (Fig. 5.1). Petroleum related products are universally used. Oil and gas are the energy that fuels the transport sector worldwide. Airplanes, vehicles, trains use petroleum products such as fuel including gasoline, diesel, kerosene etc. Natural gas (liquefied petroleum gas,

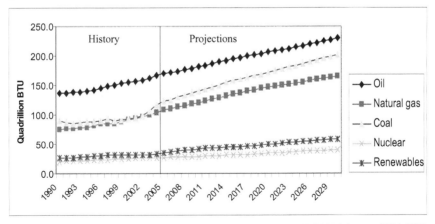

FIG. 5.1 World Energy Consumption by type, 1990-2030 (1990-2005 History; 2006-2030 Projections) (cited as Hakes, 2000)

Biological Sciences Department, Faculty of Science, Niger Delta University, Wilberforce Island, Amassoma, Bayelsa State, Nigeria, E-mail: eohimain@yahoo.com

liquefied natural gas and compressed natural gas) are used as cooking gas, fuel for gas turbines for electricity generation and heaters. Products from petroleum are also used in the chemical manufacturing sector for the production of plastics, nylons, pharmaceuticals, tyres, textiles to mention a few. Despite the advances in alternate energy, petroleum is still the energy of choice worldwide. Petroleum virtually drives the world's economy.

On the other hand, microbial biotechnology has found applications in diverse fields including agriculture, medicine, pharmaceuticals, industries etc. Microbial technology is increasingly being considered for the provision of alternative renewable energy. Despite its relevance in the energy sector, the role of microbial technology in petroleum engineering is often not appreciated. For instance, microbial enhanced oil recovery (MEOR) is now applied for the commercial production of oil from depleted reservoirs. It has also been applied to secondary or tertiary oil recovery measures such as water/chemical flooding. Microbes have also been used to selectively plug reservoirs to divert water flooding operations to less permeable zones for enhanced oil recovery. Microbes have also been applied for the repair of faulty fractures and reservoir stimulation. Microbes produce several products such as acids, gases, solvents, polymers, and surfactants, which aid hydrocarbon production. For instance, microbial production of methane by methanogenic bacteria helps to re-pressurize the reservoir. Also, the microbial production of surfactants has been applied in well bore cleaning, to enhance oil recovery and clean up of oil spills. Other methods of oil spill clean up including mechanical and chemical, are unable to completely remove/ recover all the spilled oil and in some cases residual oil after clean up could exceed permissible levels. Besides, mechanical and chemical methods may generate oily contaminants and residues that need to be managed as well. But microbiological methods, using the natural ability of microbes to biodegrade hydrocarbons, can completely mineralize oil spills. Microorganisms have also found applications in the emerging bioenergy sector. For instance, fermentative microorganisms have been used to produce fuel grade ethanol from food crops particularly sugarcane, cassava and their processing effluents. Also, biogas is currently being produced from human and agricultural wastes in many developing countries as sources of cooking gas and electricity.

However, not all activities of microorganisms are beneficial to the energy sector, some are indeed deleterious. Among the deleterious activities of microbes in the oil field environment is the microbial assisted corrosion of oil installations including flow lines, pipelines, liners, topsides and other production facilities. Some microbes such as sulphate-reducing bacteria produce hydrogen sulphide, which foul crude oil in addition to causing corrosion. Corrosion often leads to oil spillages, which can cause environmental damage. Microbes have also been implicated in the plugging of reservoirs especially, which could reduce the efficiency of water flooding operations. Uncontrolled proliferation of microbes in the reservoir could reduce the porosity and permeability of the formation. Also, the microbial degradation of drilling mud could make the mud less effective in controlling reservoir pressures during oil well drilling and work over operations, which could cause well blow up. Microbes are also involved in the degradation of petroleum and petroleum products. Following dredging operations to create access for oil exploration especially in mangrove wetlands, microbes often cause acidification through the oxidation of exposed sedimentary pyrite, which result in water and heavy metal pollution, fish and vegetation killing, corrosion of metals etc. It is therefore because of the beneficial and deleterious activities of microbes in the petroleum energy sector that underscore the need for petroleum microbiologists in the oil sector to enhance the benefits while mitigating the deleterious activities. The aim of this chapter, therefore, is to present a general overview of petroleum industrial activities starting from the origin of petroleum, through exploration, well drilling to production of petroleum but with emphasis on geomicrobial and biotechnology applications.

THE ORIGIN OF PETROLEUM

The origin of petroleum is still contentious. There are numerous theories on the origin of petroleum. Two of the most accepted theories are those of biogenic and abiogenic origin. Arguments have been advanced for each theory, though conflicting, both theories have been generally accepted.

In the biogenic (organic) theory, petroleum is said to originate from dead organic matter from plant, animal and microbial remains. These organic materials, both from land and marine origin, are transported to the marine sediment, where the action of microorganism leads to their partial decomposition and modification. In the sediment, these organic precursors mix with the mud, silt and sand. After several years of deposition and with increasing overburden, anaerobic conditions set in. Through physical, biological and chemical transformation the organic precursors are transformed to oil and gas in sedimentary rocks, which are often called source rocks. Through Earth movement and physical disturbances such as earthquakes and tectonics, oil migrate from the porous formation until their movement is stopped by impermeable cap rock overlying the porous and permeable rocks, which are also called reservoirs (Baker 1983).

In the abiotic (inorganic) theory, oil is deposited in the Earth's crust from materials incorporated into the mantle at the time of the Earth formation. Out-gassing process then transported the oil into porous and permeable sedimentary rocks. Although, the abiotic and primeval origin of petroleum is becoming more popular, Ourisson et al. (1984) through chemical analysis of varied organic sediments including coal and petroleum reveal a surprising commonality: all derive much of their organic matter from microbial lipids called bacteriohopanetetrol. Also, the presence of microorganisms in the sub-surface oil field environment and deep oil reservoirs, which have been reported by several researchers (Cord-Ruwich et al. 1987, Magot et al. 2000, Birkeland 2005, Jeanthon et al. 2005, Magot 2005, Ollivier and Cayol 2005) tend to give credence to the biotic origin of petroleum, which reinforces the importance of geomicrobiology in petroleum engineering. Besides, the possible origin of oil from an inorganic source is no longer viewed as having validity based upon the modern analysis of petroleum and sedimentary organic matter, which is today recognized as petroleum precursors.

OVERVIEW OF HYDROCARBON DEVELOPMENT

Petroleum development often commences with exploration activities in an acquired concession. Exploration is done to find hydrocarbon traps. A petroleum trap is an arrangement of rocks that contains an accumulation of hydrocarbons. The arrangement consists of an impermeable cap rock overlying the porous and permeable oil reservoir. There are different types of traps including anticlines, faulting, unconformities, salt domes or plugs (Baker 1983). Exploration is the technique used to search for oil. There are various exploration methods, seismic exploration being the most common. Seismic exploration is well documented in literature, hence this chapter shall focus on geomicrobial exploration methods only.

More often, exploratory drilling is carried out to confirm the presence of hydrocarbon in commercial quantity. If successful, more oil wells are drilled covering the entire reservoir. Oil drilling typically involves the use of drilling bits, which drill a well with the assistance of drilling mud into the formation. On drilling to the total depth, the wells are completed by installing casings and Christmas trees through cementation. Note that there are various types of well completion. The well formation is tested through well logging and testing operations. The wells are finally brought to production by perforating the completion, connecting the wells to a flow station using a flow line. The flow station receives the crude oil through flow lines and is transported through pipelines to oil terminals or refineries for export or refining respectively. Details of the seismic exploration methods, well drilling, well completion and perforation, well testing, installation of surface facilities (well heads, flow lines, flow station and pipelines) and hydrocarbon production are all beyond the scope of this chapter. Some relatively non-technical literature covering these topics are Baker (1983, 1994), Conaway (1999), Devereux (1999), Hyne (2001), Planckaert (2005), Miesner and Leffler (2006), Raymond and Leffler (2006). However in subsequent sections, the application of geomicrobial biotechnologies in petroleum development shall be the main focus, starting with geomicrobial processes triggered by the dredging of oil exploration access canal in mangrove ecosystems, followed by petroleum reservoir microbiology and application of biotechnology in oil and gas exploration.

Geomicrobiological Processes Following the Dredging of Oil Development Access Canals

Hydrocarbon accumulation is fairly distributed globally in the various regions under different climatic conditions including the arid Middle East, tundra (Russia, Alaska – USA, Canada, and North Sea) and tropical climates (Africa, Asia, South America and the Gulf of Mexico). In the tropics, most of the successfully explored hydrocarbon prospects are found in coastal wetlands especially in the Gulf of Mexico, West Africa, South America and South East Asia. Most of these wetlands are dominated by mangrove and freshwater vegetation. Generally, mangrove wetlands are spawning grounds for coastal and marine fisheries and provide feeding and nesting habitats for migratory species. Wetlands are also important in shoreline protection. However, oil exploration in mangrove areas is challenged by access difficulties, which the oil companies often overcome by carrying out dredging. Dredging in these wetlands is often carried out to create safe navigable accesses for oil and gas exploitation. During dredging, sediments and soils are removed from the right of way, placed along canal banks mostly upon fringing mangroves and abandoned. This often results in the killing of the mangroves and other biological entities that depend on it including fisheries. Several hectares of mangroves all over the tropical world fringing most of the creeks where oil exploration-related dredging has taken place have been killed likewise. There is a possible microbial role in the weathering of dredged materials that is linked to acidification and the environmental impacts. In this sub-section, the focus shall be on geomicrobiological processes involved in the formation and weathering of sulphidic dredged materials leading to wetland acidification.

Worldwide, coastal mangroves are known to contain reduced iron sulphide known as pyrite. These pyrites when undisturbed under water cover are innocuous, but their disturbance often results in severe acidification and ecosystem damage (Ohimain 2004a, 2006).

The microbial reduction of sulphate is the origin of pyrite in coastal plains (Berner 1970, van Breemen 1976, 1982, Howarth 1979, Goldhaber and Kaplan 1982, Pons et al. 1982). In many coastal regions all over the tropical world, clayey sediments are deposited over extensive areas under brackish to saline conditions. Such tidal marshes are normally covered by dense mangrove forests. Abundance of organic matter from mangrove trees, soil/sediment saturated with seawater containing high concentrations of sulphate under reducing conditions creates suitable conditions for the formation of metal sulphides mediated by sulphate-reducing bacteria. If iron is available from ferric oxides or iron-bearing silicates, much of the sulphide produced is fixed as pyrite (cubic FeS_2) and to a lesser extent iron monosulphide (FeS). The greater part of the alkalinity (HCO_3) formed during sulphate reaction is moved to the sea by tidal action leaving a potentially acid residue as pyrite. The microbial formation of pyrite may be represented by the following overall equation and presented diagrammatically in Fig. 5.2.

$$Fe_2O_{3(s)} + 4SO_{4\ (aq)}^{2-} + 8CH_2O + 1/2O_{2(aq)} \rightarrow 2FeS_{2(s)} + 8HCO^-_{3(aq)} + 4H_2O$$

| Iron III Oxide from sediments | sulphate ions from seawater | organic matter | dissolved oxygen | pyrite | carbonate |

The exposure and abandonment of pyritic dredged materials following the dredging of oil well access canals in mangrove areas, produce acidic drainage resulting from the interactions of sulphide minerals (pyrite) with oxygen, water, and bacteria particularly *Acidithiobacillus* sp. The overall stoichiometric reaction describing the oxidation of pyrite is summarized as follows:

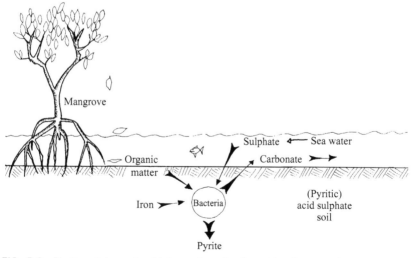

FIG. 5.2 Biogenesis/geomicrobiology of pyrite formation in coastal mangrove soils (Smith et al. 1995)

$$FeS_2(s) + 3.75O_2 + 3.5H_2O = Fe(OH)_3(s) + 2SO_4^{2-} + 4H^+$$

This reaction leads to the production of sulphuric acid causing the pH of the drainage to be low, often <3. At such a low pH, heavy metals are released and mobilized in the drainage. This low pH often leads to severe acidification with attendant consequences including heavy metal pollution, vegetation dieback, reduced plant/animal productivity, corrosion of steel, concrete, and other engineering structures, degradation of surface and groundwater quality, mortality of estuarine biota especially fishes and bioaccumulation of pollutant (Ohimain 2006, Ohimain et al. 2008a, 2008b). Figure 5.3 shows the process of microbial acid formation following the dredging and abandonment of pyritic sediments. Figure 5.4 shows ecosystem damage following the dredging of an oil flow line access canal in the Niger Delta. The pH of this site was

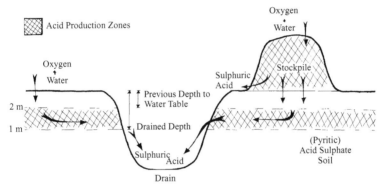

FIG. 5.3 Geomicrobial oxidation of pyrite causing the release of sulphuric acid (Smith et al. 1995)

FIG. 5.4 Vegetation damage caused by acidification from a recently dredged oil well access canal (Note: the orange coloured discharges from the backswamp indicate acidification).

<3 and both fish kill and vegetation dieback was reported at the site (Ohimain 2003). The short- and longterm impacts of acidification on fisheries are presented in Table 5.1. Dredged materials are commonly found fringing production facilities in the Niger Delta (Fig. 5.5) and elsewhere including the Gulf of Mexico (Turner and Streever 2002).

TABLE 5.1 Short-term and long-term impacts of acidification on fisheries

Short-term effects	Long-term effects
• Fish kills	• Loss of habitat
• Fish disease	• Persistent iron coating
• Mass mortalities of microscopic organism	• Alterations to water plant communities
• Increased light penetration due to lower water turbidity	• Invasion by acid-tolerant water plants
• Loss of acid-sensitive crustaceans	• Reduced spawning success due to stress
• Destruction of fish eggs	• Chemical migration barriers
• Oyster mortality	• Reduced food resources
	• Dominance by acid-tolerant plankton species
	• Growth abnormalities
	• Reduced growth rates
	• Changes in food chain and web
	• Damaged and undeveloped eggs
	• Reduced recruitment
	• Higher water temperatures due to increased light penetration
	• Increase availability of toxic elements
	• Reduced availability of nutrients
	• Poor growth of oysters and other bivalves

(Modified from Sammut and Lines-Kelly 2000)

Petroleum Reservoir Microbiology

The presence of microorganisms in the sub-surface oil field environment and deep oil reservoirs, has been extensively reported by several researchers (Cord-Ruwich et al. 1987, Bubela 1989, Sharma et al. 1993, Bhupathiraju 1999, Krumholz et al. 1999, Magot et al. 2000, Birkeland 2005, Jeanthon et al. 2005, Magot 2005, Ollivier and Cayol 2005). But what

FIG. 5.5 Abandoned dredged materials fringing oil wells and production platforms in the mangrove ecosystem of the Niger Delta.

is however not clear is whether the bacteria found in the reservoirs are indigenous or inadvertently introduced during well drilling or reservoir water flooding operations. Notwithstanding, the roles of reservoir microorganisms are now becoming clearer. However, a study carried out by Long et al. (1996) clearly demonstrated that bacteria, which exist in the sub-surface, may have originated from the Earth crust. Though, there is still the challenge of aseptically obtaining core samples from oil reservoirs, but with the recent advances in science and technology, pre-sterilized sleeves are now available, which can collect core samples aseptically at various depths for laboratory analysis (Bass and Lappin-Scott 1997). For microbes to proliferate in oil reservoirs, they may have to overcome the extreme environmental conditions prevailing in the reservoir (Table 5.2). Studies have shown that some reservoir organisms are indeed extremophiles and can grow, multiply and carry out metabolic activities even under extreme reservoir conditions such as high temperatures (>125°C), and wide pH range (1-11) and 30% salinity (Bass and Lappin-Scott 1997). Jinfeng et al. (2005) successfully carried out field trial of microbial assisted water flooding at a temperature of 73°C and salinity of 16,790 mg/l. Though the proximity of the oil–water interface has been shown to affect the composition of oil or gas, the activities of

TABLE 5.2 Extreme environmental conditions in oil reservoirs

Temperature (°C)	22-97
Porosity (%)	11-36
Permeability (mD)	10-8100 (high fractured)
Oil type	heavy asphaltic–light paraffinic
Rock type	sandstone–limestone

Source: Hitzman 1983

microbes in the reservoir often alter the mineralogical, hydrological, and geological condition of the reservoir including its rocks and fluids properties (Bubela 1989). The most significant geological changes are:

• Precipitation of dissolved minerals especially carbonates
• Change of permeability due to precipitation in pore throats
• Change in the crude oil properties
• Production of methane gas to pressurize the reservoir
• A change of porosity, either increase or decrease, depending on the equilibria of dissolved salts and products of organic acids.

Details of petroleum reservoir engineering are beyond the scope of this chapter, but can be found in Amyx et al. (1960). However, the focus of this chapter is on the role of microbes in oil reservoirs. There are both positive and negative influences of bacteria in oil reservoirs. Studies have shown that reservoir microbes could be stimulated to enhance oil production. This aspect is dealt with detail in the next section, but it is important to understand that during microbially enhanced oil recovery, several changes occur in both the fluid and rock properties. For instance, the microbial production of organic acids, surfactants and alcohols decrease the viscosity of heavy crude oil and make it easier for production. These products and the direct activities of reservoir microbes have been associated with the alteration of formation porosity and permeability (Bubela 1989, Bass and Lappin-Scott 1997). It is generally believed that the microbial modification of permeability contributes to the enhancement of oil recovery (Zekri and El-Mehaideb 2003, Jinfeng et al. 2005). Studies have also shown that methanogenic bacteria existing in the reservoir could be stimulated by nutrient addition to produce methane gas, which increased reservoir pressures by increasing the gas oil ratios (Nazina et al. 1998, 2002). In-situ microbial growth apparently resulted in chemical and petrophysical changes within the reservoir that lead to positive microbial enhancement of oil recovery. However, several studies showing the movement of bacteria through porous media/ formation have been reported (Sharman and McInerney 1994, Jenneman et al. 1985, 1986, Torbati et al. 1986, Fontes et al. 1991, Sharma et al. 1993, Bhupathiraju 1994, 1999, Krumholz et al. 1999). It is therefore evident from theoretical considerations and from experimental evidence that many factors could influence the penetration of cells in a porous matrix. These include: (1) Physical and chemical properties of the rocks, such as permeability, pore size distribution, porosity, wettability, surface charge, type of oil (e.g. polar versus non-polar), and total salinity and ionic composition of formation water. (2) Cell properties, such as shape, size, motility, type of cell growth (individually or in clumps or chains), surface charge, production of capsules and slimes, chemical reaction products (acids or gases). (3) Mode of injection, such as rate of injection, salt content of injection water and density of cell suspension. After considering all of these factors, it is essential to treat the whole reservoir as a bioreactor and study the displacement efficiency and the applicability of releasing oil by bacteria under reservoir conditions (Donaldson et al. 1989).

Also, microorganisms have caused problems in the reservoir. For example, Sulphate Reducing Bacteria (SRB) have been shown to be responsible for the production of hydrogen sulphide in the reservoir (biofouling of crude oil), thus reducing the oil quality, causing corrosion of steel/metal structures (Cord-Ruwich et al. 1987) including tubings, casings, well head, flow lines and pipelines and causing air pollution during petroleum refining. Davis (1967) reported that the concentration of hydrogen sulphide tends to increase as the oil-producing zone approached the oil–water interface.

Biotechnological Applications in the Petroleum Industry

Several problems encountered in the petroleum industry during drilling and production operations could be solved using biotechnological techniques/approaches. Such challenges include poor recovery; environmental pollution, sour crude, production of oil from shale; sewage and waste management especially waste drilling mud and cuttings. Specific geomicrobiological techniques for tackling these challenges are discussed below.

Microbial enhanced oil recovery

Typically, not all the petroleum existing underground is recoverable. Less than 50% of heavy crude oil and nearly 80% of light oil is recovered using conventional methods even under favourable circumstances. Conventional oil production occurs in three stages: primary, secondary and tertiary. In the primary stage oil extraction results due to the natural reservoir pressure, which exist within the formation causing the fluid to flow to the well bore and then to the surface. When this internal driving force diminishes perhaps due to flow resistance or declining reservoir pressures, downhole pumps or gas lifts are installed to raise oil to the surface until the energy available to drive the reservoir fluid becomes inadequate to maintain reservoir productivity. According to Amro et al. (2007), depending on the reservoir properties, only about 10-15% of the original oil in place is recoverable during primary production. When reservoir pressure diminishes to a point where production declines then the secondary recovery method is applied which involves gas re-injection and water flooding of the oil reservoirs. Oil recovery can be increased to about 40% by the application of secondary recovery methods (Lake et al. 1992, Hite et al. 2004, Amro et al. 2007). When secondary recovery is no longer economical, tertiary recovery methods (also known as enhanced oil recovery methods) involving the use of chemical or thermal energy are be considered. According to Amro et al. (2007), the major causes of poor recoveries during primary and secondary production stages are the existence of interfacial tension between oil and water, high mobility ratios and reservoir heterogeneity. Tertiary recovery methods are therefore designed to overcome these constraints. The recovery method through the use of chemical energy involves chemical and polymer flooding which includes the use of alkaline water floods, carbon dioxide flooding and polymer flooding. Thermal recovery on the other hand includes in-situ combustion and injection of steam. But, all these recovery methods suffer from technological problems and many other constraints (Bubela 1989). Hence, attention is now focused on the use of microbes for enhanced oil recovery.

Microbial enhanced oil recovery (MEOR) or microbial improved oil recovery (MIOR) in a broad sense can be described as the application of the microorganisms and/or their metabolic products to increase the production of oil from reservoirs of marginal productivity. MEOR is a secondary/tertiary oil recovery process whereby microorganisms are used to extract oil from wells through several mechanisms including:

- Increasing the solubility of residual oil in crude oil formations by biosurfactants produced by oil-degrading bacteria;
- Controlling the viscosity of flood solutions during secondary oil recovery by polymers produced by bacteria;
- Re-pressurizing and redirecting the oil flow in wells as a result of carbon dioxide produced by the growth of bacteria injected in the well;

- Reducing the viscosity of the oil by injection of microbes that can metabolize long chain alkanes (e.g. waxes) in the well.

There are several marginal wells in the Niger Delta (Fig. 5.6), which were recently acquired by marginal field operators. Many of these marginal fields were initially suspended because of several reasons such as the presence of heavy crude oil, depleted or low reservoir pressures etc. Microbial enhanced oil recovery process could be useful in the production of marginal wells.

MEOR applications use naturally occurring microorganisms which under the appropriate oil reservoir conditions, (e.g. salinity, pressure, permeability and temperature), may provide economic and environmental advantages compared to conventional recovery methods that are based on water flooding and injection of chemical agents into the well. The economic advantages result from the lower initial and follow-up costs of implementing a MEOR process and also from relatively high oil recovery rates. The environmental advantages result from the fact that conventional injection processes are more energy intensive and use chemicals such as acids (Ah-You et al. 2000). Moreover, bacteria provide a cheap, environmentally acceptable alternate to the several oilfield chemicals used in EOR. They are able to grow on low cost substrates such

2004/ 8/22 10:29am

FIG. 5.6 A marginal field in the Niger Delta

as molasses; brewery and cassava processing wastes and their small size help them penetrate deep into the porous formation where they can act in situ.

Several studies showing the movement of bacteria through porous media/formation have been reported (Sharman and McInerney 1994, Jenneman et al. 1985, 1986, Torbati et al. 1986, Fontes et al. 1991, Sharma et al. 1993, Bhupathiraju 1994, 1999, Krumholz et al. 1999). MEOR relies either on the stimulation of the bacterial population, or introduction of specific bacteria into the reservoir that has some ability to improve recovery rates. For instance, Jinfeng et al. (2005) in a field trial introduced microbes and nutrients into the reservoir to boost oil recovery. Their results indicated that both the injected and indigenous reservoir microbes were stimulated by the nutrient. These microbes enhanced oil recovery through the reduction of surface tension, production of bio-surfactants, change in oil and gas properties (density, viscosity, paraffin and asphaltene content) and increase in oil production.

The promotion of the indigenous bacteria for enhanced oil recovery has the advantage of being cheaper. There are a variety of natural products that bacteria produce which can lead to increased recovery rates. Bio-surfactants allow oil to travel more freely though the reservoir by reducing oil–water interfacial tension releasing the capillary entrapped oil. Organic acids dissolve carbonate rocks, widening fissures and channels. Microbially produced gases such as CO_2, H_2, N_2, and CH_4 can be used to re-pressurize the reservoir, and reduce oil viscosity. Finally microbial biomass and production of *in situ* biopolymers can be used to selectively plug areas of high permeability, thus diverting the water floods to sweep the fresh areas of the reservoir.

Several species of microorganisms have been found in sub-surface petroleum reservoirs at a depth of 1000 m (Grula and Sewell 1983). Stephens et al. (1999, 2000), recorded microorganisms from cores of newly drilled oil wells using an electron microscope. Such mixed microbial cultures are capable of synthesizing a large variety of biochemical products from crude oil constituents when provided with essential nutrients and favourable environmental conditions. The range of metabolic products from microbial attack of petroleum is very broad, depending on environmental conditions (pressure, temperature, salinity, pH, and the presence or absence of oxygen), supporting nutrients available for cell metabolism (nitrogen, phosphorus, etc.) and the specific bacteria interacting with the petroleum. Generally, microbial metabolic products may be gases (methane, hydrogen, carbon dioxide, hydrogen sulphide), organic acids (formic, acetic, valeric), solvents (alcohols, aldehydes, ketones), polymers (proteins, polysaccharides), surface-active compounds (poly-anionic lipids) and many other compounds ranging from simple to very complex macromolecules (Donaldson et al. 1989). The types of products and activities of microbes involved in MEOR are presented in Table 5.3. For instance in a field study, the bacterium *Clostridium tyrobutyricum* that was introduced into a reservoir along with nutrients produced hydrogen, carbon dioxide and methane, which doubled the gas oil ratio in seven months, suppressed the deleterious activities of SRB and increased hydrocarbon recovery (Baas and Lappin-Scott 1997). Specific details of the application of microbial products and/or biomass in enhanced oil recovery can be found in Hitzman (1983), Donaldson et al. (1989), Baas and Lappin-Scott (1997) and McInerney et al. (2005).

TABLE 5.3 Microbial products and activities useful in enhanced oil recovery

Products or activity	Examples	Relevance
Gases	Methane, carbon dioxide	Reservoir re-pressurization, oil viscosity reduction
Acids	Acetic acid, butyric acid, lactic acid	Increase in porosity and permeability
Solvents	Ethanol, butanol, acetone	Oil viscosity reduction, wettability alteration
Polymers	Polysaccharides, proteins, Xanthan gum	Mobility control, permeability rectification
Bio-emulsifiers	Heteropolysaccharides, proteins	Oil emulsification, wax and paraffin control
Bio-surfactants	Glycolipids, lipopeptides	Interfacial tension reduction, emulsification, wettability alteration
Hydrogen metabolism	Cleave long chain hydrocarbons, production of akylsuccinic acid	Paraffin control, viscosity reduction, methane production, reservoir re-pressurization
Biomass production	Microbial cells	Selective plugging, water flooding

Adapted from McInerney et al. (2005)

For microbes to be used for *in situ* enhanced oil recovery, they must be able to withstand extreme conditions (temperature, pressure, and salinity) in the reservoirs (Table 5.2). Lazar (1983) isolated mixed cultures of bacteria from various locations that were adapted to reservoir conditions. Species identified in mixed cultures were: *Pseudomonas,*

Escherichia, Arthrobacter, Mycobacterium, Micrococcus, Peptococcus, Bacillus, and *Clostridium.* They found that the mixed cultures were more efficient in releasing oil than pure strains. Seven oilfields were inoculated, but only two of the oilfields responded favourably with increased oil production ranging from 16 to 200%. Also, McInerney et al. (1989) successfully produced bio-surfactant from *Bacillus licheniformis* strain JF-2.

Raiders et al. (1985) reported the enhancement of oil recovery from cores by stimulating the growth and proliferation of indigenous microbes in sandstone by nutrient addition. Incremental oil recovery ranged from 10-38% of the original oil in place. Volumetric sweep efficiency was improved by the microbial selective plugging of the reservoir. In another instance (Stephens et al. 1999), reservoir microorganisms were able to enhance crude oil production and caused a sharp decline in produced water after nutrient injection. In a related field trial, Bhupathiraju et al. (1993) successfully used nutrient injection to stimulate microbial growth even in a hypersaline (6-9% sodium chloride) reservoir. Fermentation of the injected nutrient (molasses) results in the production of hydrogen, methane, ethanol and acetate and sulphide was produced as a by-product. Similarly, McInerney et al. (2000) reported increase in oil recovery following the injection of nutrients to boost microbial activities causing selective plugging of oil reservoirs in Payne County, USA. They observed selective reductions in the interwell permeability between the injected well and production wells after stimulation of the growth of indigenous reservoir microbes by nutrient injection. According to the authors, increases in alkalinity and sulphide concentrations in the produced water (brine) confirmed that microbial metabolic activity occurred as a consequence of nutrient injection. Recall that sulphide sours the crude oil thereby reducing its quality. Field trials were successfully conducted for the control of hydrogen sulphide production by the use of the bacteria, *Acidithiobacillus denitrificans* (McInerney 1996).

Microbial repair of formation damage caused by hydraulic fracturing

Typically, the production rates in all oil wells decrease as the pressure of the reservoir is depleted through oil production (Raymond and Leffler 2006). The conventional methods for stimulating the well to boost production are hydraulic fracturing and acidizing. Oil and gas well stimulation involves fracture initiation and propagation from the well bore, which is often cased, perforated and inclined with respect to the *in situ* principal stress direction. When properly performed, fracture stimulation can create a network of highly permeable flow channels that effectively increase drainage of the production zones. However the full potential of fracture stimulation is sometimes not realized because gelling agents used in the fracture fluids cause formation damage by decreasing permeability and blocking flow from the producing zone. Formation damage caused by hydraulic fracturing is one of the many problems of the petroleum industry. Recently, scientists have been involved in applying microorganisms to solve problems in the industry. Through research, they have been able to develop several biological products that can, in most cases, double existing production and eliminate the need for dangerous chemicals. A new technology developed by Micro-Bac Inc. uses a polymer product to repair costly fracture jobs that have failed or had poor results (Kunkel 2001).

Well treatments are designed to repair the formation damage caused by the polymer, restore flow and enhance the effectiveness of the fracture stimulations. Wells have been treated with microbes and their products, which remove polymer damage and restore oil flow (Bailey and Cummings 2001). Microbial repair of fractures is cheaper, less technical and a sustainable method of fracture repair.

Microbial assisted well bore cleaning

Another application of microbial technology is the production of metabolites that are useful in well bore cleaning and work over operations to enhance oil production. Typically, acidizing processes using mineral acids such as HCl are routinely used for well bore cleaning. Microbial processes are now increasingly being considered for well bore cleaning processes because they are cheap, renewable and cause less impacts on the environment. Well bore cleaning processes involve the use of drilling mud degrading (Benka-Coker and Olumagin 1995, 1996) or hydrocarbon degrading or scale removing (McInerney et al. 2005) or acid producing bacteria (Ohimain 2004a) to remove drilling muds, heavy oil, scales and other materials in the well bore.

Microbial synthesis of acids, solvents and gases helps in well bore cleaning in the following ways:

- These acids (organic and inorganic) react with various minerals, especially carbonates, and loosen clay particles and other inorganic deposits.
- Microbially produced solvents dissolve or swell the precipitated organic deposits, improving mobility of the oil.
- Gases formed by microbes include methane, carbon dioxide and hydrogen sulphide could increase well pressures, which aid in the cleaning of the well bore.

Trebbau de Acevedo and McInerney (1996) produced bio-emulsifiers from several pure cultures including thermophilic Archaea, *Methanobacterium thermoautotrophicum*, which is effective over a wide range of pH, at salt concentration up to 200 g/l and at temperatures up to 80°C. Based on the characteristics of the bio-emulsifier, the authors concluded that it is suitable for use in saline or thermophilic oil reservoirs as a mobility control agent or in well bore clean up processes.

Well stimulation and selective reservoir plugging

A number of oil wells have been suspended for decades because they have reached their economic limit of production. This suspension would allow well bore pressure build up over time. However, the wells could be stimulated within a short time of about two weeks through the injection of microbial cultures and fermentable substrates such as molasses and cassava processing wastes. Similarly, nutrients without microbial cultures are often injected into the reservoirs to stimulate microbial activity in the reservoir to mobilize residual oil or alter the flow path of a water flood (McInerney et al. 2005).

Many types of microorganisms produce polymers, biomass, biofilms and slime under reservoir conditions. The growth and production of biomass and biofilms in highly permeable zones will reduce permeability thereby causing the selective plugging and diversion of subsequent flooding through unswept areas to improve water flood sweep efficiency.

Production of cleaner fuel using microorganisms

- One of the contaminants of crude oil is sulphur. Around the world, environmental regulations require its removal from oil products, in order to reduce sulphur pollution in the atmosphere. The conventional method for removing sulphur, nitrogen and other impurities from oil involves the use of hydrogen. In hydrodesulphurization (the process of using hydrogen to remove sulphur from oil), oil, a catalyst, and hydrogen are processed at high pressure and temperatures. During this process the sulphur binds with the hydrogen and is eventually separated from the oil. This method is expensive, energy intensive and generates carbon dioxide as a by-product. Microbes are now increasingly being considered for the removal of sulphur from crude oil (biodesulphurization process).

Some of the environmental and economic advantages of the biodesulphurization process over the hydrodesulphurization process include (Ah-You et al. 2000, van Hamme et al. 2003, Killane 2005).

- Energy savings and a reduction of greenhouse gas emissions of 70-80%;
- Safer operating conditions;
- Effective removal of sulphur even in aromatic ring structures;
- Capital cost savings of approximately 50%;
- Operating cost savings of about 10-20%;
- More rapid engineering and construction time (12-18 months for BDS versus 24–36 months for HDS).

With breakthrough in genetic engineering, scientists were able to isolate the sulphur-attacking genes from bacteria. The genes were cloned (reproduced in the laboratory) and manipulated to increase their sulphur-removing abilities. They were then transferred to new bacterial hosts, in order to create a more efficient sulphur-removal process.

Bacterial engineering for sustainable oil production from shale and sandstones

North America has significant underground deposits of what is known as oil shale. This is sedimentary rock containing an organic material called kerogen which decomposes to yield oil when heated. However, oil shale is very expensive to mine, compared to the cost of recovering oil from liquid deposits. Also, the extraction process is environmentally destructive. Scientists are attempting to genetically engineer *Pseudomonas* and *Bacillus* strains of hydrocarbon degrading bacteria to develop a cost-effective, environmentally friendly oil extraction process. The goal is to modify the bacteria to simply store oil, rather than digest it. In this case, the bacteria would be used to recover oil from a largely untapped source. This is just a sample of the current and potential uses of bacteria biotechnology in the oil recovery and processing industries. As in numerous other fields, biotechnology offers solutions to improving environmental sustainability (Ag-West Biotech Inc. 1999).

Biotechnologies for remediation and pollution control in the petroleum industry

Oil exploitation, like other industrial activities has led to environmental pollution. Soils, groundwater and surface waters have become contaminated, and there are increasing pressures on oil industry operators to reduce risks associated with this contamination. The mitigation of environmental impacts is therefore a major challenge for the modern petroleum industry (Bernoth et al. 2000).

Oil spills are mostly caused by faulty equipment/mechanical failure, corrosion, sabotage and negligence. Depending on the terrain/ environment involved mechanical (booms, scooping etc.) and chemical methods (surfactants, sinking agents) are mostly used. These methods have the inherent drawback of being not able to clean up the spills to the recommended 50 ppm residual oil levels (Department of Petroleum Resources 2002). Hence, biological means offers the only possibility of reducing residual oil below the recommended limits. This could be achieved using one or a combination of the following microbial methods:

(1) Enrichment of natural organisms within the oil spill site (intrinsic bioremediation)

(2) The application of genetically modified microorganisms (engineered bioremediation)

(3) Application of microbial products (mostly enzymes).

The chemically and biologically induced changes in the composition of a polluting petroleum hydrocarbon mixture are known collectively as weathering, microbial degradation (biodegradation) plays a major role in the weathering process (Atlas 1981). Biodegradation in natural environment is complex. Luckily, hydrocarbon degrading bacteria and fungi are widely distributed in marine, freshwater and soil habitats (Atlas 1977) and could therefore catalyze the degradation of oil spills. Biodegradation can be defined as the biologically catalyzed reduction in the complexity of chemicals (mostly organic) (Alexander 1994).

Bioremediation is a family of technologies that rely on microbial processes to convert environmental pollutants to harmless products such as carbon dioxide, water and simple inorganic salts (Thomas et al. 1993). The principles underlying bioremediation processes are well understood. The role of microorganisms as catalysts in the degradation of natural and xenobiotic compounds is well known, and a wide range of habitats and environmental conditions support microbial growth and activity. Both organic and inorganic compounds can be biodegraded or transformed by microbial processes. In the most common bioremediation applications, microorganisms that occur naturally in contaminated soil or waters are encouraged to accelerate the degradation of organic contaminants such as petroleum hydrocarbons by the manipulation of environmental conditions, e.g., oxygen supply, nutrient concentrations and moisture content, temperature, and the physical state of the pollutant.

Bioremediation encompasses a range of distinct techniques for managing or reducing the health and environmental risks associated with contaminated soil and water (Table 5.4). Bioremediation processes can be applied either *in situ* or *ex situ* (e.g., after groundwater pumping or soil excavation).

TABLE 5.4 Bioremediation technologies for soil and groundwater

In situ techniques	Ex situ techniques
Bio-venting	Land farming (shallow mixed beds)
Bio-sparging	Static vented piles (biopiles)
Bio-flushing	Composting in piles or windrows
Lagoon treatments	Soil slurry reactors
Groundwater re-injection or re-infiltration	White-rot fungal processes
Natural or intrinsic remediation	

Source: Bernorth et al. 2000

In fact, microbial metabolism of groundwater pollutants is the only technology that has the potential to completely degrade pollutants *in situ* and converts them to more environmentally acceptable forms (Suflita 1989). Land farming has been frequently used by the oil industry to degrade oily wastes (Alexander 1994) and drill cuttings (Chaineau et al. 1996). This operation depends on the catabolic activities of soil microorganisms.

The science of biodegradation and bioremediation has been studied well. Excellent literature that can be consulted, such as Walker et al. (1974), Atlas (1981), Bernoth et al. (2000), van Hamme et al. (2003), Prince (2005), etc.

Microbiological prospecting of petroleum hydrocarbons

Hydrocarbons exportation often begins with the delineation and acquisition of a sedimentary basin suspected to contain oil. Typically, exploration involves a series of studies to understand sub-surface conditions/events leading up to oil accumulation, which include sedimentation, movement of petroleum from the sedimentary environment into reservoir rock and trapping of the oil. Modern petroleum exploration involves the detection of both source and reservoir rocks, traps and the drilling exploration of wells for testing and laboratory analysis. Geochemical exploration for oil is quite expensive, highly technical and time consuming, hence the consideration for geomicrobiological method for oil prospecting.

In the 1940s-70s, several researchers considered the use of microbes for oil exploration (Moilevskii and Stiehler 1940, Kluyver and Manteu 1942, Bokova 1947, Beerstecher 1954, Soli 1954, 1957, Hiztman 1960, Brisbane and Ladd 1965). Sealy (1974a, b) successfully used microbiological methods to prospect for oil in Texas with over 90% accuracy. Despite these groundbreaking findings, microbiological investigation has not kept pace with geochemical exploration of petroleum, though several patents on geomicrobial prospecting of oil and gas have been given to some researchers especially in Russia. Evidence at the Earth's surface of the presence of petroleum accumulation in the sub-surface are the primary objectives of microbial petroleum prospecting. Two important characteristics of microbes that made them suitable for oil exploitation, is the fact that microbes are able to utilize petroleum hydrocarbons and these hydrocarbon utilizing microbes are ubiquitous on Earth. During exploration, physical detection and identification of petroleum gas or petroleum seepages are often preferred, an indirect means is to detect and enumerate microorganisms that feed on seepages, particularly gases. The approach adopted includes the determination of methane or ethane or propane consuming bacteria in soil. Ethane and propane oxidizing Mycobacteria have been used for the exploration of petroleum in Russia. It should be noted that testing soil and water for gaseous hydrocarbon oxidizing bacteria and petroleum gases might have value in the reconnaissance of some areas. Microbial oil exploration, like other exploration methods (chemical, geological and geophysical), should be used in conjunction with other methods.

MICROBIAL ACTIVITIES DETRIMENTAL TO THE PETROLEUM INDUSTRY

Activities of microorganisms in oil wells are not always beneficial. Some microbes have been associated with certain deleterious activities. In considering the detrimental activities of microorganisms with respect to the petroleum industry, one of the first and major aspects is microbial corrosion of metals, particularly steel and iron. Corrosion of iron and steel by whichever cause is of interest and concern because of the vast amount of metal/steel equipment employed in petroleum industrial operations. Furthermore, corrosion of pipelines, casings, tubings and storage facilities has resulted in loss of products and pollution of the environment. Other microbes are involved in the deterioration of drilling muds, additives and other petroleum products. Yet, others are involved in the formation of hydrogen sulphide, which is toxic and also associated with the precipitation of metal sulphides, metal/steel corrosion, plugging of reservoir fouling of crude oil, contamination of fuel. Majority of the deleterious activities of microbes in oil field environment are caused by SRB (Table 5.5). Adequate knowledge of the biology of the microbes involved their mode of action, their nutritional requirements

TABLE 5.5 Sulphide generating activity of sulphate reducing bacteria that relate to petroleum microbiology

Environment	Manifestation
Solid and surface waters	Promotes corrosion or iron and steel. Precipitates iron as FeS, raises pH, promotes CaCO$_3$ precipitation promotes sulphur deposition
Sub-surface waters	Precipitates calcite in limestone reservoir promotes sulphur deposition, corrosion of oil wells: apparently have profound effects upon composition of sub-surface water in association with petroleum
Drilling muds	Deteriorated, offensive odour, corrosion of oil well casing.
Waters, particularly sea water, associated with stored petroleum fuel	Deteriorated or corrosive fuels
Metal cutting oil emulsions	"Blueing" of emulsion due to precipitate of iron as FeS, deterioration, offensive odour.
Aqueous sediment, particularly marine	Maintains low oxidation-reduction potential. Precipitates iron as FeS, raises pH

Source: Davis (1967)

and general physiological/environmental requirements may be necessary in controlling the noxious activities of these organisms.

Bio-corrosion

The subject of corrosion is a very broad inter-disciplinary study involving many aspects of chemistry, metallurgy and chemical engineering, now microbiological corrosion is gaining prominence. Corrosion is one of the major causes of oil spillage in the world. Structures which have been attacked include well casings, pipelines (exterior and interior), pumps, fuel holding tanks, and other sub-surface equipment. With the exception of cast iron, corrosion of steel and other metals by microorganisms is generally the pitting type, often leading to perforation. Sulphide stress cracking of certain steels by microbiologically produced hydrogen sulphide is well documented (Menzies 1971, Iverson 1972, Cord-Ruwisch 1987).

Microbial aspects of corrosion could occur by any or a combination of the mechanisms listed in Table 5.6.

Corrosion in aqueous environments could be regarded as an electrochemical process (Iverson 1972). A metal in solution tends to ionize.

TABLE 5.6 Mechanisms of microbial corrosion of metals in the oil and gas industry

Mechanism of corrosion	Corrosion process	Relevance to the oil industry	References
The production of corrosive metabolic products	Inorganic and organic acids are among the most corrosive products produced by microbes. Oxidation of pyrite by *Acidithiobacillus* sp. produces sulphuric acid under aerobic conditions. Extreme acidity of pH<3 is common	Acids cause corrosion of iron and steel and concrete platforms	Ohimain 2003, 2004a
	Production of H_2S by sulphate-reducing bacteria, *Desulfovibrio* under anaerobic condition	Corrosion of underground metallic structures, primarily iron and steel, and the internal corrosion of oil well casings and tubings, pipelines and flow lines, storage tanks, topsides and other production facilities	Iverson 1972, Cord-Ruwisch 1987
The production of differential aeration and concentration cells	Microbial growth and activities on metal surfaces forming tubercle cause variations in the oxygen or ion concentration on the metal surface generating electric currents, which play an important role in the corrosion of many metals. *Gallionela*, *Crenothrix* and *Leptothrix* are among the principal bacteria genera associated with pitting corrosion.	Pitting corrosion of the internal surfaces of oil and water pipes	Miller 1971, Iverson 1972
Depolarization of cathodic processes	By utilizing the hydrogen protection on metal surfaces, sulphate-reducing bacteria depolarize the metal and cause it to corrode.	Corrosion of pipelines and flow lines	Iverson 1972, Cord-Ruwisch 1987
Disruption of natural and protective films	Microorganisms on the metal surface may prevent natural oxide film formation from taking place or removing these films once they have formed. Protective films of iron sulphide formed on steel could be detached by sulphate-reducing bacteria.	Various applied protective organic coatings used to coat buried pipes (polyester, waxy, or asphalt coatings) have been observed to become attacked by soil organisms.	Iverson 1972
The breakdown of corrosion inhibitors	Microorganisms breakdown corrosion inhibitors and predispose the metal to corrosion. Nitrite and nitrate that are used to inhibit corrosion are attached by *Pseudomonas* sp.	Corrosion of cooling tower during in petroleum refinery	Iverson 1972

$$M \rightarrow M^{z+} + Ze \tag{1}$$

Removal of electrons by one or more mechanisms causes the reaction to go to the right, thus leading to corrosion. Two principal mechanisms for the removal of electron are:

$$2H^+ + 2e \rightarrow 2H \rightarrow H_2 \tag{2}$$

$$H_2O + \tfrac{1}{2} O_2 + 2e \rightarrow 2 (OH) \tag{3}$$

The oxidizing agents (also referred to as depolarizing agents) involved in these mechanisms are hydrogen ions and oxygen, respectively. If a plentiful supply of an oxidizing agent were present, the corrosion process would be expected to go unimpeded, were it not for the formation of films of oxidation products which will slow down one or more of the above reactions (1, 2 and 3). Bio-corrosion mainly involves the stimulation of these electrochemical processes (Iverson 1972). For instance, by utilizing the hydrogen produced on metal surface by SRB can tilt the above electrochemical reactions to the right, thus depolarizing and corroding the metal. This also has the tendency of either disrupting or preventing the formation of natural protective layer/films on metal surface. There are reported cases of microbial attack and breakdown of corrosion inhibitors (Iverson 1972). Also the microbial production of raw sulphuric acid through pyrite oxidation is capable of causing etching corrosion of oil field installations, since pyrite containing dredged materials are typically abandoned close to oil installations (Ohimain et al. 2004, 2008a, 2008b).

Microbial souring of oil and sulphide toxicity

The production of hydrogen sulphide in the reservoir by sulphate-reducing bacteria has been associated with the souring of oil and gas reservoirs. This reduces the quality, price and acceptability of the petroleum product. Hydrogen sulphide production by bacteria may occur in the reservoir and after production during storage and transportation of oil containing water.

Studies have shown that indigenous SRB are present in oil and gas reservoirs (Cord-Ruwisch 1987, Baas and Lappin-Scott 1997). Injection of seawater (for water flooding operations), which typically contain high concentrations of sulphate, could stimulate the growth SRB and the production of hydrogen sulphide in the reservoir. In addition, SRB cause metal corrosion, produce biofilm, which plug the reservoir.

Biogenically produced hydrogen sulphide is highly toxic when exposed to workers. Hydrogen sulphide is extremely toxic if inhaled. It easily escapes from contaminated waters and may accumulate under poorly ventilated conditions. It is usually recognized by its distinctive, unpleasant odour, but high concentrations anaesthetize the sense of smell (Cord-Ruwisch et al. 1987). Notwithstanding, studies have shown that a certain bacterium, *Clostridium tyrobutyricum,* isolated from an oil reservoir is capable of suppressing the activities of SRB (Wagner et al. 1995).

Microbial reservoir plugging

Water flooding of petroleum reservoirs is a common, effective means for the secondary recovery of oil and microbiology is an integral part of this application. Floodwaters employed in secondary oil recovery vary depending upon the availability and formation of water disposal requirement. Floodwaters may consist of recycled water, water supplied from other sources including river water, and/or both. Saline produced water may be recycled for the dual purpose of re-pressuring the oil reservoir and circumventing the problem of disposal of the salt water without polluting the recipient environment. Uncontrolled uses of floodwater often lead to the contamination of the reservoir by microbes. However, in the Niger Delta, the source of water for reservoir flooding is not from seawater, but from dedicated boreholes, hence they are less likely to be prone to SRB contamination. The proliferation of noxious microbes results in formation damage causing reduction in the permeability of reservoir rocks in various ways (Donaldson et al. 1989):

- The formation of microbial films clog the liners and reservoir pore spaces (plugging due to biofilm formation)
- Bacteria (especially dead forms) also cause particulate plugging without necessarily forming films
- The precipitation of ferrous sulphide by SRB results in the plugging of the well
- The bacterial reduction of sulphate in formation can lead to the precipitation of calcite
- The activities of iron bacteria (*Lepthothrix, Crenothrix, Sphaerotis, Gallionella* and *Acidithiobacillus*) in oil wells results in the precipitation of ferric hydroxide and ferric sulphate, which could cause plugging in addition to the bacteria themselves.

Though microbial reservoir plugging is hazardous, it has also been used positively to enhance water flooding operations.

Decomposition of drilling fluids and additives

In drilling processes, turning the bit is accompanied by the all-important function of the drilling fluid in effectively removing the cuttings produced. The viscosity and other properties of the mud are often adjusted to perform this function under variable conditions of rock composition, temperature, pressure and water encountered. Oil well drilling and the application of drilling fluids in the drilling process are now highly developed technologies. The microbiology of drilling fluids is a relatively small aspect of the overall technology involved, but even today the actual economic importance of microbial activity in modifying the composition of drilling fluids has not been fully appreciated. In the Niger Delta, Benka-Coker and Olumagin (1995, 1996) isolated drilling mud utilizing bacteria. While these bacteria have the advantage of treating drilling wastes biologically, they could also cause the breakdown of drilling mud.

A drilling fluid performs at least five functions, namely, (1) bringing the borehole cuttings to the surface, (2) cooling and lubricating the drill bit, (3) walling up and stabilizing the borehole, (4) counteracting oil, gas and water pressures and (5) reducing friction between the drill pipe and the borehole. Additives are typically added to drilling mud in quantities just enough to withstand the well bore pressures. Too much mud concentration could lead to formation damage, while too little could lead to well blow out. Organic additives such as starch, natural gums, carboxymethyl cellulose (CMC) or chrome lignosulphonates are added to the drilling fluids to prevent or control water loss into porous formations which otherwise would be excessive due to the higher pressure differential exerted by the column of drilling fluid in the borehole. Organic additives in drilling fluids have various other purposes including; dispersant (viscosity) effects and their emulsifying effects. Both natural and synthetic organic materials are often added to drilling muds. Unfortunately, several microbes isolated from the oil field environment have the capability of degrading most of these additives. Microbial degradation of such organic additives is detrimental to their desired properties and the drilling programme may become risky and unnecessarily expensive. Several fungi and bacteria have been isolated which degrade CMC, starch, lignosulphonates and natural gum.

Microbial contamination deterioration of petroleum products

Microbial utilization of hydrocarbons is beneficial in the areas of petroleum prospecting, microbial enhanced oil recovery and biodegradation of petroleum wastes. The ability of microbes to utilize petroleum, however, has its detrimental aspects, particularly with respect to the deterioration of certain manufactured fuel and lubricants. Microbes have been reported to cause serious damage to petrol, fuel oils, kerosene, jet-aircraft oil, lubricants and other products. The attack of jet fuel tanks by the fungus *Cladosporium* has caused serious problems to the aviation industry. Also, the microbial attack on crude oil pipelines and storage facilities has been implicated in the loss of products and spillage to the environment.

At least four microbial mechanisms exist for the deterioration of fuels, particularly gasoline and jet-aircraft fuels.

1. Microbial growth upon the fuel hydrocarbon components. This occurs to a great extent with the higher molecular weight hydrocarbons of kerosene (jet aircraft) fuels and results in formation of a microbial sludge.
2. Microbial growth upon organic additives in fuels with consequent microbial sludge development.
3. Production of hydrogen sulphide in associated waters by sulphate-reducing bacteria. The hydrogen sulphide reacting with fuel components results in highly corrosive organic sulphides.
4. Microbial deterioration of organic storage tank linings. Resulting in both microbial sludge development and eventual serious metal corrosion of the tanks.

FUTURE PERSPECTIVES AND CONCLUSION

Oil exploration activities have intensified globally owing to the rising cost of crude oil, which is currently above US $100/bbl. Like other fossil fuels, the global reserves of crude oil are declining as production intensifies. Therefore, attention is now focused on other less conventional sources of oil particularly shale in North America. In Nigeria, previously abandoned marginal fields are re-opening. Some wells that had lost their primary energy drives are being considered for secondary and tertiary production. Environmental impacts arising from the consumption of petroleum continue to be the major focus of world debate. While a country like Nigeria, which up till now still flare her associated gas, is now embarking on massive associated gas gathering projects for domestic power supply and export. Even non-associated gas wells are being developed to meet gas demands from other countries. More liquefied natural gas (LNG) plants are being built, while the capacity of the existing LNG plants is being expanded by the addition of more production trains. More compressor stations and gas plants are being installed. Other forms of gas management are being considered including the conversion of gas to liquids (GTL), production of natural gas liquids (NGL) and compressed natural gas (CNG). Apart from the installation of more gas gathering facilities and pipelines, regional efforts have led to the construction of the West African Gas Pipeline, while the proposed Trans Saharan Gas Pipe from Africa to Europe is under consideration.

Attention has also focused on alternative sources of energies including biogas, biodiesel and bioethanol. These forms of energy have some advantages over crude oil since they are renewable sources and produce less carbon. But advancements in bio-energies have caused some problems including global food crises, poverty and hunger in developing countries.

The intensification of oil exploration and development of alternative energies have created new opportunities for microbiologists. For instance, the presence of microbes the traditional role of microbiologist in the petroleum industry is in the area of environmental and waste management especially oil spill clean up/bioremediation, sewage monitoring and water testing. But the occurence of microbes in petroleum reservoirs and their beneficial and deleterious activities has created opportunities in the oil and gas sector for reservoir microbiologists. Generally, the future forseen is that there will be an increasing role for microbiologists in the oil sector. New job titles will likely emerge including petroleum microbiologists, petroleum reservoir microbiologists, production microbiologists, geomicrobiologists, petroleum quality control microbiologists, corrosion control microbiologists and biogas and bioethanol process/production microbiologists etc.

REFERENCES

Ag-West Biotech Inc. 1999. Bacteria Biotech: Developing Environmental Alternatives in the Oil Industry (Part II), AgBiotech Infosource 43: 1-3.

Ah-You, K., M. Suleiman, and J. Jaworski. 2000. Biotechnology and Cleaner Production in Canada. Life Sciences Branch, Industry Canada.

Alexander, M. 1994. Biodegradation and Bioremediation. Academic Press, San Diego, USA.

Amro, M.M. and M.A. Al-Mobarky. 2007. Society of Petroleum Engineers. *In:* E.S. Al-Homadhi and S. King [eds.]. Improved Oil Recovery by Application of Ultrasound Waves to Water Flooding. Paper SPE 105370, presented at the 15th Society of Petroleum Engineers Middle East Oil and Gas Show and Conference held in Bahrain International Exhibition Center, Kingdom of Bahrain. March, 11-14.

Amyx, J.W., D.M Bass Jr., and R.L. Whiting. 1960. Petroleum Reservoir Engineering. McGraw-Hill. London.

Atlas, R.M. 1977. Stimulated petroleum biodegradation. Critical Review Microbiology 5: 371-386.

Atlas, R.M. 1981. Microbial degradation of petroleum hydrocarbons: An environmental perspective. Microbiology Review 45: 180-209.

Bailey, S. and B. Cummings. 2001. Biotechnology repairs frac damage. Hart's E&P. January 2001: 84.

Baker, R. 1983. The Production Story. Petroleum Extension Service, University of Texas, Houston.

Baker, R. 1994. A Primer of Oil Well Drilling. 5th ed. Petroleum Extension Service, University of Texas, Houston.

Bass, C. and H. Lappin-Scott. 1997. The Bad Guys and the Good Guys in Petroleum Microbiology. Oil Field Review. University of Exeter, England.

Benka-Coker, M.O. and A. Olumagin. 1995. Effects of waste drilling fluid on bacterial isolation from a mangrove swamp oilfield location in the Niger Delta of Nigeria. Bioresource Technology 55: 175-179.

Benka-Coker, M.O. and A. Olumagin. 1996. Waste drilling-fluid-utilising microorganism in a tropical mangrove swamp oilfield location. Bioresource Technology 53: 211-215.

Berner, R.A. 1970. Sedimentary pyrite formation American Journal of science 15: 81-103.

Bernoth, L., I. Firth, P. McAllister, and S. Rhodes. 2000. Biotechnologies for remediation and pollution control in the mining industry. Minerals Metallurg. Processing 17: 105-111.

Bhupathiraju, V.K., M.J. Mc Inerney, and R.M. Knapp. 1993. Pre-test studies for a microbially enhanced oil recovery field in hypersaline oil reservoir. Geomicrobiological Journal 11: 19-34.

Bhupathiraju, V.K., A. Oren, P.K. Sharma, R.S. Tanner, C.R. Woese, and M.J. Mc Inerney. 1994. *Haloamacrobium salsugo* sp. nov., A moderately halophilic, anaerobic bacterium from a subterranean brine. Journal of Systematic Bacteriology 44: 565-572.

Bhupathiraju, V.K., M.J. McInerney, C.R. Woese, and R.S. Tanner. 1999. *Aloanaerobium kushneri* sp. nov., an obligately halophilic, anaerobic bacterium from an oil brine. International Journal of Systematic Bacteriology 49: 953-960.

Birkelands, N.K. 2005. Sulfate-reducing bacteria and archaea. *In:* B. Ollivier and M. Magots [eds.]. Petroleum Microbiology. ASM Press, Washington, DC. pp. 35-54.

Brown, L.B. and A.A. Vadie. 1992. Microorganisms, microbial by-products and oil-bearing formation mineral. US. DOE Progress Rev 64: 117-136.

Bubela, B. 1989. Geobiology and microbiologically enhanced oil recovery. *In:* E.C. Donaldson, G.V. Chilingarian, and T.F. Yen [eds.]. Microbial Enhanced Oil Recovery. Development in Petroleum Science 22. Elsevier, Amsterdam. pp. 75-98.

Bukova, E.N. 1947. Oxidation of gaseous hydrocarbons by bacteria as a basis of microbiological prospecting for petroleum. Docklady Akad. Nauk SSSR, 56: 755-757.

Chaineau, C.H., J.L. Movel, and J. Oudot. 1996. Land treatment of oil based drill cuttings in an agricultural soil. J. Environ. Qual. 25, pp. 858-876.

Crolet, J.L. 2005. Microbial corrosion in the oil industry: A corrosionist's view. *In:* B. Ollivier and M. Magot [eds.]. Petroleum Microbiology. ASM Press, Washington, DC. pp. 143-169.

Conaway, C.F. 1999. The petroleum industry: A Non-technical Guide. Pennwell Corporation, Tulsa, Oklahoma.

Cord-Ruwisch, R., W. Kleinitz, and F. Widdel. 1987. Sulfate-reducing bacteria and their activities in oil production. Journal of Petroleum Technology 1: 97-106.

Davis, J.B. 1967. Petroleum Microbiology. Elsevier Publishing, Amsterdam.

Devereux, S. 1999. Drilling Technology in Non-technology Language. Pennwell Corporation, Tulsa, Oklahoma.

Donaldson, E.C., R.M. Knapp, T.Y. Yen, and G.V. Chilingarian. 1989. The subsurface environment. *In:* E.C. Donaldson, G.V. Chilingarian, and T.F. Yen [eds.]. Microbial Enhanced Oil Recovery. Development in Petroleum Science 22. Elsevier, Amsterdam. pp. 15-35.

DPR. 2002. Environmental Guidelines and Standard for the Petroleum Industry in Nigeria. Department of Petroleum Resources, Lagos.

Fontes, D.E., A.L. Mills, G.M. Hornberger, and J.S. Hermen. 1991. Physical and chemical factors influencing transport of microorganism through porous media. Applied and Environmental Microbiology 57: 2473-2481.

Goldhaber, M.B. and I.R. Kaplan. 1982. Control and consequences of sulphate reduction in recent marine sediments. *In:* J.A. Kittrick, D.S. Fanning, and L.R. Hossner [eds.]. Acid Sulphate Weathering SSSA special Publication No. 10. pp. 1-18. Soil Science Society of America, Madison, Wisconsin. pp. 19-36.

Grula, M.M. and G.W. Sewell. 1983. Microbial interactions with polyacrylamide polymers. *In:* E.C. Donaldson and J.B. Clark [eds.]. Proceedings 1982 International Conference on Microbial Enhancement of Oil Recovery. NTIS, Springfield, Va., pp. 219-134.

Hakes, J.E. 2000. International Energy Outlook 2000 with Projections to 2020. Energy Information Administration. www.eia.doe.gov. Downloaded 20 September 2007.

Hite, J.R., S.M. Avasti, and L.B. Paul. 2004. Planning EOR Project. Paper SPE 92006, Presented at the 2004 SPE International Petroleum Conference, Puebla, Mexico, November 8-9.

Hitzman, D.O. 1960. Comparison of geomicrobiological prospecting methods used by various investigators. Development in Industrial Microbiology 2: 33-42.

Hitzman, D.O. 1983. Petroleum microbiology and the history of its role in enhanced oil recovery. *In:* E.C. Donaldson and J.B. Clark [eds.]. Proceedings 1982 International Conference on Microbial Enhancement of Oil Recovery. NTIS, Springfield, Va., pp. 162-218.

Huang, H. and S. Larter. 2005. Biodegradation of petroleum in subsurface geological reservoir. *In:* B. Ollivier and M. Magot [eds.]. Petroleum Microbiology. ASM Press, Washington, DC. pp. 91-121.

Hyne, N.J. 2001. Non-technical Guide to Petroleum Geology, Exploration, Drilling and Production. 2nd ed, Pennwell Corporation, Tulsa, Oklahoma.

Inverson, W.P. 1972. Biological corrosion. *In:* M.G. Funtana and R.W. Stachle [eds]. Advances in Corrosion Science and Technology, Vol. 2. Plenum Press, New York. pp. 1-43.

Javaheri, M., G.E. Jennrman, M.J. McInerney, and R.M. Knapp. 1985. Anaerobic production of a biosurfactant by *Bacillus licheniformis* JF-2. Applied and Environmental Microbiology 50: 698-700.

Jeanthon, C., O. Nercessian, E. Corre and A. Grabowski-lux. 2005. Hyperthermopholic and methanogenic archaea in oil field. *In:* B. Ollivier and M. Magot [eds.]. Petroleum Microbiology. ASM Press, Washington, DC. pp. 55-69.

Jenneman, G.E. 1989. The potential for in-situ microbial applications. *In:* E.C. Donaldson, G.V. Chilingarian and T.F. Yen [eds]. Microbial Enhanced Oil Recovery. Development in Petroleum Science 22. Elsevier, Amsterdam. pp. 15-35.

Jenneman, G.E., P.D. Moffitt, G.A. Bala, and R.H. Webb. 1996. Field demonstration of sulphide removal in brine by bacteria indigenous to a Canadian reservoir. SPE paper, 38768.

Jenneman, G.E., M.J. McInerney, and R.M. Knapp. 1985. Microbial penetration through nutrient–saturated Berea sandstone. Applied and Environmental Microbiology 50: 383-391.

Jenneman, G.E., M.J. McInerney, M.E. Crocker, and R.M. Knapp. 1986. Effect of sterilization by dry heat or autoclaving on bacterial penetration through Berea sandstone. Applied and Environmental Microbiology 51: 39-43.

Jinfeng, L., M. Lijun, M. Bozhong, L. Rulun, N. Fangtian, and Z. Jiaxi. 2005. The field pilot of microbial enhanced oil recovery in a high temperature petroleum reservoir. Journal of Petroleum Science and Engineering 48: 265-271.

Killane, J.J. 2005. Biotechnological upgrading of petroleum. *In:* B. Ollivier and M. Magot [eds.]. Petroleum Microbiology. ASM Press, Washington, DC. pp. 239-255.

Kluyver, A.J. and A. Manteu. 1942. Some observation on the metabolism of bacteria oxidation hydrogen. Antonie Van Leeuwenhoek. Journal of Microbiology Serol. 8: 71-85.

Krumholz, L.R., S.H. Harris, S.T. Tay, and J.M. Suflita. 1999. Characterization of two subsurface H_2-utilizing bacteria, *Desulfomicrobium hypiogeium* sp. nov. and *Acetobacterium psammolithicum* sp. nov., and their ecological roles. Applied and Environmental Microbiology 65: 2300-2306.

Kunkel, B. 2001. Intelligent tools increase oil recovery. Hart's E&P. *In:* S.I. Kuznetsov [ed]. 1961. Geologic Activity of Microorganisms (translated from Russian by Consultants Bureau, New York, 1962). Tr. Inst. Mikrobiol. Akad. Nauk SSSR, pp. 45-47.

Kuznetsov, S.I., M.V. Ivanov, and N.N. Lyalikova. 1962. Introduction to Geological Microbiology (Translated from Russian by P.T. Broneer and C.H. Oppenheimer). Introduction to Geological Microbiology. McGraw-Hill, New York, N.Y.

Lake, L.W., R.L. Schmidt, and P.B. Venuto. 1992. A Niche for Enhanced Oil Recovery in the 1990s Oilfield Review, January, 1992, p 55-61.

Lazar, I. 1983. Microbial enhancement of oil recovery in Romania. *In:* E.C. Donaldson and J.B. Clark [eds.]. Proceedings 1982 International Conference on Microbial Enhancement of Oil Recovery. NTIS, Springfield, Va., pp. 140-148.

Long P.E., T.C. Onstott, J.K. Fredrickson, T.O. Stevens, G. Goa, B.N. Bjorstad, D.R. Boone, R. Griffiths, R.B. Hallett, and J.C. Lorenz. 1996. Origin of Subsurface Microorganisms: Evidence from a Volcanic Thermal Aureole presented at the international symposium on Subsurface Microbiology, Davos, Switzerland, September, 15-21.

Magot, M. 2005. Indigenous microbial communities in oil fields. *In:* B. Ollivier and M. Magot [eds.]. Petroleum Microbiology. ASM Press, Washington, DC. pp. 21-33.

Magot, M., B. Ollivier, and B.K.C. Patel. 2000. Microbiology of petroleum reservoir. Antonie Leewenhock, 77: 103-116.

McInerney, M.J., P.D. Nagle, and M.R. Knapp. 2005. Microbially enhanced oil recovery: Past, present and future. *In:* B. Ollivier and M. Magot [eds.]. Petroleum Microbiology. ASM Press, Washington, DC. pp. 215-237.

McInerney, M.J. and T. Sublette. 1997. Petroleum Microbiology: Biofouling, souring and improved oil recovery. *In:* T. Burst. [ed]. Manual of Environment Microbiology. American Society for Microbiology Press, Washington, D.C. pp. 600-607.

McInerney, M.J., M. Javaheri, and D.P. Nagle Jr. 1989. Properties of the biosurfactant produced by *Bacillus licheniformis* strain J.F. 2. Journal of Industrial Microbiology 4: 1-7.

McInerney, M.J., N.Q. Wofford, and K.L. Sublette. 1996. Microbial control of hydrogen sulphide production in a porous medium. Applied Biochemistry and Biotechnology 57/58: 933-944.

McInerney, M.J., R.M. Knapp, L.J. Chisholm, V.K. Bhupathiraju, and J.D. Coates. 2000. Use of indigenous or injected microorganism for enhanced oil recovery. *In:* C.R. Bell, M. Brylinksky, and P. Johnson-Green [eds.]. Microbial Biosystems: New Frontier Atlantic Canada Society for Microbial Ecology. Halifax Canada. Proceedings of the 8th international symposium on Microbial Ecology.

Menzies, I.A. 1971. Introductory corrosion. *In:* J.D.A. Miller [ed]. Microbial Aspects of Metallurgy. Medical & Technical Publishing Co. England. pp. 35-60.

Miesner, T.C. and W.L. Leffler. 2006. Oil and Gas Pipelines in Non-technical Languages. Pennwell Corporation, Tulsa, Oklahoma.

Miller, J.D.A. 1971. Microbial Aspects of Metallurgy. MTP Ltd., England.

Mogilevskii, G.A. 1940. The bacterial method of prospecting for oil and natural gases. Razvedka Nerd 12: 32-33.

Nazina, T.N., A.E. Ivanova, V.S. Ivoilov, Y.M. Miller, G.F. Kandaurova, R.R. Ibatullin, S.S. Belyaev, and M.V. Ivanov. 1999. Results of the trail of the microbiological method for the enhancement of oil recovery at the carbonate collector of the Romashkinskoe oilfield: Biogeochemical and productional characteristics. Microbiology 68(2): 222-226.

Nazina, T.N., A.A. Grigoryan, Y.F. Xue, D.S. Sokolova, E.V. Novikova, T.P. Tourova, A.B. Poltaraus, S.S. Belyaev, and M.V. Ivanov. 2002. Phylogenetic diversity of aerobic saprotrophic bacteria isolated from the Daqing oilfield. Microbiology 71(1): 91-97.

Ohimain, E.I. 2003. Environmental impacts of oil mining activities in the Niger Delta mangrove ecosystem. Proceedings of the International Mine Water Association (IMWA) conference held in Johannesburg, South Africa. *In:* D. Armstrong, A.B. de Villiers, R.L.P. Kleinmann, T.S. McCarthy and P.J. Norton [eds.]. Mine Water and the Environment. International Mine Water Association (IMWA), Sandton, South Africa. pp. 503-517.

Ohimain, E.I. 2004a. Environmental impacts of dredging in the Niger Delta; options for sediment relocation that will mitigate acidification and enhance natural mangrove restoration. Terra et Aqua 97: 9-19.

Ohimain, E.I. 2004b. Available options for the bioremediation and restoration of abandoned pyritic dredge spoils causing the death of fringing mangroves in the Niger delta. *In:* M. Tsezos, A. Hatzikioseyian, and E. Remoudaki [eds.]. Biohydrometallurgy: A sustainable technology in evolution. National Technical University of Athens, Zografou, Greece, pp. 475-481.

Ohimain, E.I., W. Andriesse, and M.E.F. van Mensvoort. 2004. Environmental Impacts of Abandoned Dredged Soils and Sediments: Available Options for their Handling, Restoration and Rehabilitation. Journal of Soils and Sediments 4(1): 59-65.

Ohimain, E.I., T.O.T. Imoobe, and D.D.S. Bawo. 2008a. Changes in water physico-chemical properties following the dredging of an oil well access canal in the Niger Delta. World Journal of Agricultural Sciences, 4(6): 752-758.

Ohimain, E.I., G. Jonathan and S.O. Abah. 2008b. Variations in heavy metal concentrations following the dredging of an oil well access canal in the Niger Delta. Advances in Biological Research 2(5-6): 97-103.

Ohimain, E.I. 2006. Indicators of wetland acidification and their relevance to impact assessment: CBBIA small grants program, 2004SGP04. Final report.

Ollivier, B. and J.L. Cayol. 2005. Fermentative iron reducing and nitrate reducing microorganism. *In:* B. Ollivier and M. Magot. [eds.]. Petroleum Microbiology. ASM Press, Washington, DC. pp. 71-88.

Ourisson, G., P. Albrecht, and M. Rohmer. 1984. The microbial origin of fossil fuels. Scientific American 251(2): 44-51.

Plankaert, M. 2005. Oil reservoir and oil production. *In:* B. Ollivier and M. Magot [eds.]. Petroleum Microbiology. ASM Press, Washington, DC. pp. 3-19.

Pons, L.J., N. Van Breemen, and P.M. Driessen. 1982. Physiography of coastal sediments and development of potential soil acidity. *In:* J.A. Kittrick, D.S. Fanning, and L.R. Hossner [eds.]. Acid sulfate weathering, SSSA special publication No. 10. Madison Wisconsin, pp. 1-18.

Prince, R.C. 2005. The microbiology of marine oil spill. *In:* B. Ollivier and M. Magot [eds.]. Petroleum Microbiology. ASM Press, Washington, DC. pp. 317-335.

Raiders, R.A., D.C. Freeman, G.E. Janneman, R.M. Knapp, M.J. McInerney, and D.E. Menzie. 1985. The use of microorganism to increase the recovery of oil from cores. Paper presented at the 6th annual technical conference and exhibition of the Society of Petroleum Engineers, Las Vegas.

Raymond, M.S. and W.L. Leffler. 2006. Oil and Gas Production in Non-technical Language. Pennwell Corporation, Tulsa, Oklahoma.

Sammut, J. and R. Lines-Kelly. 2000. An Introduction to Acid Sulphate Soils. Natural Heritage Trust, Australia Seafood Industry and Environment Australia.

Sanders, F.P. and J.P. Sturman. 2005. Biofouling in the oil industries. *In:* B. Ollivier and M. Magot. [eds.]. Petroleum Microbiology. ASM Press, Washington, DC. pp. 171-198.

Sealy, J.R. 1974a. A geomicrobial method of prospecting for oil: Oil and Gas Journal. 8: 142-146.

Sealy, J.R. 1974b. A geomicrobial method of prospecting for oil: Oil and Gas Journal. 15: 98-102.

Sharma, P.K., M.J. McInerney, and R.M. Knapp. 1993. In situ growth and activity and modes of penetration of *Escherichia coli* in unconsolidated porous materials. Appl. Environ. Microbiol. 59(11): 3686-3694.

Sharma, P.K. and M.J. McInerney. 1994. Effect of grain size on bacterial penetration, reproduction and metabolic activity in porous glass bead chambers. Applied and Environmental Microbiology 60: 1481-1486.

Smith, B., R. Bushand, and J. Sammut. 1995. Acid Sulfate soil in the Nnosa River catchments. Noosa Council Australia.

Soli, G.G. 1954. Geomicrobiological prospecting. Bulletin of the American Associated of Petroleum Geologist 3: 2555-2558.

Soli, G.G. 1957. Microorganism and geochemical methods of oil prospecting. Bulletin of the American Associated of Petroleum Geologist 41: 134-140.

Stephens, J.O., L.R. Brown, and A.A. Vadie. 1999. MEOR – A low cost solution for enhanced water flood performance DOE Contract Report No. DE-FC 22-94 BC 14962.

Stephens, J.O., L.R. Brown, and A.A. Vadie. 2000. The Utilization of the Microflora Indigenous to and Present in Oil-bearing Formations to Selectively Plug the More Porous Zones thereby Increasing Oil Recovery during Water Flooding. DOE/BC/14962 – 24. Hughes Eastern Corporation, Mississippi.

Suflita, J.M. 1989. Microbiological principles influencing the biorestoration of aquitors. Transport and fate of contaminants in the subsurface. United States Environmental Protection Agency (USEPA). Seminan Publication No. EPA/625/4–89/019. Ohio, pp. 85-99.

Sunde, E. and T. Torsvik. 2005. Microbial control of hydrogen sulfide production oil reservoir. *In:* B. Ollivier and M. Magot [eds.]. Petroleum Microbiology. ASM Press, Washington, DC. pp. 201-213.

Thomas, C.P., G.A. Bala and M.L. Durall. 1993. Surfactant-based enhanced oil recovery mediated by naturally occurring microorganisms. SPE paper, 22844.

Torbati, H.M., R.A. Raiders, E.C. Donaldson, M.J. McInerney, G.E. Jannerman, and R.M. Knapp. 1986. Effort of microbial growth on pore entrance size distribution in sandstones cores. Journal of Industrial Microbiology pp. 227-234.

Trebbau de Acevedo, G. and M.J. McInerney. 1996. Emulsifying activitity in thermophilic and extremely thermophilic microorganism. Journal of Industrial Microbiology 16: 1-7.

Turner, R.E. and B. Streever. 2002. Approaches to Coastal Wetland Restoration: Northern Gulf of Mexico. SPB Academic Publishing, The Hague, The Netherlands.

van Breemen, N. 1976. Genesis and solution chemistry of acid sulphate soils in Thailand. Report 848. Centre for Agricultural Publishing and Documentation, Wageningen.

Van Hamme, J.D., A. Singh, and O.P. Ward. 2003. Recent advances in petroleum microbiology. Microbiology and Molecular Biology Reviews 67: 503-549.

Vence, I. and D.R. Thrasher. 2005. Reservoir souring: Mechanisms and prevention. *In:* B. Ollivier and M. Magot [eds.]. Petroleum Microbiology. ASM Press, Washington, DC. pp. 143-169.

Wagner, M., D. Lungerhausen, H. Murtada, and G. Rosenthal. 1995. Development and Application of a new Biotechnology of the Molassess In-situ Method: Detailed Evaluation Selected Wells in the Romashkino Carbonate Reservoir. *In:* Proceedings of the 5th US Department of Energy International Conference on Microbial Enhanced Oil Recovery and Related Technology for Solving Environmental Problems. Dallas, Texas, September 11-14, pp. 153-173.

Walker, J.D. and R.R. Colwell. 1976. Biodegradation rates of component of petroleum. Canadian Journal of Microbiology 22: 1209-1213.

Zajic, J.E. and E.C. Donaldson (eds.). 1985. Microbes and Oil Recovery. Bioresources Publications, El Paso, Texas.

Zajic, J.E., T.R. Jack, and E.A. Sullivan. 1997. Chemical and microbially assisted leaching of anthabasca oil sands coke. *In:* W. Schwartz [ed.]. Conference on Bacterial Leaching. Braunschweig–Stockhiem.

Zekri, A.Y. and R. El-Mehaideb. 2003. Steam/bacteria to treatment of asphaltene deposition in carbonate rocks. Journal of Petroleum Science and Engineering 37: 111-121.

ZoBell, C.E. 1947. Bacterial release of oil from oil-bearing materials, Parts I and II. World Oil 126(13): 36-47(1); 127(1): 35-41 (II).

Bio/Geomarkers and Mineralizing Hydrothermal Settings: Selected Earth Analogs, Microbiological Ecosystems and Astrobiological Potential

Jesús Martínez-Frías[1]*, Ester Lázaro[2] and Abraham Esteve-Núñez[3]

INTRODUCTION

A fundamental problem related to the astrobiological study of terrestrial (e.g. early Earth rocks and minerals) and planetary materials (e.g. asteroidal and Mars meteorites, Martian rocks) is not only recognizing and quantifying carbon-related compounds that may be present, but also differentiating those molecules formed abiotically from those generated by extinct or extant life. Whereas biological markers or '*biomarkers*' are molecular fossils (Simoneit 2004a, b), there is not a clear and official definition of the term '*geomarker*', and it has been ambiguously used to refer to different topics, not always following a formal concept. Biomarkers are defined as complex organic compounds, which originated from formerly living organisms and which are composed of carbon, hydrogen, and other elements. Abiotic organic compounds are not biomarkers per se because they do not originate from biosynthesis

Centro de Astrobiología, CSIC/INTA, Associated to the NASA Astrobiology Institute, Ctra de Ajalvir. Km. 4, 28850 Torrejón de Ardoz, Madrid, Spain, E-mails: [1]martinezfj@inta.es, [2]lazarolm@inta.es, [3]estevena@inta.es
*Corresponding author

(Simoneit 2004a, b). Biomarkers occur in many different types of materials and geological settings (sediments, rocks, crude oils, etc.) and normally display negligible or no structural changes from their parent organic molecules in living organisms (see Eglinton and Calvin 1967, Summons and Powell 1986, Summons and Walter 1990, Becker et al. 1997, Simoneit et al. 1998, Brocks et al. 1999, Brocks 2001, Simoneit 2002, 2004a, b, Brocks and Summons 2004, Brocks and Pearson 2005, Zhang et al. 2005). In accordance with Peters et al. (2006) three principal characteristics permit biomarkers to be distinguished from many other organic compounds: "a) biomarkers have structures composed of repeating subunits, indicating that their precursors were components in living organisms; b) each parent biomarker is common in certain organisms, and c) these organisms can be abundant and widespread". Taking into account these concepts, and considering that hydrothermal systems have been proposed as logical candidates for the location of the emergence of life (an assumption that is supported by genetic evidence that modern hyperthermophilic organisms are closer to a common ancestor than any other forms of life) (Shock 1997, Russel et al. 2005), the present contribution aims to offer a synthetic overview of a) the main geological (mineralizing hydrothermal) scenarios regarding geo and biomarkers; b) the main microbiological features regarding hyperthermophilic bacteria and thermophilic viruses. This information is extremely useful for the understanding of hydrothermal ecosystems, both past and present (Barns et al. 1997), and for the possible extrapolation of such knowledge to the astrobiological exploration of Mars and other planetary bodies.

BIO AND GEOMARKERS

One of the most outstanding aims of geobiology is to rebuild the co-evolution of life and environment throughout Earth history (Brocks and Pearson 2005). Hornek (2000) states that in any astrobiological search for extant life, more data are required on the geology (paleolakes, volcanism, hydrothermal vents, carbonates), climate (hydrosphere, duration of phases which allow liquid water) and radiation environments. As Dr. Jack Farmer says (Zabala 2001): *"The selection of landing sites is a crucial step for implementing the strategy for Mars astrobiology. If we land in the wrong place, we will not be able to access the most favorable past or present environments for life"... "The two most important geological environments for biosignature preservation are: 1) sites of rapid mineral precipitation (e.g. mineralizing springs, evaporative lakes, etc.) and 2) low energy, anoxic lake environments where fine-grained, clay-rich sediments (e.g. shales, water-lain volcanic ash, etc.) were deposited. Virtually all fossil microbial biosignatures found in the Precambrian record on Earth are found in these settings"*. Recent contributions tackling the use of biomarkers for the astrobiological exploration of Mars and other planetary bodies comprise diverse topics: a) studies about what the death of cryptoendolithic microorganisms leaves behind as inorganic traces of microbial life (Wierzchos et al. 2003, 2006); b) the effect of ionizing radiation on the preservation of amino acids (as potential biomarkers) on Mars (Kminek and Bada 2006); c) the development of instrumentation for the detection of biomarkers (Dickensheets et al. 2000, Parro et al. 2005, Sims et al. 2005, Rull and Martinez-Frias 2006, Villar and Edwards 2006); the evaluation of the use of morphology as an indicator of biogenicity (Allen et al. 2000, Garcia-Ruiz et al. 2002, Westall and Folk 2003).

Our planet and its biosphere have evolved together, and the geobiological history of ecosystems is recorded in sedimentary rocks through billions of years (Brocks and Pearson 2005). Hence, in order to identify potential biomarkers regarding the type of microbes which lived (or still live) at the surface of Mars or other planets, we will need to use the minerals, rocks (and other geological features as geomarkers to understand the geological and environmental contexts (Martínez-Frías et al. 2006), to comprehend the often intangible link between the 'bio- and geo- (planetary)-worlds'. A significant handicap in reaching this objective is the differential development and in many cases conceptual and epistemological 'gaps' between the studies regarding 'bio- and geomarkers'. But while biomarkers are perfectly defined, what are geomarkers? A references search in the principal scientific database (ISI-Web of Science) shows nearly 14,000 references including the term biomarker(s), but there are only 5 publications in which the term 'geomarker(s)' occur(s) explicitly. Until very recently (Martínez-Frías et al. 2007a, b) there was not a clear and official definition of the term, and it had been ambiguously used to refer to different topics, not always following a formal concept. The specific works in which the term 'geomarker' are particularly quoted deal with a) the use of benzohopanes as geomarkers of sediments and petroleums (Hussler et al. 1984) the characterization of oil shales, using the alkane geomarkers as indicators of the immature character of sediments and the nature and environmental features of the source (Vayisoglu et al. 1997), the utilization of spectroscopic techniques (Raman spectroscopy) for planetary exploration in the near future to search for extinct or extant life signals (Villar and Edwards 2006), the use of some minerals (graphite) to carry out paleogeographic reconstractions (Dissanayake et al. 2000, Sreeraj et al. 2000, and e) the use of mineralogical textures, crystal-chemical features and isotopic values as geomarkers in methanogenic and hydrothermal chimneys (Martínez-Frías et al. 2004, 2007a, b, Merinero et al. 2006, Gonzalez et al. 2007, Merinero et al. 2008).

Regarding the specific relationship with life, it is important to note that the term 'geosignature' had already been proposed (Schulze-Makuch et al. 2002, Schulze-Makuch and Irwin 2004). This review must be taken with caution, because the non-use of the term 'geomarker' does not necessarily mean that the geological aspects and indicators regarding the history and evolution of a specific geological environment have not been tackled, but it shows a spasmodic development that we want to stress in the present chapter. Thus, a definition of the term 'geomarker' was necessary, and it was proposed (Martínez-Frías et al. 2007a, b) that *Geomarkers can be defined as any geological (i.e. mineralogical, geochemical, metallogenetic, sedimentological, petrological, tectonic) feature or set of features, which can be used as proxy indicators of the physical, chemical and/or biological characteristics of the environment in which they occur, and/or of the process which formed them.*

HYDROTHERMAL SYSTEMS, MINERALIZATION AND ASTROBIOLOGICAL SIGNIFICANCE

Recently, the origin and early evolution of life has undergone an unprecedented development. New theories concerning the origin of life such as cosmic (e.g. cometary, asteroidal) sources and types of organics, the possible role of marine (and also terrestrial) hydrothermal systems on the chemistry of Early Earth and the postulate of an 'RNA' (and other) world have given new perspectives to: 1) tackle this problem from different angles, and 2) understand their potential implications in the astrobiological exploration of other planetary bodies. The discovery of deep-sea hydrothermal vents, in the late 1970s, opened a window into an unknown and unexplored geosphere and biosphere. Submarine hydrothermal vents are singular sites where hot, metal-bearing hydrothermal solutions that have been convected through newly formed volcanic (mainly basaltic) crust are exhaled onto the sea floor. Upon exhalation, these fluids interact with sediments and rocks and precipitate their metallic load to form a wide variety of edifices; mounds and venting structures displaying a wide typological variety of minerals (base and precious metal sulfides and sulfosalts, carbonates, sulfates, oxides, oxi-hydroxides, etc.) and complex parageneses (Rona and Scott 1993, Herzig and Hannington 1995, Humphris et al. 1995, Barnes 1997, Herzig et al. 2002, Rona 2003, among others). More than 30 years after their discovery we know that modern hydrothermal vent environments (and hydrocarbon seeps) are located at characteristic geotectonic, geochemical and biological interfaces where H_2S and CH_4 enriched fluids are discharged at the seafloor, sustaining abundant chemosynthetic ecosystems (ChEss program, http://www.noc.soton.ac.uk/chess/).

It is not the objective of the present contribution to specifically tackle the controversial topic of the origin of life and its possible relationship with all likely genetic scenarios, but there is significant geological, geochemical and biological (genetic) evidence that hydrothermal systems should be considered extraordinary sites for the initiation and development of life on Earth (see Russell and Hall 1997, Cairns-Smith et al. 1992, Ferris 1992, Hennet et al. 1992, Russell et al. 1994, MacLeod et al. 1994, Marshall 1994, Walter 1996, McClendon 1999, McCollom et al. 1999, Reysenbach et al. 1999, Zierenberg et al. 2000, Blake et al. 2001, Line 2002, Martin and Russell 2003, Russell and Martin 2004, Russell et al. 2005, among others). Turcotte (1980) points to the fact that crustal heat flows were much higher during the Archean, and volcanism (and hydrothermalism) were more extensive. Komatiitic lavas are among the most ancient lavas of our planet and they represent some of the oldest ultramafic magmatic rocks preserved in the Earth's crust, at 3.5 Ga. Komatiitic volcanism is mainly restricted to Archean terranes indicating that average crustal temperatures must have been much higher at that time. Recently, the astrobiological significance of these volcanic rocks has been emphasized (Nna-Mvondo and Martínez-Frías 2005, 2006). In broad terms, hydrothermal systems were important in the differentiation and early evolution of Earth because they linked the global lithospheric, hydrologic, and atmospheric cycles of the elements (Des Marais 1996).

Regarding hydrothermalism, if following a similar database search, the term 'hydrothermal' is added as a key word, it is interesting to note that no references regarding geomarkers appear, whereas there are almost 100 regarding biomarkers. Essentially they cover the following topics:

- studies of hydrothermal simulation experiments (Bushnev et al. 2004, Zarate del Valle et al. 2006) and biomarkers consisting of hopanes, gammacerane, tricyclic terpanes, carotane and its cracking products, steranes, and drimanes (hydrothermal bitumen) (Zarate del Valle and Simoneit 2005);
- characterization of prokaryote activity in supergene reactions within submarine hydrothermal sulfidic sediments (Glynn et al. 2006);
- the use of biomarkers as indicators of the biogeochemical cycling by Archaea and Bacteria (Summons and Powell 1986, Campbell 2006), worms (Pond et al. 2002, Phleger et al. 2005), fishes (Guerreiro et al. 2004), shrimps (Pond et al. 1997), gastropods (Pranal et al. 1996) and sea stars (Howell et al. 2003) related, or not related, with hydrothermal vents;

- isotopic and mineralogical studies of carbon, and textural and geochemical features indicative of microbially mediated alteration of basaltic glass in the Achaean (Barberton Greenstone Bell) (Banerjee et al. 2006);

- thermal alteration of organic matter in terrestrial impact craters; biomarker ratios in the rocks that contained pre-existing liquid hydrocarbons are used to show the variation in the degree of heating across the structure (Parnell et al. 2005);

- characterization of hydrocarbons (terpane distribution) trapped in fluorite deposits (Guilhaumou et al. 2001) and biogeochemical studies of organic matter in ore deposits, their host rocks and sediments, and lignites (Simoneit 1992, Guezennec et al. 1998, Spangenberg et al. 1999, Stefanova et al. 1999, Logan et al. 2001, Bao et al. 2005);

- analyses of deep-sea hydrothermal sub-vent core samples (biomarkers) (Takano et al. 2005), vent sediments (Rushdi and Simoneit 2002, Yamanaka and Sakata 2004) and warm (65°C) deep-sea hydrothermal waters and shales (Brocks et al. 2003, Kenig et al. 2003);

- scientific constraints to find a good definition of the term 'biomarkers' as molecular fossils and geochemical indicators of life (Simoneit 2002, 2004a, b);

- morphological studies (e.g., petrographic identification of body fossils) as the best method to identify organisms present in ancient hot-spring accumulations (Guidry and Chafetz 2003);

- the application of mapping, optical and electron microscopy, digital image analysis, micro-Raman spectroscopy and other geochemical techniques in questioning the evidence for Earth's oldest fossils, reinterpreting the microfossil-like structure, as secondary artefacts formed from amorphous graphite within multiple generations of metalliferous hydrothermal vein chert and volcanic glass (Brasier et al. 2002);

- the use of the oxygen isotope ratios of PO_4 as an inorganic indicator of enzymatic activity and P metabolism, and a new biomarker in the search for life (Blake et al. 2001);

- studies of organic geochemistry of hydrothermal petroleum associated with submarine volcanism (Yamanaka et al. 1999);

- characterization of biomarkers (pristane, phytane, regular isoprenoids paraffin, methyl-heptadecyl) in pyritic stromatolites (Xia et al. 2000);

- comparative study of extracelular polymeric substances and biofilms from hot springs, deep-seas, volcanic lakes, and shallow marine/littoral environments as possible biomarkers (Westall et al. 2000).

An important issue related to astrobiological exploration and hydrothermalism is that hydrothermal environments often exhibit high rates of mineralization, which favors microbial fossilization. Thus, hydrothermal deposits are often rich storehouses of paleobiologic information (Farmer 2000). Likewise, as Hornek (2000) indicates in her study of the microbial world and the case for Mars exploration, microbial prokaryotes have flourished on Earth for more than 3.5 billion years. They dominated the Earth's biosphere during the first 2 billion years of its history before the first unicellular mitotic eukaryotes appeared. Therefore, in a search for extant life beyond the Earth, microorganisms are the most likely candidates for a putative biota extra-terrestrial habitat.

THREE SELECTED EARTH ANALOGS

Among the exclusive mineralizing hydrothermal models, which have been proposed as interesting Earth analogs for the study of geo- and biomarkers (9°50'N and 21°N East Pacific Rise, Gorda, Juan de Fuca and Explorer Ridge, Lost City, Gulf of Cadiz, Guaymas Basin, Lucky Strike, etc. (see for instance, Vents Program http://www.pmel.noaa.gov/vents/)) we think it appropriate to mention here for astrobiological interest: a) a fossil hydrothermal system in SE Iberian (Martínez-Frías et al. 2004, Grymes and Briggs 2005, Martínez-Frías et al. 2006); b) some selected carbonate-precipitating hot springs (Allen et al. 2000); and c) selected areas displaying currently active hydrothermal and mineralizing processes (altered rocks and precipitates) in Iceland (Geptner et al. 2005).

(a) Fossil hydrothermal system, SE Iberia: The Jaroso Hydrothermal System is a volcanism-related mineralizing hydrothermal system, of Upper Miocene age, whose characteristics can aid to the geological and astrobiological exploration of Mars (Martínez-Frías et al. 2001, 2004). The Mars Exploration Rover Opportunity's Moessbauer spectrometer showed in 2004 the presence of an iron-bearing sulfate called jarosite in the area 'El Capitan' on Mars' Meridiani Planum (Christensen et al. 2004, Klingelhofer 2004, Madden et al. 2004, Squyres et al. 2004). Jarosite is one of the most significant pieces of evidence for the existence of liquid water in Mars' past. It was characterized for the first time in the Jaroso Ravine, which is the 'world type locality' (Amar de la Torre 1852; Martinez-Frias 1999), and the best outcrop where the mineralization and alteration associated with the JHS have attained the maximum surface expression (Martínez-Frías et al. 2004). All mineral deposits originated from the JHS make up a metallogenetic belt of hydrothermal mineralization which extends roughly 50 km SW-NE, from the Cabo de Gata region (Almeria province) to the Aguilas area (Murcia province). Morphologically, the deposits are polymetallic veins and hydrothermal breccias hosted in Permian-Triassic basement and locally in Neogene volcanic edifices and stratabound ores hosted in Upper Miocene, shallow-marine sandy marls. Paleobathymetry data (Montenat and Seilacher 1978) for the time of emplacement of the hydrothermal fluids indicate an approximate depth of 200–300 m below sea level. The JHS includes oxy-hydroxides (e.g. hematite), gold, silver, Hg-Sb, and base-metal sulfides and different types of sulfosalts (mainly rich in Ag and Sb) (Martínez-Frías et al. 1989, Martínez-Frías 1991). Hydrothermal sulfuric acid weathering of the ores has generated large volumes of oxide and sulfate minerals of which jarosite is the most abundant. It is important to stress that some extinct shallow-marine hydrothermal vent structures, which are associated with the mineralizing process of the JHS, are still preserved 'in situ' in the sandy Mars substrate, constituting perfect targets for carrying out detailed isotopic analysis (fluid geomarkers) (Fig. 6.1). It has been proposed and generally accepted that the JHS is genetically linked with the late episodes of Upper Miocene calc-alkaline and shoshonitic volcanism in the area. The conjunction of the metallogenetic peculiarities of the JHS, its geodynamic setting, its mineralogy and paragenetic sequences and the isotopic results obtained from the study of vent chimneys give information about the fluid–rock interactions and indicate that the Jaroso Mars analog comprises many interesting and diverse geological keys which can be used as geomarkers for the study of similar ancient processes on Mars.

(b) Selected carbonate-precipitating hot springs: Hot spring deposits have been cited as prime locations for the search for life on Mars. Allen et al. (2000) completed research on microscopic physical biomarkers. Samples were selected from four hot springs similar in temperature, water composition, and precipitate mineralogy. They were examined to document physical biomarkers at the scale of microorganisms — submillimeter to submicrometer. The sampling sites included Le Zitelle in the Iterbo region of Italy, Narrow Gauge in the Mammoth Hot Springs complex of Yellowstone National Park, Wyoming, Jemez Springs on the slopes of the Valles caldera, New Mexico and and the Hot Springs

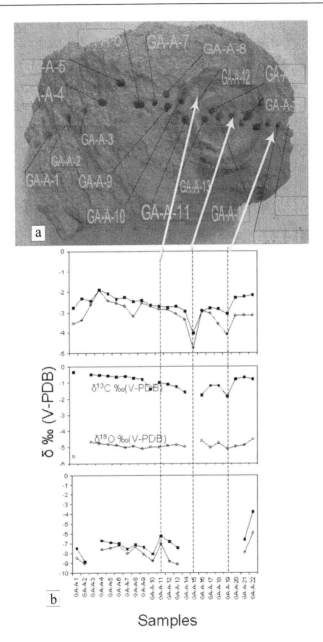

FIG. 6.1 Extinct low-temperature hydrothermal chimney from the Jaroso Mars Analog (SE Spain) (Martínez-Frías et al. 2006). Sampling and stable isotope variations from the central orifice to the outer rim. Note that siderite is not present where ankerite (more stable phase) occurs. The paleoenvironmental interpretation carried out from the isotopic data emphasizes the importance of the stable isotopes as fluid geomarkers.

Color image of this figure appears in the color plate section at the end of the book.

National Park, Arkansas. Their study included microbes that live in such environments and the preservation of microbial forms, biofilms, mineral precipitates, and petrographic fabrics indicative of life in hot spring deposits. An extremely significant result was that microorganisms at all four sites produce microscopic physical biomarkers. The authors found: a) a range of markers which provide either equivocal or unequivocal indication of the presence of microbial life in carbonate hot springs; b) detected that these biomarkers are rapidly destroyed or altered but that others can be incorporated in the geologic record; and c) that a fossil hot spring deposit might preserve such biomarkers on the desiccated Martian surface.

(c) Active hydrothermal and mineralization areas in Iceland: Recently, Geptner et al. (2005) studied the fossilization of biota and the formation of low- and high-temperature hydrothermally altered rocks in solfataric fields, artificial hot lakes, in natural hot springs, and on a heated beach within the present-day rift zone at the Reykjanes, Nesjavellir, Geysir, Landmannalaugar, Námafjall, and Öxarfjörður geothermal areas. They studied hydrothermally altered rocks which were formed in low- and high-temperature settings in solfataric fields, hot springs, and artificial hot lakes. The following mineral types were found to exist: (1) smectite assemblages with iron oxides and hydroxides; (2) smectite assemblages with sulfides (pyrite); (3) kaolinite-metahalloysite assemblages with sulfates, anatase, and boehmite; and (4) siliceous assemblages. As the authors stated (Geptner et al. 2005), the occurrence of microbial species mineralized with clay, Fe-bearing minerals, and silica at recent hot springs and in solfataric fields indicates that possible life existed in the high-temperature hydrothermal process. Given the geological and metallogenetic similarities of these areas with the JHS, the mineralogical, geochemical and microbiological study of selected samples has been initiated in the Complutense University of Madrid, the Centro de Astrobiología and the University of Valladolid (Spain), in the context of their geological and astrobiological interest as a potential 'active Mars analog'.

HYPERTHERMOPHILIC MICROORGANISMS

Microorganisms thriving in high temperature terrestrial and deep-sea hydrothermal systems have stimulated new theories of life's origins. In these extreme environments, the microbial and geochemical interactions are well coupled, providing many of the basic constituents for the primordial synthesis of organic molecules and for the evolution of fundamental metabolic processes. Genetic evidence based on the sequence of 16s rDNA suggests that hyperthermophilic microorganisms are closer to the common ancestor than any other form of life (Holm et al. 1992), thus hydrotermal environments become the most adequate candidates for studying the emergence of life.

This first hyperthermophilic microorganism was isolated from Yellowstone National Park (Brock 1978), since then more than 30 genus of bacteria and archaea have been identified with the ability to grow at temperatures higher than 80°C. These microorganisms are found in diverse environments related with volcanism and/or hydrothermalism.

Most hyperthermophiles exhibit a chemolithoautotrophic mode of nutrition, thus molecular hydrogen, sulfide, sulfur and ferrous ion are common electron donors used as energy sources (chemolithotrophic), while CO_2 is the only carbon source required to build organic cell material (autotrophic). In terms of the use of electron acceptors, most of hyperthermophiles are anaerobic or facultative aerobes able to use oxygen at low concentrations. This is in agreement with the environmental conditions given that the hydrothermal fluid, suitably cooled by mixing with oceanic seawater, still remains reduced because of the high content of hydrogen sulfide. Among the electron acceptors used by hyperthermophiles one can find nitrate, sulfate, sulfur, CO_2, and Fe (III).

Terrestrial Hyperthermophiles

Hot oxygen environments in solfataric fields harbor acidophilic hyperthermophile archaea from the genus *Sulfolobus* (strict aerobe), *Metallosphaera* and *Acidianus* (Stetter 2002). Among neutrophilic hyperthermophiles from mildly acidic hot spring the genera *Pyrobaculum*, *Thermoproteus*, *Thermofilum*, *Desulfurococcus*, *Sulfophobococcus*, *Methanothermus* and *Thermosphaera* are found (Stetter 2002) .

Marine Hyperthermophiles

A variety of hyperthermophiles are adapted to the high salinity of seawater and are represented by the archaeal genera *Pyrolobus*, *Pyrodictium*, *Hyperhtermus*, *Stetteria*, *Thermodiscus*, *Igneococcus*, *Staphylothermus*, *Aeropyrum*, *Pyrobaculum*, *Methanopyrus*, *Pyrococcus*, *Thermococcus*, *Archaeglobus* and *Ferroglobus*. In the last five years, novel archeae genera like *Ignicoccus* (Huber et al. 2002), *Geoglobus* (Kashefi et al. 2002), *Palaeococcus* (Takai et al. 2000), *Methanothermococcus* (Takai et al. 2002), *Methanocaldococcus* (L'Haridon et al. 2003), *Methanotorris,* and Geogemma (Kashefi and Lovley 2003) have been described.

Aquifex and *Thermotoga* have been the bacterial genera most studied in marine hydrothermal environments (Reysenbach et al. 2001a; Reysebanch et al. 2001b). However, recent studies revealed the novel and interesting genera *Deferribacter* (Miroshnichenko et al. 2003a), *Thermodesulfobacterium* (Moussard et al. 2004), *Geothermobacter* (Kashefi et al. 2003) or *Caldithrix* (Miroshnichenko et al. 2003b).

Although a number of different archea and bacteria has been identified by molecular tools, the isolation of these hyperthermophilic microorganisms is required in order to perform a study under controlled conditions, especially those with biogeochemical reaction, leading to products which can be truly used as geomarkers in high-temperature (>80°C) hydrothermal ecosystems.

Fe (III)-Reducing Hyperthermophiles

Among the different types of microorganisms described in hyperthermophilic environments, Fe (III)-reducing microorganisms are especially relevant because of their wide distribution (Lovley 2004, Holden and Feinberg 2005). They also has astrobiological relevance because it has been proposed that the first membrane system capable of electron transport and energy conservation through a chemiosmotic mechanism began by catalyzing the oxidation of hydrogen with the reduction of Fe (III). According to that hypothesis, bacterial and archaea hyperthemophiles closely related to the last common ancestor of modern life were tested for hydrogen oxidation coupled to Fe (III) reduction. These sulfur reducers, and even *Thermotoga maritime,* previously considered to have only a fermentative metabolism, could grow as a respiratory organism when Fe (III) was provided as an electron acceptor. These results provide microbiological evidence that Fe (III) reduction could have been an important process on Early Earth and suggest that microorganisms might contribute to Fe (III) reduction in modern hot biospheres (Vargas et al. 1998). In addition, hydrogen-iron redox biogeocoupling was also found in the strain 121 an archea belonging to the novel genus *Geogemma*, isolated from an active 'black smoker' hydrothermal vent at the Juan de Fuca Ridge, which expanded the upper temperature limit for life to 121°C (Kashefi and Lovley 2003).

These Fe (III)-reducing microorganisms are able to use Fe (III) as respiratory substrate conserving energy in the reduction process (Lovley 2004). Although these bacteria can use CO_2 as a carbon source, some of them like *Geoglobus ahangari* and *Ferroglobus placidus* are also able to couple the oxidation of acetate, long-chain fatty acids and aromatic compounds to Fe (III) reduction that completes the carbon cycle that begins with CO_2 fixation by methanogens and other autotrophs in hyperthermophilic environments (Tor et al. 2001). This ability suggests for the first time that the complete oxidation of complex organic matter back to carbon dioxide may be possible in hot microbial ecosystems (Lovley 2004).

Geosignatures from Microbial Fe (III)-Reduction

Hyperthermophiles reduce poorly cystalline Fe (III)-oxide as ferryhydrite to ultra-fine grained crystals of that magnetite (Kashefi and Lovley 2000). The magnetite is very similar to the that produced and accumulated out of the cell by mesophilic Fe (III)-reducers as *Geobacter metallireducens* (Lovley et al. 1987), which produces a high amount of suparamagnetic magnetite with just 4% of it showing a single domain (Moskowitz et al. 1989). However, a recent study has reported that a unique form of tabular, single-domain magnetite can be obtained when *G. metallireducens* is cultured under low CO_2 conditions (Fig. 6.2). This novel magnetite has a distinct crystal habit and magnetic properties. Thus, it could be used as a biosignature to distinguish between biological and abiotic Fe (III)-reducing processes in hyperthermophilic environments. It is important to point out that this magnetite is not the one forming part of the intracellular magnetosomes of magnetotactic bacteria, which led to inconclusive evidence for the presence of life in the Martian meteorite ALH84001 (McKay et al. 1996). In contrast, magnetite produced by microbial Fe (III)-reduction is extracellular and 5000-fold more abundant than the one produced by an equivalent biomass of magnetotactic bacteria (Lovley et al. 1987).

The possibility that Fe (III)-reducers might isotopically fractionate iron ($^{56}Fe/^{54}Fe$) during Fe (III)-reduction is an additional tool for identifying presence of life (Johnson and Beard 2005). Unfortunately there is no reported information on magnetite-iron fractionation by hyperthermophiles, but recent findings demonstrate that Fe isotopes record aqueous/solid-phase Fe redox cycling during microbial Fe (III)-reduction (Crosby et al. 2005). The observation that microbial Fe(III)-reduction produces Fe (II)aq that has low $\delta^{56}Fe$ values might be taken as a biological geosignature for this type of iron metabolism (Johnson et al. 2005, Johnson and Beard 2005), but it could also be geological/ mineralogical.

Another possible biomarker for microbial Fe(III)-reduction is the content of structural phosphorous associated to magnetite. Abiotic magnetites produced in various environments incorporate structural

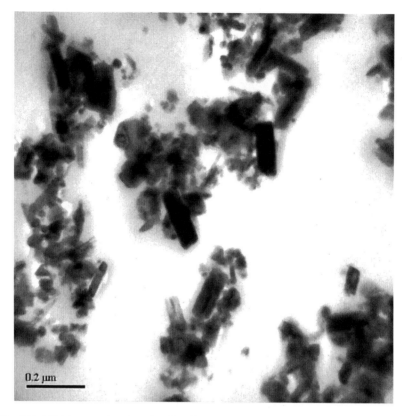

FIG. 6.2 TEM image showing the disc-like shape of tabular magnetite at 45 degrees tilting. Courtesy Dr. H.Vali

phosphorous, but some well-characterized biological magnetites seem to contain no phosphorus. Thus, it was concluded that natural magnetites containing occluded P are unlikely to be formed by microorganisms (Jurado et al. 2003).

Other Potential Biological Geomarkers

Studies with a wide range of microorganisms indicate that wolfram (W) is typically preferred over molybdenum with increasing growth temperatures above 60°C. The preference for W at the higher temperatures appears to be due to the greater thermal stability of W complexes. For instance, *Pyrococcus furiosus* appears incapable of utilizing Mo when cells are limited in W (Mukund and Adams 1996). There also seem to be a correlation between the amounts of W present and the environments where hyperthermophiles are found, but it is still not well defined if W attracts hyperthermophiles or whether the W is accumulated and deposited by the microorganisms (Holden and Adams 2003).

THE WORLD OF HEAT LOVING VIRUSES

Viruses constitute the simplest biological entities in nature. They consist of one or several nucleic acid molecules covered by several protective layers. Usually there is a protein layer, the capside, which can be enclosed by a lipid envelope. The number of proteins encoded by the viral genomes is rather low and clearly insufficient to provide the metabolic activities necessary to live independently. The success of viruses to survive depends on their ability to infect cells and to exploit the cellular metabolism for their own benefit. This means that they must be able to use the cellular resources to replicate their genomes, synthesize their proteins, assemble new virus particles and develop strategies that permit the liberation of the progeny viruses outside of the cell.

Despite their structural simplicity, viruses have been able to adapt to the wide diversity of the cellular world adopting an extraordinary diversity of strategies. There are approximately 5100 viruses characterized which infect organisms belonging to the three domains of life (Bacteria, Eukarya and Archaea). They can use DNA or RNA to store the genetic information, and it is worth noting that viruses and viroids (not reviewed in this chapter) are the only biological systems in which the genomic material can be constituted by RNA. This feature has been hypothesized to be a remnant of the RNA world and, according to this interpretation RNA viruses can be considered molecular fossils of the primitive replicator molecules that were present in the prebiotic Earth (Weiner and Maizels 1999). Viruses have also explored diverse ways to replicate their genomes, including some that are exclusive to them (RNA replication and retrotranscription). Both processes take place with high error rate, which confers to RNA viruses a wide genetic diversity and adaptive capacity. Finally, it should be mentioned that the modifications in the capside structure and in the external morphology are factors that condition the cellular types which can be infected and the stability of viruses in the external environment.

Taking into account this extraordinary adaptive potential, it is expected to find that viruses are able to survive and infect organisms living in extreme hydrothermal environments. The interest in the search for viruses in these ecosystems relies not only on obtaining knowledge of the biodiversity present at high temperatures. In the same way that in the decade of 1960 bacteriophages were a very valuable tool to understand the mechanisms of genome replication and the transmission of the genetic information, thermophilic and hyperthermophilic viruses could serve as model systems to understand the molecular basis required for adaptation to high temperatures. In addition, the study of these viruses could help to understand how life originated in the extreme conditions of primitive Earth, especially if the hypothesis of a hot origin for the first forms of life is considered. The biotechnological applications that could derive from the isolation of viral enzymes that work efficiently at high temperatures is another relevant question to explore.

VIRUSES IN HYDROTHERMAL ENVIRONMENTS. AN UNEXPECTED DIVERSITY OF MORPHOTYPES

Of the three domains of life, organisms belonging to the crenarchaeota phyla of the archaea domain are the most abundant in hot, acidic springs (Takai and Sako 1999). The hosts for most hyperthermophilic viruses isolated up to now are members of the crenarchaeota genera *Sulfolobus, Acidianus, Thermoproteus* and *Pyrobaculum* (Prangishvili and Garret 2005). The first two genera comprise extreme acidophiles, whereas members of the last two genera are neutrophiles and obligate anaerobes. A thermal archaeal virus has also been isolated from a hyperthermophilic marine euryarchaeote (Geslin et al. 2003). Bacteriophages infecting thermophilic bacteria have not been studied extensively. They have been isolated from hot springs, mud spots, and solfataric fields (Rachel et al. 2002). Recently, the isolation of two lytic bacteriophages of thermophilic bacteria which were purified from deep-sea hydrothermal fields in the Pacific was reported for the first time (Liu et al. 2006).

The low number of viruses found in hydrothermal environments (≈25) contrasts with the approximately 5100 identified viruses distributed within biosphere. One reason for this large difference is that isolation of viruses from extreme environments usually presents many difficulties associated with the cultivability of the host strains. This restriction has probably introduced a bias in the type of viruses that have been isolated, which could represent only a minor part of a much wider virus diversity. Another difficulty lies on the considerable lower concentration of virus particles in hot springs than in 'temperate' marine environments, probably due to the lifestyle of hyperthermophilic viruses mainly based in non-lytic infections to reduce the exposure of the viral particles to the extreme conditions (Snyder et al. 2003). Recently, new protocols for detecting virus-like particles (VLPs) directly from hot springs have been developed. These methods are based on direct observation by epifluorescence microscopy and in the detection of conserved genes in comparative genomics studies (Wiedenheft et al. 2004). The possibility of detecting VLPs independently of the cultivability of the host organisms opens new perspectives to the study of the spatial and temporal distribution of the viruses present at high temperature. The role played by viruses in the ecology and geochemistry of hydrothermal environments could also be explored with a deeper insight.

Crenarchaeal hyperthermophilic viruses have been classified into five approved families and two other families are under consideration (Prangishvili and Garret 2005). The classification is based on the virion morphology and the characteristics of the virus genome. There is a sharp contrast between the low number of morphotypes found in viruses which infect mesophilic bacteria and euryarchaea with the high morphologic diversity of the viruses infecting hyperthermophilic crenarchaeota organisms. Ninety seven percent of non-thermophilic viruses are composed of an icosahedral head and a helical tail (head-and-tail viruses). The remaining 3% are tail-less icosahedra, filaments or pleomorphic particles. In contrast to this scenario, the isolation of hyperthermophylic viruses has shown an extraordinary diversity of morphotypes, some of them never seen before (Rice et al. 2001, Rachel et al. 2002). Next, a brief description of the virus morphologies found in high temperature environments is provided:

(a) Spindle-shaped particles: They have in common a single short tail with fibers that facilitate the attachment of the virion to the host membrane. Most viruses in this group infect *Sulfolobus* spp. (Stedman et al. 2003). They comprise the Fuselloviridae family that includes SSVI — the first crenarchaeal virus isolated (Fig. 6.3) — the unclassified STSV1 virus, and the two-tailed virus (ATV) of the Bicaudaviridae family which infects the genus *Acidianus*. The last virus can develop a very long tail at each one of its pointed ends after being released from its host in the absence of any exogenous energy source or cofactors (Häring et al. 2005). This is the first example of a virus with host-independent extracellular activity. The process only occurs when the virus is outside its host and exposed to the harsh environment, and probably represents a special adaptation to facilitate host recognition, reducing the time of exposure to the high temperatures. The spindle-shaped morphology is the most frequent among hyperthermophile crenarchaeota viruses and is also present in the only hyperthermophilic virus isolated from an euryarchaeota organism (Geslin et al. 2003).

(b) Droplet-shaped particles: Viruses included in this group possess a large number of densely packed thin filaments protruding from the pointed end of the virion (Arnold et al. 2000). This morphotype is only found in the SNDV virus.

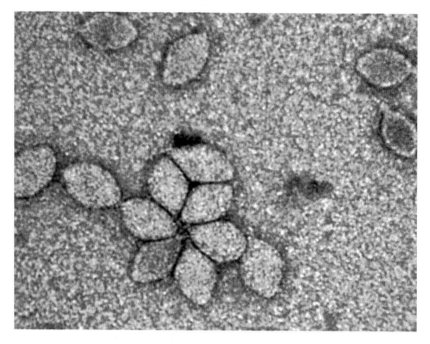

FIG. 6.3 Electronic micrograph of SSV1, the first crenarchaeal virus isolated (courtesy Dr. Kenneth Stedman).

(c) Bottle-shaped particles: The only example in this group is the enveloped virion ABV of the Ampullaviridae family (Häring et al. 2005) that resembles a bottle, the narrow end of which is likely involved in cellular adsorption.

(d) Stiff rod-shaped particles: This group comprises viruses with similar morphology to some RNA plant viruses (Prangishvili et al. 1999).

(e) Filamentous enveloped particles: (Janekovic et al. 1983).

(f) Icosahedral and globular particles: Some of the viruses with these morphologies resemble RNA viruses of the Paramyxoviridae family, which infects vertebrates (Rice et al. 2004)

It is generally accepted that these unusual morphotypes constitute adaptations for survival at high temperatures. However, it is unclear how they contribute to the stability of virus particles in hot, acidic environments.

All the viruses described in this chapter have double-stranded DNA genomes that can be linear or circular. Similar to their hosts, they have a low GC content, indicating that the higher genetic stability of GC pairing with respect to AT pairing has not been exploited. The genomes of most crenarchaeal viruses isolated have been sequenced, which has permitted to obtain information about the genetic diversity and to carry out comparative genomic studies. In general, the search for sequence similarities among genes in hyperthermophilic viruses does not yield significant matches with genes from bacteria or eukarya. There are only some homologies with genes of other crenarchaeal viruses (Prangishvili et al. 2006). Viruses with similar morphologies and genome structures show more genetic similarity than viruses with different morphologies. Unique genes observed in related viruses might reflect different evolutionary histories or particular adaptations for a specific hot environment. The wide distribution of hot springs and hydrothermal vents all around the world opens up questions concerning the relationships between the biogeochemical features of specific locations and the virus diversity found in them.

HYPERTHERMOPHILIC VIRUSES. A TOOL TO ELUCIDATE THE PHYLOGENY OF ARCHAEA?

Most cellular characteristics of archaea resemble those of bacteria and, according to this, one might expect that archaeal viruses would be similar to bacteriophages. This is true for the viruses found in the archaeal kingdom euryarchaeota, which mainly infect mesophilic or moderately thermophilic organisms. The similarities between bacteriophages and euryarchaeota viruses are manifested in many features, such as the morphology of the virions, the high homology degree found in many genes and the lytic cycles adopted by most euryarchaeal viruses. This situation contrasts with the great and unusual variety of virus shapes found in the archaeal kingdom Crenarcheota, many of which are exclusive to them. Other differences are the presence of an envelope in most chrenarchaeal viruses and, with the only known exception of ATV, the adoption of non-lytic cycles. Crenarchaeal viruses cause persistent infections, meaning that there is a continuous virus production without apparent interference to cellular replication.

Comparative-genomic studies show that euryarchaeal and crenarchaeal viruses strongly differ in genome organization and gene content. Whereas euryarchaeal viruses share many genes with bacteriophages, crenarchaeal viruses are almost unrelated to any other viruses (Prangishvili et al. 2006). These biological differences between crenarchaeal and euryarchaeal viruses might be explained by distinct evolutionary histories of the two kingdoms or by the exposure to different selective pressures. Most studies suggest that the high temperatures at which hyperthermophilic viruses have to adapt constitutes a strong selective pressure that can be responsible for many of the differences reported between crenarchaeal and euryarchaeal viruses.

HYPERTHERMOPHILIC VIRUSES AS BIO/GEOMARKERS

In general, viruses have a great influence in the transference of matter and energy in ecosystems, which is manifested by changes in the biogeochemical structure (Fuhrman 1999). Lytic infections convert cells into viruses plus cellular debris, which is made of dissolved molecules plus colloids and cell fragments. Most of these products become available to other bacteria, although some cellular debris may resist degradation. Released polymers modify the characteristics of water, influencing many biological and microscopic physical chemical processes.

As it has been previously pointed out, most crenarchaeal hyperthermophilic viruses are non-lytic. Consequently, there is little liberation of cellular debris to the external environment. This fact probably limits the influence of viruses in the geology and microbial diversity of hot ecosystems. However, there are other alternative ways to modify the external environment. Although viruses do not have the metabolism or extracellular activity, they can alter the geology or mineralogy of the environment through interference with the metabolism of their hosts. Further studies are necessary to determine the interaction between geology/mineralogy in a given environment and the virus diversity which is present in it.

ASTROBIOLOGICAL INTEREST OF THE STUDY OF HYPERTHERMOPHILIC VIRUSES

The astrobiological relevance of the study of viruses adapted to high temperatures is clear. These viruses are a very valuable tool to understand the biochemical and molecular mechanisms responsible for adaptation to hot environments similar to those of the primitive Earth.

There are a number of arguments indicating that viruses could have existed in the prebiotic phase of evolution (Balter 2000). The16S rRNA-based phylogenetic tree of life places the hosts of hyperthermophilic viruses closest to the root of the tree. Taking this fact under consideration, an important role for hyperthermophilic viruses at the earliest stages of evolution has been suggested.

All archaeal hyperthermophilic viruses found until now have DNA genomes. If we take into account that RNA viruses are able to adapt much faster to adverse conditions and environmental changes than DNA viruses, it would be expectable to find RNA viruses living in hot environments. The application of culture-free techniques might facilitate the isolation of hyperthermophilic viruses with RNA genomes, which could lead to new insights into the role played by RNA in the early evolution of life (Ortmann et al. 2006).

FUTURE PERSPECTIVES

It is broadly accepted that the subaerial and submarine mineralizing hydrothermal episodes: 1) contributed, through the evolution of the Earth, to nearly continuous fluid–rock interaction processes; 2) were particularly remarkable in relation to the behavior and geochemical cycles of some elements and minerals (oxides and oxi-hydroxides, sulfides, sulfates), mainly iron and metal sulfides (Schoonen et al. 2004); and 3) revealed a new world related to the origin of life and biomineralization processes (Fortin et al. 1998). These new findings have changed our viewpoints about fluid geodynamics, building mechanisms of new subaerial and submarine structures, microbial metabolism and survivability under extremophilic conditions, opening extraordinary future perspectives for the study of potential hydrothermal system on Mars and other planetary bodies. It is a fact that they have introduced new questions about the physical and chemical limits to life (Prieur et al. 1995), and how this knowledge can be used to find terrestrial analogs for searching extinct or extant extra-terrestrial life.

Very recently (Squyres et al. 2008) interpreted the Martian silica-rich materials identified by Spirit to have originated under hydrothermal conditions, as they occur in association with volcanic materials and, in some cases, are closely mixed with ferric sulfates that are also probably of hydrothermal origin (see, for instance, Rodriguez-Losada et al. 2000, Bustillo and Martínez-Frías 2003, Geptner et al. 2005). As the authors point out, hydrothermal environments that generated the Martian opal could be extremely significant from the astrobiological point of view.

Thus, if one wants to correctly use the term 'bio/geo' or 'geo/bio', it is essential that the studies cover both aspects from an ambivalent perspective. Only through the combination of both types of markers (biomarkers and geomarkers) will one have a complete scientific panorama, which will not only contribute to the understanding of ancient environments and their associated life, but will also help in the recognition and exploration of new potential extinct or extant extraterrestrial ecosystems. The case of mineralizing hydrothermal systems has extraordinary astrobiological potential.

ACKNOWLEDGEMENTS

We wish to acknowledge the institutional support of the Centro de Astrobiologia. Special thanks to Prof. R. Lunar (Complutense University, Madrid), Prof. F. Rull (Valladolid University and CAB), Dr. A. Delgado (Estacion Experimental del Zaidin, Granada), Prof. L. Vazquez (Complutense University and CAB) and Dr. J. Gomez-Elvira (CAB). Also thanks to Prof. Dr. Sudhir Kumar Jain for his extremely kind invitation to write the present chapter, his constant help and editorial supervision. Finally, thanks to Dr. Andrew Hill for revision of the English version.

REFERENCES

Allen, C.C., F.G. Albert, H.S. Chafetz, J. Combie, C.R. Graham, T.L. Kieft, S.J. Kivett, D.S. McKay, A. Steele, A.E. Taunton, M.R. Taylor, K.L. Thomas-Keprta, and F. Westall. 2000. Microscopic physical biomarkers in carbonate hot springs: Implications in the search for life on Mars. Icarus 147(1): 49-67.

Amar de la Torre, R. 1852. Descripción de los minerales, algunos de ellos nuevos, que constituyen el fil´on del Barranco Jaroso de Sierra Almagrera, por el caballero profesor el doctor Augusto Breithaupt, de Freiberg. Revista Minera 3: 745-754.

Arnold, H.P., U. Ziwsu, and W. Zillig. 2000. SNDV, a novel virus of the extremely thermophilic and acidophilic archaeon Sulfolobus. Virology 272: 409-416.

Balter, M. 2000. Evolution on life's fringes. Science 289: 1866-1867.

Banerjee, N.R., H. Furnes, K. Muehlenbachs, H. Staudigel, and M. de Wit. 2006. Preservation of similar to 3.4-3.5 Ga microbial biomarkers in pillow lavas and hyaloclastites from the Barberton Greenstone Belt, South Africa. Earth Planet. Sci. Lett. 241, 3-4: 707-722.

Bao, Z.W., Z.H. Zhao, and J. Guha 2005. Organic geochemistry of sedimentary rock-hosted disseminated gold deposits in southwestern Guizhou Province, China. Acta Geol. Sin.-Engl. Ed. 79-1: 120-133.

Barnes, H.L. (ed). 1997. Geochemistry of Hydrothermal Ore Deposits. Wiley, New York.

Barns, S.M., C.F. Delwiche, J.D. Palmer, S.C. Dawson, K.L. Hershberger, and N.R. Pace. 1997. Phylogenetic perspective on microbial life in hydrothermal ecosystems, past and present. The CIBA Foundation, http://www.novartisfound.org.uk/catalog/202abs.htm

Becker, L., D.P. Glavin, and J.L. Bada. 1997. Polycyclic aromatic hydrocarbons (PAHs) in Antarctic Martian meteorites, carbonaceous chondrites, and polar ice. Geochim. Cosmochim. Acta 61(2): 475-481.

Blake, R.E., J.C. Alt, and A.M. Martini. 2001. Oxygen isotope ratios of PO4: An inorganic indicator of enzymatic activity and P metabolism and a new biomarker in the search for life. Proc. Natl. Acad. Sci. 98-5: 2148-2153.

Brasier, M.D., O.R. Green, A.P. Jephcoat, A.K. Kleppe, M.J. Van Kranendonk, J.F. Lindsay, A. Steele, and N.V. Grassineau. 2002. Questioning the evidence for Earth's oldest fossils. Nature 416: 6876-6881.

Brock, T.D. 1978. Thermophilic Microorganisms and Life at High Temperatures. Springer-Verlag, New York.

Brocks, J.J. 2001. Molecular fossils. *In:* McGraw-Hill Yearbook of Science & Technology. McGraw Hill, London. pp. 252-255.

Brocks, J.J. and R.E. Summons. 2004. Sedimentary hydrocarbons, biomarkers for early life. *In:* H.D. Holland and K. Turekian [eds.]. Treatise in Geochemistry Elsevier-Pergamon. Oxford, pp. 65-115.

Brocks, J.J. and A. Pearson. 2005. Building the Biomarker Tree of Life. Reviews in Mineralogy and Geochemistry 59: 233-258.

Brocks, J.J., G.A. Logan, R. Buick, and R.E. Summons. 1999. Archean molecular fossils and the early rise of Eukaryotes. Science 285: 1033-1036.

Brocks, J.J., G.D. Love, C.E. Snape, G.A. Logan, R.E. Summons, and R. Buick. 2003. Release of bound aromatic hydrocarbons from late Archean and Mesoproterozoic kerogens via hydropyrolysis. Geochim. Cosmochim. Acta 67-8: 1521-1530.

Bushnev, D.A., N.S. Burdel'naya, S.N. Shanina, and E.S. Makarova. 2004. Generation of hydrocarbons and hetero compounds by sulfur-rich oil shale in hydrous pyrolysis. Pet. Chem., 44-6: 416-425.

Bustillo, M.A. and J. Martínez-Frías. 2003. Green opals in hydrothermalized basalts (Tenerife Island, Spain): Alteration and ageing of silica pseudoglass. Journal of Non-crystalline Solids 323: 27-33.

Cairns-Smith, A.G., N.G. Holm, R.M. Daniel, J.P. Ferris, R.J. Hennet, E.L. Shock, B.R. Simoneit, and H. Yanagawa. 1992. Marine hydrothermal systems and the origin of life: Future research. Orig Life Evol Biosph. 22(1-4): 181-242.

Campbell, K.A. 2006. Hydrocarbon seep and hydrothermal vent paleoenvironments and paleontology: Past developments and future research directions. Paleogeogr. Paleoclimatol. Paleoecol. 232: 2-4, 362-407.

Christensen, P.R., M.B. Wyatt, T.D. Glotch, A.D. Rogers, S. Anwar, R.E. Arvidson, J.L. Bandfield, D.L. Blaney, C. Budney, W.M. Calvin, A. Fallacaro, R.L. Fergason, N. Gorelick, T.G. Graff, V.E. Hamilton, A.G. Hayes, J.R. Johnson, A.T. Knudson, H.Y. McSween, Jr., G.L. Mehall, L.K. Mehall, J.E. Moersch, R.V. Morris, M.D. Smith, S.W. Squyres, S.W. Ruff, and M.J. Wolff. 2004. Mineralogy at Meridiani Planum from the Mini-TES Experiment on the Opportunity Rover. Science 306 (5702): 1733-1739.

Crosby, H.A., C.M. Johnson, E.E. Roden, and B.L. Beard. 2005. Coupled Fe(II)-Fe(III) electron and atom exchange as a mechanism for Fe isotope fractionation during dissimilatory iron oxide reduction. Environ. Sci. Technol. 39: 6698-6704.

Des Marais, D.J. 1996. Stable light isotope biogeochemistry of hydrothermal systems. *In:* G. Bock and J. Goode [eds.]. Evolution of Hydrothermal Ecosystems on Earth (and Mars?). John Wiley & Sons, New York. 83-93: 273-299.

Dickensheets, D.L., D.D. Wynn-Williams, H.G.M. Edwards, C. Schoen, C. Crowder, and E.M. Newton. 2000. A novel miniature confocal microscope/Raman spectrometer system for biomolecular analysis on future Mars missions after Antarctic trials. J. Raman Spectrosc. 31-7: 633-635.

Dissanayake, C.B., R. Chandrajith, and J.P. Boudou. 2000. Graphite as a geomarker—application to continental reconstruction of Pan-African Gondwana terrains. Gondwana Res. 3: 405-413.

Eglington, G. and M. Calvin. 1967. Chemical fossils. Sci. Amer., 261: 32-43.

Farmer, J. 2000. GSA Today. V. 10, No. 7, July 2000 http://www.geosociety.org/pubs/gsatoday/gsat0007.htm

Ferris, J.P. 1992. Chemical markers of prebiotic chemistry in hydrothermal systems. Origins of Life and Evolution of the Biosphere 22: 109-134.

Fortin, D., F.G. Ferris, and S.D. Scott. 1998. Formation of Fe-silicate and Fe-oxides on bacterial surfaces in samples collected near hydrothermal vents on the Southern Explorer Ridge in the northeast Pacific Ocean. American Mineralogist 83: 1399-1408.

Furhman, J.A. 1999. Marine viruses and their biogeochemical and ecological effects. Nature 399: 541-548.

Garcia Ruiz, J.M., A. Canerup, A.G. Christy, N.J. Welham, and S.T. Hyde. 2002. Morphology: An ambiguous indicator of biogenicity. Astrobiology 2-3: 353-369.

Geptner, A.R., T.A. Ivanovskaya, and E.V. Pokrovskaya. 2005. Hydrothermal Fossilization of Microorganisms at the Earth's Surface in Iceland. Lithology and Mineral Resources 40-6: 505-520.

Geslin, C., M.L. Romancer, G. Erauso, M. Gaillard, G. Perrot, and D. Prieur. 2003. PAV1, the first virus-like particle isolated from a hyperthermophilic euryarchaeote, *Pyrococcus abyssi*. J. Bacteriol. 185: 3888-3894.

González, F.J., L. Somoza, R. Lunar, J. Martínez-Frías, J.A. Martín Rubí, and V. Díaz del Río. 2007. Fe–Mn nodules associated with hydrocarbon seeps: The new discovery of the Gulf of Cadiz (eastern Central Atlantic). Episodes 30-3: 187-196.

Grymes R. and R. Briggs. 2005. Astrobiological explorations in the NASA Astrobiology Institute: Earth analogues for Mars and beyond. Geophysical Research Abstracts, 7-09899, 2005 SRef-ID: 1607-7962/gra/EGU05-A-09899.

Guerreiro, V., L. Narciso, A.J. Almeida, and M. Biscoito. 2004. Fatty acid profiles of deep-sea fishes from the Lucky Strike and Menez Gwen hydrothermal vent fields (mid-Atlantic ridge) Cybium 28-1. Suppl. S: 33-44.

Guezennec, J., F. Rocchiccioli, B. Maccaron-Gomez, N. Khelifa, J. Dussauze, and A. Rimbault. 1998. Occurrence of 3-hydroxyalkanoic acids in sediments from the Guaymas basin (Gulf of California). FEMS Microbiol. Ecol. 26-4: 335-344.

Guidry, S.A. and H.S. Chafetz. 2003. Depositional facies and diagenetic alteration in a relict siliceous hot-spring accumulation: Examples from Yellowstone National Park, USA. J. Sediment. Res. 73-5: 806-823.

Guilhaumou, N., N. Ellouz, T.M. Jaswal, and P. Mougin. 2001. Genesis and evolution of hydrocarbons entrapped in the fluorite deposit of Koh-i-Maran, (North Kirthar Range, Pakistan). Mar. Pet. Geol. 17-10: 1151-1164.

Glynn, S., R.A. Mills, M.R. Palmer, R.D. Pancost, S. Severmann, and A.J. Boyce. 2006. The role of prokaryotes in supergene alteration of submarine hydrothermal sulfides. Earth Planet. Sci. Lett. 244, 1-2: 170-185.

Häring, M., R. Rachel, X. Peng, R.A. Garrett, and D. Prangishvili. 2005. Viral diversity in hot springs of Pozzuoli, Italy, and characterization of a unique archaeal virus, *Acidianus*, bottle-shaped virus, from a new family, the Ampullaviridae. J. Virol. 79: 9904-9911.

Häring, M., G. Vestegaard, R. Rachel, L. Chen, R.A. Garret, and D. Prangishvili. 2006. Independent virus development outside a host. Nature 436: 1101-1102.

Hennet, R.J-C., N.G. Holm and M.H. Engel. 1992. Abiotic synthesis of amino acids under hydrothermal conditions and the origin of life: A perpetual phenomenon? Naturwissenschaften 79: 361-365.

Herzig, P.M. and M.D. Hannington. 1995. Polymetallic massive sulfides at the modern seafloor: A review. Ore Geology Reviews 10: 95-115.

Herzig, P.M., M.D. Hannington, and S. Petersen. 2002. Technical requirements for exploration and mining of seafloor massive sulphide deposits and cobalt-rich ferromanganese crusts. *In*: Polymetallic Massive Sulphides and Cobalt-rich Ferromanganese Crusts: Status and Prospects, International Seabed Authority Technical Study 2, Report on the UN Workshop on Seafloor Mineral Resources 2000, Kingston, Jamaica, 91-100.

Holden, J.F. and M.W.W. Adams. 2003. Microbe–metal interaction in marine hydrothermal environments. Current Opinion in Chemical Biology 7: 160-165.

Holden J.F. and L.F. Feinberg. 2005. Microbial iron respiration near 100°C. *In*: R.B. Hoover, G.V. Levin, A.Y. Rozanov, and G.R. Gladstone [eds.]. Astrobiology and Planetary Missions. Proceedings from SPIE Vol. 5906. The International Society for Optical Engineering, Bellingham, Washington, US. pp. 57-67.

Holm, N.G., A.G. Cairns-Smith, R.M. Daniel, J.P. Ferris, R.J. Hennet, E.L. Shock, B.R. Simoneit, and H. Yanagawa. 1992. Marine hydrothermal systems and the origin of life: Future research. Orig Life Evol Biosph. 22: 181-242.

Horneck, G. 2000. The microbial world and the case for Mars. Planet Space Sci. 48-11: 1053-1063.

Howell, K.L., D.W. Pond, D.S.M. Billett, and P.A. Tyler. 2003. Feeding ecology of deep-sea seastars (Echinodermata: Asteroidea): A fatty-acid biomarker approach. Mar. Ecol.-Prog. Ser. 255: 193-206.

Huber, H., M.J. Hohn, R. Rachel, T. Fuchs, V.C. Wimmer, and K.O. Stetter. 2002. A new phylum of Archaea represented by a nano-sized hyperthermophilic symbiont. Nature 417: 63-67.

Humphris, S.E., P.M. Herzig, D.J. Miller and Leg 158 Shipboard Scientific Party. 1995. The internal structure of an active sea-floor massive sulphide deposit. Nature 377: 713-716.

Hussler, G., J. Connan, and P. Albrecht. 1984. Novel families of tetra- and hexacyclic aromatic hopanoids predominant in carbonate rocks and crude oils. Org. Geochem. 6: 39-49.

Janekovic, D., S. Wunderl, I. Holz, W. Zillig, A. Gierl, and H. Neumann. 1983. TTV1, TTV2, and TTV3, a family of viruses of the extremely thermophilic, anaerobic, sulphur reducing archaebacterium *Thermoproteus tenax*. Mol. Gen. Genet. 192: 39-45.

Johnson, C.M and B.L. Beard. 2005. Biogeochemical cycling of iron isotopes. Science 309: 1025-1027.

Johnson, C.M., E.E. Roden, S.A. Welch, and B.L. Beard. 2005. Experimental constraints on Fe isotope fractionation during magnetite and Fe carbonate formation coupled to dissimilatory hydrous ferric oxide reduction. Geochim. Cosmochim. Acta 69: 963-993.

Jurado, M.J., V. Barron, and J. Torrent. 2003. Can the presence of structural phosphorus help to discriminate between abiogenic and biogenic magnetites? J Biol Inorg Chem. 8-8: 810-814.

Kashefi, K. and D.R. Lovley. 2000. Reduction of Fe(III), Mn(IV), and toxic metals at 100 degrees C by *Pyrobaculum islandicum*. Appl Environ Microbiol. 66: 1050-1056.

Kashefi, K. and D.R. Lovley. 2003. Extending the upper temperature limit for life. Science 301: 934.

Kashefi, K., J.M. Tor, D.E. Holmes, C.V. Gaw Van Praagh, A.L. Reysenbach, and D.R. Lovley. 2002. *Geoglobus ahangari* gen. nov., sp. nov., a novel hyperthermophilic archaeon capable of oxidizing organic acids and growing autotrophically on hydrogen with Fe(III) serving as the sole electron acceptor. Int J Syst Evol Microbiol 52: 719-728.

Kashefi, K., D.E. Holmes, J.A. Barros, and D.R. Lovley. 2003. Thermophily in the Geobacteraceae: *Geothermobacter ehrlichii* gen. nov., sp. nov., a novel member of the Geobacteraceae from the ldquoBag Cityrdquo hydrothermal vent. Appl Environ Microbiol 69: 2985-2993.

Kenig, F., D.J.H. Simons, D. Crich, J.P. Cowen, G.T. Ventura, T. Rehbein-Khalily, T.C. Brown, and K.B. Anderson. 2003. Branched aliphatic alkanes with quaternary substituted carbon atoms in modern and ancient geologic samples. Proc. Natl. Acad. Sci. 100-22: 12554-12558.

Klingelhöfer, G., R.V. Morris, B. Bernhardt, C. Schröder, D.S. Rodionov Jr., P.A. de Souza, A. Yen, R. Gellert, E.N. Evlanov, B. Zubkov, J. Foh, U. Bonnes, E. Kankeleit, P. Gütlich, D.W. Ming, F. Renz, T. Wdowiak, S.W. Squyres, and R.E. Arvidson. 2004. Jarosite and Hematite at Meridiani Planum from Opportunity's Mössbauer Spectrometer Science 306, 5702: 1740-1745.

Kminek, G. and J.L. Bada. 2006. The effect of ionizing radiation on the preservation of amino acids on Mars. Earth and Planetary Science Letters 245, 1-2, 1-5.

Line, M.A. 2002. The enigma of the origin of life and its timing. Microbiology 148: 21-27.

Liu, B., W. Suijie, Q. Song, X. Zhang, and Xie, L. 2006. Two novel bacteriophages of thermophilic bacteria isolated from deep-sea hydrothermal fields. Curr. Microbiol. 53: 163-166.

Logan, G.A., M.C. Hinman, M.R. Walter, and R.E. Summons. 2001. Biogeochemistry of the 1640 Ma McArthur River (HYC) lead-zinc ore and host sediments, Northern Territory, Australia. Geochim. Cosmochim. Acta 65-14: 2317-2336.

Lovley, D.R. 2004. Potential role of dissimilatory iron reduction in the early evolution of microbial respiration. *In:* J. Seckbach [ed.]. Origins, Evolution and Biodiversity of Microbial Life. Kluwer, The Netherlands. pp. 301-313.

Lovley, D.R., J.F. Stolz, G.L. Nord Jr., and E.J.P. Phillips. 1987. Anaerobic Production of Magnetite by a Dissimilatory Iron-reducing Microorganism. Nature 330 (6145): 252-254.

Macleod, G., C. McKeown, A.J. Hall, and M.J. Russell. 1994. Hydrothermal and oceanic pH conditions of possible relevance to the origin of life. Origins of Life and Evolution of the Biosphere 23: 19-41.

Madden, M.E.E., R.J. Bodnar, and J.D. Rimstidt. 2004. Jarosite as an indicator of water-limited chemical weathering on Mars. Nature 431: 821-823.

Marshall, W.L. 1994. Hydrothermal synthesis of amino acids. Geochimica et Cosmochimica Acta 58: 2099-2106.

Martin, W. and M.J. Russell. 2003. On the origin of cells: A hypothesis for the evolutionary transitions from abiotic geochemistry to chemoautotrophic prokaryotes, and from prokaryotes to nucleated cells. Philosophical Transactions of the Royal Society B 358: 59-85.

Martínez-Frías, J. 1991. Sulphide and sulphosalt mineralogy and paragenesis from the Sierra Almagrera veins (Betic Cordillera). Est. Geol. 47 (5-6): 271-279.

Martínez-Frías, J. 1999. Mining vs. Geological Heritage: The Cuevas del Almanzora Natural Area (SE Spain), AMBIO: A Journal of the Human Environment 28-2: 204-207.

Martínez-Frías, J., J. García Guinea, J. López Ruiz, J.A. López, and R.Benito. 1989. Las mineralizaciones epitermales de Sierra Almagrera y de la cuenca de Herrerías (Cordilleras Béticas). Rev. Soc. Esp. Min. 12: 261-271.

Martínez-Frías, J., R. Lunar, J. Mangas, A. Delgado, G. Barragan, E. Sanz-Rubio, E. Diaz, R. Benito, and T. Boyd. 2001. Evaporitic and hydrothermal gypsum from SE Iberia: Geology, geochemistry, and implications for searching for life on Mars. Geological Society of America (GSA) Annual Meeting, Boston, Massachusetts, November 5-8.

Martínez-Frías, J., R. Lunar, J.A. Rodríguez-Losada, and A. Delgado. 2004. The volcanism-related multistage hydrothermal system of El Jaroso (SE Spain): Implications for the exploration of Mars Earth. Planets and Space 56: 5-8.

Martínez-Frías, J., G. Amaral, and L. Vázquez. 2006. Astrobiological significance of minerals on Mars surface environment. Reviews in Environmental Science and Bio/technology 5: 219-231.

Martínez-Frías, J., A. Delgado-Huertas, F. García-Moreno, E. Reyes, R. Lunar, and F. Rull. 2007a. Oxygen and carbon isotopic signatures of extinct low temperature hydrothermal chimneys in the Jaroso Mars analog. Planetary & Space Science 55: 441-448.

Martínez-Frías, J., E. Lázaro, and A. Esteve-Núñez. 2007b. Geomarkers versus biomarkers: Paleoenvironmental and astrobiological significance. AMBIO: Journal of the Human Environment 36-5: 425-427.

Merinero, R., R. Lunar, J. Martínez–Frías, L. Somoza, and V. Díaz-del-Río. 2006. Iron-rich coccoidal microcrystals and framboids in submarine, methane-derived carbonate chimneys (Gulf of Cadiz, SW Iberian Peninsula): Mineralogy, textures and astrobiological relevance. Geophysical Research Abstracts, Vol. 8, 01382, 2006 SRef-ID: 1607-7962/gra/EGU06-A-01382.

Merinero, R., R. Lunar, J. Martínez–Frías, L. Somoza, and V. Díaz-del-Río. 2008. Iron oxide and sulphide mineralization in hydrocarbon seep-related carbonate submarine chimneys, Gulf of Cadiz (SW Iberian Peninsula). Marine and Petroleum Geology (in press). DOI: 10.1016/J.MARPETGEO.2008.03.005.

McKay, D.S., E.K. Gibson Jr., K.L. Thomas-Keprta, H. Vali, C.S. Romanek, S.J. Clemett, X.D. Chillier, C.R. Maechling, and R.N. Zare. 1996. Search for past life on Mars: Possible relic biogenic activity in Martian meteorite ALH84001. Science. 273(5277): 924-30.

McClendon, J.H. 1999. The origin of life. Earth-Science Reviews 47: 71-93.

McCollom, T.M., G. Ritter, and B.R.T. Simoneit. 1999. Lipid synthesis under hydrothermal conditions by Fischer-Tropsch-Type reactions. Origins of Life and Evolution of the Biosphere 29: 153-166.

Miroshnichenko, M.L., A.I. Slobodkin, N.A. Kostrikina, S. LrsquoHaridon, O. Nercessian, S. Spring, E. Stackebrandt, E.A. Bonch-Osmolovskaya, and C. Jeanthon. 2003a. *Deferribacter abyssi* sp. nov.—a new anaerobic thermophilic bacterium from the deep-sea hydrothermal vents of the Mid-Atlantic Ridge. Int J Syst Evol Microbiol 53: 1637-1641.

Miroshnichenko, M.L., N.A. Kostrikina, N.A. Chernyh, N.V. Pimenov, T.P. Tourova, A.N. Antipov, S. Spring, E. Stackebrandt, and E.A. Bonch-Osmolovskaya. 2003b. *Caldithrix abyssi* gen. nov., sp. nov., a nitrate-reducing, thermophilic, anaerobic bacterium isolated from a Mid-Atlantic Ridge hydrothermal vent, represents a novel bacterial lineage. Int J Syst Evol Microbiol 53: 747-752.

Montenat, C. and A. Seilacher. 1978. Les turbidites messiniennes à Helminthoides et Paleodictyon du bassin de Vera (Cordillère bétiques orientales). Indications Paleobathymetriques: Bull. Soc. Géol. Fr.,7: T-20.

Moskowitz, B.M., R.B. Frankel, D.A. Bazylinski, H.W. Jannasch, and D.R. Lovley. 1989. A Comparison of Magnetite Particles Produced Anaerobically by Magnetotactic and Dissimilatory Iron-Reducing Bacteria. Geophys Res Lett. 16(7): 665-668.

Moussard, H., S. LrsquoHaridon, B.J. Tindall, A. Banta, P. Schumann, E. Stackebrandt, A.-L. Reysenbach, and C. Jeanthon. 2004. *Thermodesulfatator indicus* gen. nov., sp. nov., a novel thermophilic chemolithoautotrophic sulfate-reducing bacterium isolated from the Central Indian Ridge. Int J Syst Evol Microbiol 54: 227-233.

Mukund, S. and M.W. Adams. 1996. Molybdenum and vanadium do not replace tungsten in the catalytically active forms of the three tungstoenzymes in the hyperthermophilic archaeon *Pyrococcus furiosus*. J Bacteriol. 178(1): 163-167.

Nakagawa, S., K. Takai, K. Horikoshi, and Y. Sako. 2004. *Aeropyrum camini* sp. nov., a strictly aerobic, hyperthermophilic archaeon from a deep-sea hydrothermal vent chimney. Int J Syst Evol Microbiol 54: 329-335.

Nna-Mvondo, D. and J. Martínez-Frías. 2005. Volcanic Source for Abiotic Organics and Inorganics on Early Earth and Mars Eos Trans. AGU, 86(52), Fall Meet. Suppl. B31B-0988.

Nna-Mvondo, D. and J. Martínez-Frías. 2006. Komatiitic volcanoes as possible suitable sites for abiotic chemistry and biological processes on early Earth. Geophysical Research Abstracts. Vol. 8, 04311, SRef-ID: 1607-7962/gra/ EGU06-A-04311.

Ortmann, A.C., B. Wiedenheft, T. Douglas, and M. Young. 2006. Hot crenarchaeal viruses reveal deep evolutionary connections. Nature 4: 520-528.

Parnell, J., G.R. Osinski, P. Lee, P.F. Green, and M.J. Baron. 2005. Thermal alteration of organic matter in an impact crater and the duration of postimpact heating. Geology 33-5: 373-376.

Parro, V., J.A. Rodriguez-Manfredi, C. Briones, C. Compostizo, P.L. Herrero, E. Vez, E. Sebastian, M. Moreno-Paz, M. Garcia-Villadangos, P. Fernandez-Calvo, E. Gonzalez-Toril, J. Perez-Mercader, D. Fernandez-Remolar, and J. Gomez-Elvira. 2005. Instrument development to search for biomarkers on mars: Terrestrial acidophile, iron-powered chemolithoautotrophic communities as model systems Planetary & Space Science 53-7: 729-737.

Peters, K.E., C.C. Walters and J.M. Moldowan. 2006. The Biomarker Guide. Cambridge University Press, Cambridge.

Phleger, C.F., M.M. Nelson, A.K. Groce, S.C. Cary, K. Coyne, J.A.E. Gibson, and P.D. Nichols. 2005. Lipid biomarkers of deep-sea hydrothermal vent polychaetes — *Alvinella pompejana*, *A. caudata*, *Paralvinella grasslei* and *Hesiolyra bergii*, Deep-Sea Res. Part I-Oceanogr. Res. Pap. 52-12: 2333-2352.

Pond, D.W., D.R. Dixon, M.V. Bell, A.E. Fallick, and J.R. Sargent. 1997. Occurrence of 16:2(n-4) and 18:2(n-4) fatty acids in the lipids of the hydrothermal vent shrimps *Rimicaris exoculata* and *Alvinocaris markensis*: Nutritional and trophic implications. Mar. Ecol.-Prog. Ser. 156: 167-174.

Pranal, V., A. FialaMedioni, and J. Guezennec. 1996. Fatty acid characteristics in two symbiotic gastropods from a deep hydrothermal vent of the west Pacific. Mar. Ecol.-Prog. Ser. 142, 1-3: 175-184.

Prangishvili, D. and R.A. Garret. 2005. Viruses of hyperhermophilic Crenarchaea. Trends Microbiol. 12: 535-542.

Prangishvili, D., H.P. Arnold, D. Gotz, U. Ziese, I. Holz, J.K. Kristjansson, and W. Zillig. 1999. A novel virus family, the Rudiviridae: Structure, virus-host interactions and genome variability of the Sulfolobus viruses SIRV1 and SIRV2. Genetics 152: 1387-1396.

Prangishvili, D., R.A. Garret, and E.V. Koonin. 2006. Evolutionary genomics of archaeal viruses: Unique viral genomes in the third domain of life. Virus Res. 117: 52-67.

Prieur, D., C. Jeanthon, and G. Erauso. 1995. Hyperthermophilic life at deep-sea hydrothermal vents. Planet Space Science 43: 115-122.

Rachel, R., M. Bettstetter, B.P. Hedlund, M. Haring, A. Kessler, K.O. Stetter, and D. Prangishvili. 2002. Remarkable morphological diversity of viruses and virus-like particles in terrestrial hot environments. Arch. Virol. 147: 2419-2429.

Reysenbach, A.-L. 2001a. Thermotogales. *In*: D.R. Boone and G.M. Garrity Bergeyrsquos [eds.]. Manual of Systematic Bacteriology. Vol. 1, 2nd ed. Springer, Berlin, Heidelberg, New York. pp. 369-387.

Reysenbach, A.-L. 2001b. Aquificales. *In*: D.R. Boone and G.M. Garrity Bergeyrsquos [eds.]. Manual of Systematic Bacteriology. Vol. 1, 2nd ed. Springer, Berlin, Heidelberg New York. pp. 359-367.

Reysenbach, A.-L., S. Seitzinger, J. Kirshtein, and E. McLaughlin. 1999. Molecular constraints on a high-temperature evolution of early life. Biol. Bull. 196: 367-372.

Reysenbach, A.-L., A. Banta, P. Messner, P. Schumann, E. Stackebrandt, and C. Jeanthon. 2003. *Methanocaldococcus indicus* sp. nov., a novel hyperthermophilic methanogen isolated from the Central Indian Ridge. Int J Syst Evol Microbiol 53: 1931-1935.

Rice, G., K. Stedman, J. Snyder, B. Wiedenheft, D. Willits, S. Brumfield, T. McDermott, and M.J. Young. 2001. Viruses from extreme thermal environments. Proc. Natl. Acad. Sci. USA 98: 13341-13345.

Rice, G., L. Tang, K. Stedman, F. Roberto, J. Spuhler, E. Gillitzer, J.E. Johnson, T. Douglas, and M. Young. 2004. The structure of a thermophilic archaeal virus shows a double-stranded DNA viral capside type that spans all domains of life. Proc. Natl. Acad. Sci. USA 101: 7716-7720.

Rodríguez-Losada, J.A., J. Martínez-Frías, M.A. Bustillo, A. Delgado, A. Hernandez-Pacheco, and J.V. De la Fuente Krauss. 2000. The hydrothermally altered ankaramite basalts of Punta Poyata (Tenerife, Canary Islands). Journal of Volcanology and Geothermal Research 103 (1-4): 367-376.

Rona, P.A. 2003 Resources of the Seafloor. Science 299, 5607: 673-674.

Rona, P.A. and S.D. Scott. 1993. A special issue on sea-floor hydrothermal mineralization: New perspectives. Econ. Geol., 88, 8: 1935-1975.

Rull, F. and J. Martinez-Frias. 2006. Raman spectroscopy goes to Mars. Spectroscopy Europe 18-1: 18-21.

Rushdi, A.I. and B.R.T. Simoneit. 2002. Hydrothermal alteration of organic matter in sediments of the Northeastern Pacific Ocean: Part 1. Middle Valley, Juan de Fuca Ridge. Appl. Geochem. 17-11: 1401-1428.

Russell, M.J. and A.J. Hall. 1997. The emergence of life from iron monosulphide bubbles at a submarine hydrothermal redox and pH front. Journal of the Geological Society of London 154: 377-402.

Russell, M.J. and W. Martin. 2004. The rocky roots of the acetyl-CoA pathway. Trends Biochem Sci 29: 358-363.

Russell, M.J., R.M. Daniel, A.J. Hall, and J. Sherringham. 1994. A hydrothermally precipitated catalytic iron sulphide membrane as a first step toward life. Journal of Molecular Evolution 39: 231-243.

Russell, M.J., A.J. Hall, A.J. Boyce, and A.E. Fallick. 2005. On Hydrothermal Convection Systems and the Emergence of Life. 100th Anniversary Special Paper. Econ. Geology, 100-3: 419-438.

Schoonen, M., A. Smirnov, and C. Cohn. 2004. A Perspective on the Role of Minerals in Prebiotic Synthesis. AMBIO: A Journal of the Human Environment 33, 8: 539-551.

Shock, E.L. 1997. Hydrothermal systems as environments for the emergence of life. The CIBA Foundation, http://www.novartisfound.org.uk/catalog/202abs.htm

Schulze-Makuch, D. and L.N. Irwin. 2004. Signatures of Life and the Question of Detection, Advances in Astrobiology and Astrophysics. Life in the Universe. Springer. Berlin, Heidelberg. pp. 149-172.

Schulze-Makuch, D., L.N. Irwin, and H. Guan. 2002. Search parameters for the remote detection of extraterrestrial life. Planetary and Space Sciences 50: 675-683.

Simoneit, B.R.T. 1992. Aqueous organic geochemistry at high temperature/high pressure. Orig Life Evol Biosph 22: 43-65.

Simoneit, B.R.T. 2002. Molecular indicators (Biomarkers) of Past Life. The Anatomical Record 268: 186-195.

Simoneit, B.R.T. 2004a. Prebiotic organic synthesis under hydrothermal conditions: An overview. Adv. Space Res. 33-1: 88-94.

Simoneit, B.R.T. 2004b. Biomarkers (molecular fossils) as geochemical indicators of life. Adv. Space Res., 33-8: 1255-1261.

Simoneit, B.R.T., R.E. Summons, and L.L. Jahnke. 1998. Biomarkers as tracers for life on early Earth and Mars. Orig. Life Evol. Biosph. 28: 4-5, 475-483.

Sims, M.R., D.C. Cullen, N.P. Bannister, W.D. Grant, O. Henry, R. Jones, D. McKnight, D.P. Thompson, and P.K. Wilson. 2005. The specific molecular identification of life experiment (SMILE). Planet Space Sci. 53-8, 781-791.

Snyder, J., K. Stedman, G. Rice, B. Wiedenheft, J. Sphuler, and M.J. Young. 2003. Viruses of hyperthermophilic Archaea. Res. Microbiol. 154: 474-482.

Spangenberg, J.E., L. Fontbote, and S.A. Macko. 1999. An evaluation of the inorganic and organic geochemistry of the San Vicente Mississippi Valley-type zinc-lead district, central Peru: Implications for ore fluid composition, mixing processes, and sulphate reduction. Econ. Geol. Bull. Soc. Econ. Geol. 94-7: 1067-1092.

Squyres, S.W., J.P. Grotzinger, R.E. Arvidson, J.F. Bell III, W. Calvin, P.R. Christensen, B.C. Clark, J.A. Crisp, W.H. Farrand, K.E. Herkenhoff, J.R. Johnson, G. Klingelhöfer, A.H. Knoll, S.M. McLennan, H.Y. McSween, J.W. Morris, R.V. Rice, Jr., R. Rieder, and L.A. Soderblom. 2004. In situ evidence for an ancient aqueous environment at Meridiani Planum. Mars, Science, 306-5702: 1709-1714.

Squyres, S.W., R.E. Arvidson, S. Ruff, R. Gellert, R.V. Morris, D.W. Ming, L. Crumpler, J.D. Farmer, D.J. Des Marais, A. Yen, S.M. McLennan, W. Calvin, F.J. Bell, III, B.C. Clark, A. Wang, T.J. McCoy, M.E. Schmidt, and P.A. de Souza, Jr. 2008. Science, 320-5879: 1063-1067.

Sreeraj, K., H. Wada, and M. Santosh. 2000. Graphite as geomarker and fluid index in east Gondwana terrains. Gondwana Res. 3: 560-561.

Stedman, K.M., Q. She, H. Phan, H.P. Arnold, I. Holz, R.A. Garrett, and W. Zillig. 2003. Relationships between fuselloviruses infecting the extremely thermophilic archaeon Sulfolobus: SSV1 and SSV2. Res. Microbiol. 154: 295-302.

Stefanova, M., S.P. Marinov, and C. Magnier. 1999. Aliphatic biomarkers from Miocene lignites desulphurization. Fuel 78-12: 1395-1406.

Stetter, K.O. 2002. Hyperthermophilic microorganisms. *In*: G. Horneck and C. Baumstark-Khan [eds.]. Astrobiology. The Quest for the Conditions of Life. Springer, Berlin, Heidelberg, New York. pp. 169-184.

Summons R.E. and T.G. Powell. 1986. Chlorobiaceae in Palaeozoic seas—Combined evidence from biological markers, isotopes and geology. Nature 319: 763-765.

Summons, R.E. and M.R. Walter. 1990. Molecular fossils and microfossils of prokaryotes and protists from Proterozoic sediments. Am. J. Science 290-A: 212-244.

Suttle, C.A. 2005. Viruses in the sea. Nature 437: 356-361.

Takai, K. and Y. Sako. 1999. A molecular view of archaeal diversity in marine and terrestrial hot water environments. FEMS Microbiol. Ecol. 28: 177-188.

Takai, K., A. Sugai, T. Itoh, and K. Horikoshi. 2000. *Palaeococcus ferrophilus* gen. nov., sp. nov., a barophilic, hyperthermophilic archaeon from a deep-sea hydrothermal vent chimney. Int J Syst Evol Microbiol 50: 489-500.

Takai, K., A. Inone, and K. Horikoshi. 2002. *Methanothermococcus okinawensis* sp. nov, thermophilic methane-producing archaeon isolated from a western Pacific deep-sea hydrothermal vent. Int J Syst Evol Microbiol 52: 1089-1095.

Takai, K., K.H. Nealson, and K. Horikoshi. 2004. *Methanotorris formicicus* sp. nov., a novel extremely thermophilic, methane-producing archaeon isolated from a black smoker chimney in the Central Indian Ridge. Int J Syst Evol Microbiol 54: 1095-1100.

Takano, Y., Y. Edazawa, K. Kobayashi, T. Urabe, and K. Marumo. 2005. Evidence of sub-vent biosphere: Enzymatic activities in 308 degrees C deep-sea hydrothermal systems at Suiyo seamount, Izu-Bonin Arc, Western. Earth Planet. Sci. Lett. 229, 3-4: 193-203.

Tor, J.M., K. Kashefi, and D.R. Lovley. 2001. Acetate oxidation coupled to Fe(III) reduction in hyperthermophilic microorganisms. Appl Environ Microbiol. 67: 1363-1365.

Turcotte, D.L. 1980. On the thermal evolution of the Earth. Earth Planet. Sci. Lett. 48: 53-58.

Vargas, M., K. Kashefi, E.L. Blunt-Harris, and D.R. Lovley. 1998. Microbiological evidence for Fe(III) reduction on early Earth. Nature 395-6697: 65-67.

Vayisoglu, E.S., O.B. Harput, B.R. Johnson, B. Frere, and K.D. Bartle. 1997. Characterization of oil shales by extraction with N-methylpyrrolidone. Fuel, 76-4: 353-356.

Villar, S.E.J. and H.G.M. Edwards. 2006. Raman spectroscopy in astrobiology. Anal. Bioanal. Chem. 384-1: 100-113.

Walter, M.R. 1996. Ancient hydrothermal ecosystems on Earth: A new palaeobiological frontier. *In:* G. Bock and J. Goode [eds.]. Evolution of Hydrothermal Ecosystems on Earth (and Mars?). John Wiley & Sons, New York. pp. 112-127.

Weiner, A.M. and N. Maizels. 1999. The genomic tag hypothesis: Modern viruses as molecular fossils of ancient strategies for genomic replication, and clues regarding the origin of protein synthesis. Biol Bull. 196: 327-328.

Westall, F. and R.L. Folk. 2003. Exogenous carbonaceous microstructures in Early Archaean cherts and BIFs from the Isua Greenstone Belt: Implications for the search for life in ancient rocks. Precambrian Res., 126, 3-4: 313-330.

Westall, F., A. Steele, J. Toporski, M. Walsh, C. Allen, S. Guidry, D. McKay, E. Gibson, and H. Chaftez. 2000. Polymeric substances and biofilms as biomarkers in terrestrial materials: Implications for extraterrestrial samples. Journal of Geophysical Research 105-24: 511-527.

Wiedenheft, B., K. Stedman, F. Roberto, D. Willits, A.K. Gleske, L. Zoeller, J. Snyder, T. Douglas, and M. Young. 2004. Comparative genomic analysis of hyperthermophilic archaeal Fuselloviridae viruses. J. Virol. 78: 1954-1961.

Wierzchos, J., C. Ascaso, L.G. Sancho, and A. Green. 2003. Iron-rich diagenetic minerals are biomarkers of microbial activity in Antarctic rocks. Geomicrobiol. J. 20: 15–24.

Wierzchos, J., C. Ascaso, F.J. Agar, I. Garcia-Orellana, A. Carmona-Luque, and M.A. Respaldiza. 2006. Identifying elements in rocks from the Dry Valleys desert (Antarctica) by ion beam proton induced X-ray emission. Nucl. Instrum. Methods Phys. Res. Sect. B-Beam Interact. Mater. Atoms, 249: 571-574.

Yamanaka, T. and S. Sakata. 2004. Abundance and distribution of fatty acids in hydrothermal vent sediments of the western Pacific Ocean. Org. Geochem. 35-5: 573-582.

Yamanaka, T., C. Mizota, T. Murae, and J. Hashimoto. 1999. A currently forming petroleum associated with hydrothermal mineralization in a submarine caldera, Kagoshima Bay, Japan. Geochem. J. 33-6: 355-367.

Yamanaka, T., J. Ishibashi, and J. Hashimoto. 2000. Organic geochemistry of hydrothermal petroleum generated in the submarine Wakamiko caldera, southern Kyushu, Japan. Org. Geochem. 31-11: 1117-1132.

Xia, X.H., C.T. Liu, and Z.M. Li. 2000. Hot-water deposition of pyritic stromatolite and its relation to biomineralization. Acta Geol. Sin.-Engl. Ed. 74-3: 529-533.

Zabala, V.A. 2001. Meet the Scientist: Dr. Jack Farmer. The Martian Chronicles, 9. http://chapters.marssociety.org/youth/mc/issue9/mts.php3

Zarate-del Valle, P.F. and B.R.T. Simoneit. 2005. Hydrothermal bitumen generated from sedimentary organic matter of rift lakes—Lake Chapala, Citala Rift, western Mexico. Appl. Geochem. 20-12: 2343-2350.

Zarate-del Valle, P.F., A.I. Rushdi and B.R.T. Simoneit. 2006. Hydrothermal petroleum of Lake Chapala, Citala Rift, western Mexico. Bitumen compositions from source sediments and application of hydrous pyrolysis. Appl. Geochem. 21-4: 701-712.

Zhang, C.L., Z.Y. Huang, J. Cantu, R.D. Pancost, R.L. Brigmon, T.W. Lyons, and R. Sassen. 2005. Lipid biomarkers and carbon isotope signatures of a microbial (eggiatoa) mat associated with gas hydrates in the Gulf of Mexico. Appl. Environ. Microbiol 71-4: 2106-2112.

Zierenberg, R.A., M.W.W. Adams, and A.J. Arp. 2000. Life in extreme environments: Hydrothermal vents. Proc. Natl. Acad. Sci. USA 97: 12961-12962.

CHAPTER 7

Spirulina Biotechnology

Hiren Doshi[1], Arabinda Ray[1*] and I.L. Kothari[2]

INTRODUCTION

Spirulina is a blue–green filamentous spiral-shaped marine alga, widely found in ocean and sea waters. Botanists classify it as microalga belonging to Cyanophyceae; but according to bacteriologists it is a bacterium due to its prokaryotic structure. The two most important species of *Spirulina* are alga *S. maxima* and *S. platensis*. Because of its high nutrition content characterized by more than 70% amino acid content, together with essential minerals, vitamins, a whole spectrum of natural mixed carotene, fatty acids *S. platensis* is gaining more attention. *Spirulina* is an inexpensive attractive alternative food source for poor countries very often devastated by natural calamities. It is a traditional food for some Mexican and African people (http://www.javeriana.edu.co/universitas_scientiarum/vol8n1/bernal.htm) *Spirulina* has another unique application. It is a very efficient biosorbent, *Spirulina* sp., was found to contain detectable levels of mercury and lead (Slotton et al. 1989) when grown under the contaminated condition, implying that it can take up toxic metals from the environment. *Spirulina* can thus be cultivated in wastewater to improve water quality.

CHEMICAL COMPOSITION

About 70% of protein is present in *Spirulina*, making it an excellent source of protein (Switzer 1980). It has essential amino acids, such as leucine

[1]Department of Chemistry, Sardar Patel University, Vallabh Vidyanagar-388 120, Gujarat, India, E-mails: drhirendoshi@yahoo.co.in, arabinda24@yahoo.co.in

[2]Department of Bioscience, Sardar Patel University, Vallabh Vidyanagar-388 120, Gujarat, India, E-mail: ilkothari@yahoo.com

*Corresponding author

(~11%), valine (~7.5%) and isoleucine (~7.0%). A high provitamine-A concentration is also present in *Spirulina* (Belay 1997). It is interesting to note that an excessive dose of β-carotene may be toxic, but *Spirulina* when consumed as a source of vitamin A, only the required amount of vitamin-A will be obtained from *Spirulina*. This is because provitamin-A is converted into vitamin A by organisms present in human body, depending on the requirement of the individual. Hence *Spirulina* is a harmless source of vitamin A (Henrikson 1994). It has about 14% carbohydrates, mainly glucose, rhamnose, mannose, xylose and galactose (Shekharam et al. 1987). *Spirulina* is rich in vitamin B_{12} (the world's richest source of vitamin B_{12}), has about 4-7% and contains essential fatty acids, viz. linoleic and γ-linoleic acids (http://www. javeriana.edu.co/universitas_scientiarum/vol8n1/bernal.htm).

Absence of cellulose in its cell wall makes *Spirulina* an ideal food for persons with intestinal problems (Richmond 1992). *Spirulina* is consumed by certain types of flamingoes found in Africa. There are natural pigments in *Spirulina* and it has been suggested that some of these pigments are responsible for imparting colours to these flamingoes (Ciferri 1983; Henrikson 1994).

SPIRULINA: AN EDIBLE ALGA

The history of *Spirulina* for human consumption is not of recent origin. In the early 16th century, cake made of *Spirulina* was used as food in today's Mexico city. P. Dangeard (http://www.javeriana.edu.co/universitas _scientiarum/vol8n1/bernal.htm) reported that a cake called dihe obtained from *Spirulina* was consumed by Kanembu, an African tribe residing in the sub-desert area of Kanembu. As already stated, this blue-green alga, *Spirulina platensis* has a high protein content, in addition to essential amino and fatty acids and essential vitamins and minerals (Campanella et al. 1998). These make *Spirulina* an attractive inexpensive alternative food source with a very high nutritional value.

Spirulina is available commercially as tablets and powder and is recommended as food supplements for persons having vitamin and protein deficiency. Reports on the medicinal and nutritional use of *Spirulina* in Thailand, India and Peru are available (Saxena et al. 1983).

SPIRULINA AS PHARMACEUTICAL MATERIAL

Spirulina is also being examined as a source of potential pharmaceuticals. So far exciting results have been obtained from the preliminary study on its ability to inhibit viral replication, strengthen the immune system and inhibit cancer. One of the most effective substances to counteract cancer causing free radicals is β-carotene and *Spirulina* as indicated above is a very good source of it. It has been found in the laboratory that *Spirulina* sp. could inhibit, shrink and destroy oral cancer cells (Momotaj and Iftikhar 2001). A group of medical scientists has published studies on a purified water extract from *Spirulina* named calcium *Spirulina* which is polymerized sugar molecule unique to *Spirulina* and contains both sulphur and calcium. This molecule has been found to inhibit the replication of HIV-1 (Richard and Ronald: http://www.*Spirulina*.com/ SPL...). *Spirulina* has some therapeutic values in treating cases of arsenicosis. A study conducted in Bangladesh reported that 3 gm *Spirulina* sp. per day per person given for a period of 3 months showed evidence of improvement by diminishing the visible manifestation of arsenicosis (Momotaj and Iftikhar 2001). It has been registered that *Spirulina* consumption for 4 weeks reduces 45% serum cholesterol levels in human beings. *Spirulina* has been recommended as an adequate source of iron for anaemic pregnant women because iron in *Spirulina* is better absorbed than ferrous sulphate. United States is the largest producer of *Spirulina*, followed by Thailand, India and China.

SPIRULINA FOR BIOREMEDIATION

Because of their non-biodegradability, incremental accumulation of toxic metals poses a great threat to the ecosystem. Many researchers have addressed this problem and search for effective treatment is still going on. Conventional methods such as chemical precipitation, ion exchange, etc. to remove toxic metals/pollutants are not very effective and have certain disadvantages (Volesky 1990). These disadvantages are particularly apparent at low metal ion concentrations encountered in wastewater.

In search of an effective treatment for removal of toxic metal ions, attention has been focused on the metal binding capacities of biological materials. These include quoting a few: peat moss (Ho and Mckay 1998), yeast (Clemens et al. 1999), algae (Jennett et al. 1982, Doshi et al. 2006), bacteria (Norberg and Rydin 1984), and various other aquatic floras (Gardea-Torresdey et al. 1998a, b, Cetinkaya et al. 1999, Dokken et al. 1999, Hussein et al. 2004). The use of biosorbents to remove toxic metal ions is becoming increasingly popular due to its high adsorbent capacity, low cost, and regeneration of the absorbent (Heitzer and Sayler 1994, Eccles 1999), and such finding opens up the possible use of such organisms for large-scale removal of harmful metals from the environment. Algae have been used for pharmaceutical reasons for detoxification of heavy metals in the human body due to very efficient adsorption of the toxic ions and this phenomenon has also been used to remove heavy metals from industrial wastewater (David and Volesky 1998).

Spirulina is an interesting object for studies of different metal ions adsorption and accumulation. The cell wall of this alga consists of polysaccharides, proteins and lipids having carboxyl and phosphate groups. Thus *Spirulina* has moieties that can bind with metal ions. It is thus a serious candidate for efficient removal of metal ions particularly from wastewater. Accumulation and sorption of CO^{2+}, Cu^{2+}, Zn^{2+} and Sr^{2+} by *Spirulina platensis* was studied by varying pH, initial metal concentration, temperature and uptake kinetics were investigated in detail that might lead to the cost-efficient continuous removal of these toxic metal ions from industrial effluents. Maximum loading capacities were found to be 29.4, 172.5, 53.4 and 22.3 mg/g of dry weight for CO^{2+}, Cu^{2+}, Zn^{2+} and Sr^{2+}, respectively (http://www.aseanbiotechnology. info/scripts.). *Spirulina* (*Arthrospira*) *platensis* TISTR 8217 was employed to remove Cd^{+2} from wastewater at low concentrations (less than 100 mg/l) of cadmium (Rangsayatorn et al. 2002). It was found that the chemical composition of *Spirulina* sp. cells had a strong influence on their adsorption capacity. The maximum adsorption capacities for Pb^{+2} and Cd^{+2} were 172.4 and 54.0 mg/g of cells when cells exhibited the higher polysaccharide content, while Cr^{+6} adsorption is more when protein content of the cells is higher (Hernandez and Olguín 2002). More literatures are available on the use of *Spirulina* sp. to remove Pb^{2+} (Doke et al. 2005, Hong and Shan-Shan, 2005), Cu^{2+} (Parmeggiani and Masini, 2003), and Cd^{2+} (Augusto da Costa and de Franca 1998, 2003). Chojnacka et al. 2005, found that Lyphilizate of *Spirulina* sp. had the highest biosorption capacity towards Cr^{3+} (185 mg/g) and Cu^{2+} (196 mg/g), but photoautotrophic form takes up Cd^{2+} (159 mg/g). Recently kinetics, microscopic and spectroscopic studies on the sorption of Cd^{2+} (Doshi et al. 2007b), Ni^{2+}, Cu^{2+}, Cr^{3+}, $Cr_2O_7^{2-}$ (Doshi et al. 2007a) and AsO_4^{-3} (Doshi et al. 2008) have been reported. In this chapter the focus is on the bioremediation potential of *Spirulina*.

There are reports about the use of *Spirulina* to remove toxic metal ions from water. Many of these have been cited in the earlier sections. The potential of *Spirulina* for bioremediation is described in detail with reference to certain metal ions in the discussion that follows. The use of kinetic data, Langmuir and Freundelich adsorption isotherms, images from SEM and Fluorescence Microscope and IR frequencies for qualitative and quantitative understanding of metal ions uptake is elaborately discussed. It is hoped that this might provide a fairly high quality working knowledge about the sorption studies.

EXPERIMENTAL

Preparation of Biomass

Spirulina sp., the cyanobacterium was obtained from the algal bloom collected locally. The materials thus collected were washed with water a number of times to remove dirt particles. *Spirulina* sp. was separated from the bloom using standard method and cultured under ideal conditions (Allen 1968). Chemicals used for the growth medium are shown in Table 7.1. The live and dead sample of *Spirulina* sp. was characterized with an image analyzer (Carl Zeiss KS 300) with integrated camera. A small sample of the live biomass was placed on a slide and for the dry sample a few drops of water were added to moisten it. Photographs taken are shown in Plate 7.1. The image shows the presence of pure *Spirulina* sp. This live biomass was used for sorption studies. The live biomass was sun dried for 48 hours. The powdered dried materials were also used as adsorbent.

TABLE 7.1 Composition of Growth Medium

Constituents	Concentration (g/L)	Constituents	Concentration (g/L)
EDTA	0.08	$FeSO_4 \cdot 7H_2O$	0.01
$CaCl_2 \cdot 2H_2O$	0.04	$MgSO_4 \cdot 7H_2O$	0.20
NaCl	1.00	K_2SO_4	1.00
$NaNO_3$	2.50	K_2HPO_4	0.50
$NaHCO_3$	16.8		

SORPTION AND KINETICS STUDIES

The sorption experiments were conducted in 50 ml flasks containing 25 ml of salt solutions with initial concentrations ranging from 0.05 to 0.5 g/25 ml. Dry biomass of 0.3 g was added to each of the flasks. Concentration of the ions, viz. Ni^{+2}, Cr^{+3} and $Cr_2O_7^{-2}$, in solution was estimated spectrophotometrically using UV-160A, Shimadzu at 30 minutes, time interval for four to six hours in all the sorption experiments. To obtain the concentration, a calibration curve was prepared for each of the ions. Generally, 0.2 ml solution was taken and diluted to 2 ml with water. The concentration of the metal ions in solutions thus obtained was determined from the absorption at λ_{max} 390 nm for Ni^{+2}, 428 nm for Cr^{+3} and 440 nm for $Cr_2O_7^{-2}$. Cu^{+2} were estimated as Cu^{+2}-EDTA complex at λ_{max} 720 nm. For all the sorption experiments, equilibrium concentration of each metal ion in solution was determined after about 16 or 18 hours. In case of Cd^{+2} and AsO_4^{-2} adsorption by *Spirulina*, the concentration of metal ions in solution was obtained from Atomic Absorption Spectral data. Similar experiments were carried out for the live *Spirulina* sp. also.

OTHER ANALYTICAL MEASUREMENTS

IR spectra of biomass and metal treated biomass were obtained on Perkins Elmer FTIR (spectrum GX). The SEM photographs (taken on model-ESEM EDAX XL-30, Philips make) of the said samples are shown in Plate 7.2. Fluorescence microscopic images (taken on Olympus BX50 trinocular microscope) of the samples are shown in Plate 7.3.

RESULTS AND DISCUSSION

Adsorption Studies

Two well-known adsorption isotherms Langmuir (Kuppasamy et al. 2004) and Freundlich (Kuppasamy et al. 2004) were considered to identify the isotherm that describes better the equilibrium adsorption of metal ions onto the biomass. These isotherms expressing the equilibrium metal uptake (q_e mg/g) and the concentration of metal ions in solution at equilibrium (C_e mg/L) are as follows:

Langmuir isotherm

$$q_e = \frac{q_{max} b C_e}{1 + b C_e} \quad (1a) \qquad \text{or} \qquad \frac{1}{q_e} = \frac{1}{q_{max}} + \frac{1}{q_{max} b C_e} \quad (1b)$$

Freundelich isotherm

$$q_e = K_F C_e^{1/n} \quad (2a) \qquad \text{or} \qquad \ln q_e = \ln K_F + 1/n \ln C_e \quad (2b)$$

where, q_{max} (mg/g) is the maximum uptake by the biomass, b, K_F and n are the constants corresponding to the respective isotherms. The sorption process on to both live and dead biomass by Cd^{+2} ions follows Langmuir fairly well (Fig. 7.1) (Doshi et al. 2007b). Langmuir adsorption isotherm provides a fairly good knowledge about the Cd^{+2} uptakes by *Spirulina* sp. The maximum uptake for Cd^{+2} (Table 7.2) as derived from Langmuir is 355 mg/g (obtained from best fit) by dead *Spirulina*, while for live, this value is 625 mg/g. When compared with uptake of cadmium by other biosorbents such as 35 and 40 mg Cd/g by Yeast Cells *S. cerevisiae* and *K. fragilis* (Table 7.2), 70 mg Cd/g by *Saccharomyces cerevisiae* (Table 7.2), about 100 mg Cd/g by biomass of *Ascophyllum nodosum* (Table 7.2), it appears that both dead and live *Spirulina* sp. are excellent biosorbent for cadmium. An important assumption in Langmuir adsorption isotherm is the independence of the available adsorption sites, and deviation from Langmuir isotherm may be traced to the failure of this assumption. A suggestion has been put forward that energetically most favourable sites are occupied. It has been observed that adsorption from solution follows Freundlich adsorption isotherm better (Fig. 7.2). The plots (Doshi et al. 2007b) of Lnq_e vs. LnC_e for the sorption of Cd^{+2} ions on to both live and dead *Spirulina* are found to be a straight line. The values of n obtained from Freundlich isotherm are 2.3 and 0.86 respectively for live and dead *Spirulina* sp. for the uptake of Cd^{+2}. It has been suggested (Kadivelu and Namasivayam 2000) that n values between 1 and 10 represent beneficial adsorption absent.

Table 7.2 Comparison of metal ions biosorption by different biosorbents on the basis of maximum uptake capacity (q_{max}: mg/g).

A: Cr^{+3}

Biosorbent Species	q_{max}
Synechococcus sp. *PCC* 7942 (Gardea et al. 1998a)	5
Alfalfa Biomass (Dokken et al. 1999)	16
Creosote Bush Stems (Gardea et al. 1998b)	52
Dead *Spirulina* (Doshi et al. 2007a)	167
Live *Spirulina* (Doshi et al. 2007a)	304

B: $Cr_2O_7^{-2}$

Biosorbent Species	q_{max}
Synechococcus sp. *PCC* 7942 (Gardea et al. 1998a)	0.8
Chlorella vulgaris (Cetinkaya et al. 1999)	33
Non-living *Pseudomonas* (Hussein et al. 2004)	111
Dead *Spirulina* (Doshi et al. 2007a)	143
Live *Spirulina* (Doshi et al. 2007a)	333

C: Cu^{+2}

Biosorbent Species	q_{max}
Synechococcus sp. *PCC* 7942 (Gardea et al. 1998a)	11
Desulfovibrio desulfuricans (Doshi et al. 2006)	16
Ulva reticulata (Doshi et al. 2006)	74
Dead *Spirulina* (Doshi et al. 2007a)	100
Non-living *Pseudomonas* (Hussein et al. 2004)	163
Chlorella sp. (Doshi et al. 2006)	220
Live *Spirulina* (Doshi et al. 2007a)	389

Table 7.2 Contd.

Table 7.2 Contd.

D: Ni^{+2}

Biosorbent Species	q_{max}
Synechococcus sp. PCC 7942 (Gardea et al. 1998a)	3
Creosote Bush Leaves (Gardea et al. 1998b)	5
Chlorella sp. (Doshi et al. 2006)	122
Lyngbya taylorii (Doshi et al. 2006)	163
Dead *Spirulina* (Doshi et al. 2007a)	515
Non-living *Pseudomonas* (Hussein et al. 2004)	556
Live *Spirulina* (Doshi et al. 2007a)	1378

E: AsO_4^{-3}

Biosorbent Species	q_{max}
Hematite (Hansen et al. 2006)	0.4
Activated alumina (Hansen et al. 2006)	15.5
Penicillium purpurogenum (Hansen et al. 2006)	35.6
Lessonia nigrescens (Hansen et al. 2006)	45.2
DSP (Doshi et al. 2008)	402.0
LSP (Doshi et al. 2008)	525.0

E: Cd^{+2}

Biosorbent Species	q_{max}
S. cerevisiae (Hadi B et al. 2003)	35
K. fragilis (Hadi B et al. 2003)	40
Saccharomyces cerevisiae (Volesky B. et al. 2004)	70
A. nodosum (Holan Z. et al. 2004)	100
DSP (Doshi et al. 2007b)	355
LSP (Doshi et al. 2007b)	625

The sorption of Cu^{+2}, $Cr_2O_7^{-2}$, Cr^{+3}, Ni^{+2} (Doshi et al. 2007a) and AsO_4^{-3} (Doshi et al. 2008) for both live and dead biomass also follows Langmuir fairly well (Figure 7.1). The maximum uptake for Cu^{+2}, $Cr_2O_7^{-2}$, Cr^{+3} and Ni^{+2} as derived from Langmuir is 130, 100, 167 and 515 mg/g respectively by dead *Spirulina*, while for live, these values are 389, 333, 304 and 1378 mg/g (Table 7.2). A comparison of metal ions taken by different adsorbents is presented in Table 7.2. It is evident from this table that the uptake capacity of live *Spirulina* is more compared to other biomasses so far investigated. It also appears (Table 7.2) that both dead and live *Spirulina* sp. are excellent biosorbent for nickel.

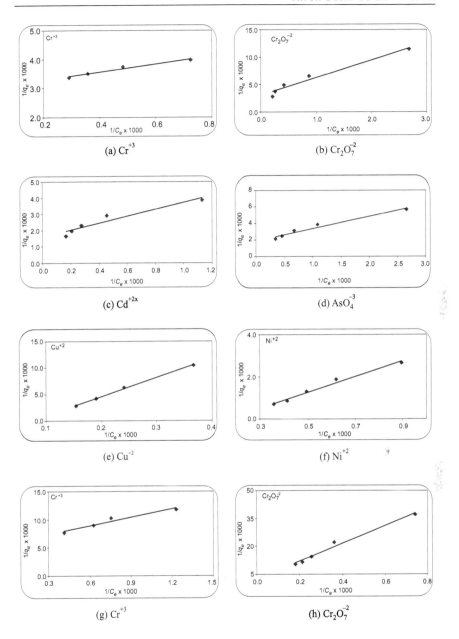

(a) Cr^{+3}

(b) $Cr_2O_7^{-2}$

(c) Cd^{+2x}

(d) AsO_4^{-3}

(e) Cu^{-2}

(f) Ni^{+2}

(g) Cr^{+3}

(h) $Cr_2O_7^{-2}$

Fig. 7.1 Contd.

Fig. 7.1 Contd.

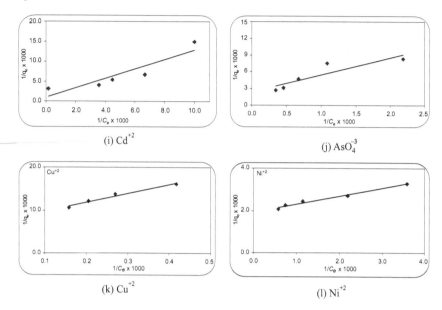

(i) Cd^{+2}

(j) AsO$_4^{-3}$

(k) Cu^{+2}

(l) Ni^{+2}

FIG. 7.1 Langmuir isotherm for the sorption of metal ions on live (a to f) and dead (g to l) *Spirulina* sp.

The Langmuir plots (Fig. 7.1) of the adsorption of arsenic by live and dead *Spirulina* are also straight lines (Doshi et al. 2008). The maximum uptake of arsenic as obtained from Langmuir plot is 525 mg/g for live *Spirulina* sp. and 402 mg/g for dead *Spirulina* sp. The importance of live *Spirulina* sp. to take up AsO$_4^{-3}$ from aqueous bodies is evident from Table 7.2, where As (V) uptake by different sorbents along with *Spirulina* sp. is shown.

The sorption of arsenic by both live and dead *Spirulina* sp. *also* follows Freundlich isotherm. The plot (Fig. 7.2) of ln q$_e$ vs. ln C$_e$ shows straight lines (Doshi et al. 2008) interestingly, the values of n for live and dead *Spirulina* are 2.05 and 1.51 respectively, indicating that the adsorption of As (V) by *Spirulina* sp. is beneficial.

It thus appears that live *Spirulina* sp. is a better option than the dead one for sorption of the metal ions investigated so far. Due to stress-induced enzymatic activities, chemicals are secreted when live *Spirulina* sp. is treated with pollutants. The chemicals thus secreted will aid to the sorption of ions (Jiunn-Tzong Wu et al. 1998, Siripornadulsil et al. 2002). Such activities in live *Spirulina* sp. possibly aid in sorption process, while in dead *Spirulina* sp. such activities are absent. This probably results in greater uptake by live *Spirulina* sp. (Doshi et al. 2007a, b, 2008).

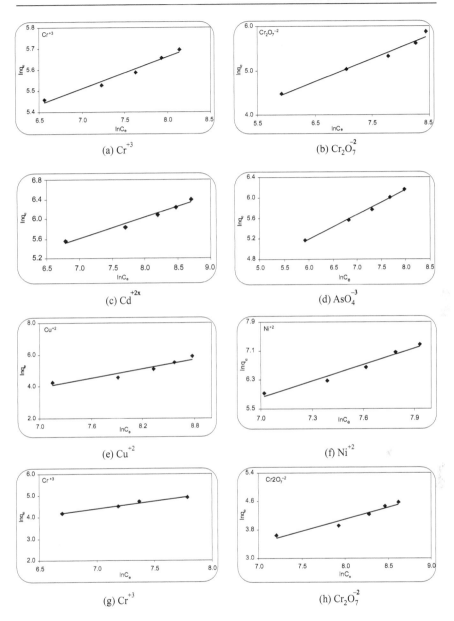

(a) Cr^{+3}

(b) $Cr_2O_7^{-2}$

(c) Cd^{+2x}

(d) AsO_4^{-3}

(e) Cu^{+2}

(f) Ni^{+2}

(g) Cr^{+3}

(h) $Cr_2O_7^{-2}$

Fig. 7.2 Contd.

Fig. 7.2 Contd.

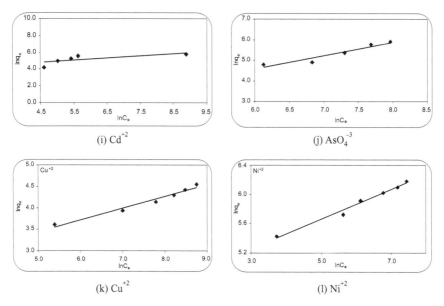

(i) Cd^{+2} (j) AsO$_4^{-3}$

(k) Cu^{+2} (l) Ni^{+2}

FIG. 7.2 Freundlich isotherm for the sorption of metal ions on live (a to f) and dead (g to l) *Spirulina* sp.

KINETIC STUDIES

Two simple kinetic models were employed to analyze sorption rates of metal ions on live and dead *Spirulina* sp. Both pseudo first order equation-3 (Ho and Mckay, 1998, Kuppasamy et al. 2004) and pseudo second order, equation-4 (Ho and Mckay, 1998, Kuppasamy et al. 2004) kinetic equations were employed

$$\text{Log } (q_e - q_t) = \text{Log}q_e - k_1t/2.303 \tag{3}$$

$$t/q_t = 1/(k_2q_e^2) + t/q_e \tag{4}$$

Here q_e is the same as indicated in Langmuir equation, q_t is the amount of pollutants adsorbed at time t for per unit weight of adsorbent (mg/g). k_1 is first and k_2 is second order rate constant.

Kinetics of Cr^{+3}, Cu^{+2}, Ni^{+2} and Cr$_2$O$_7^{-2}$ adsorption

The sorption of Cu^{+2}, Ni^{+2} and Cr^{+3} on to dead *Spirulina* sp. biomass follows pseudo first order rate law. The plots are shown in Fig. 7.3. The calculated values of q_e agree fairly well with the experimental q_e (Table 7.3) and this fact lends support to the validity of first order kinetic for this sorption. However, the sorption of Cr$_2$O$_7^{-2}$ on dead *Spirulina* sp.

does not follow the pseudo first order kinetics as the calculated q_e differs widely from the experimental q_e. This sorption process follows pseudo second order kinetic rate equation. The uptake of all the metal ions by live *Spirulina* sp. follows second order kinetics. The graphs are shown in Fig. 7.3.

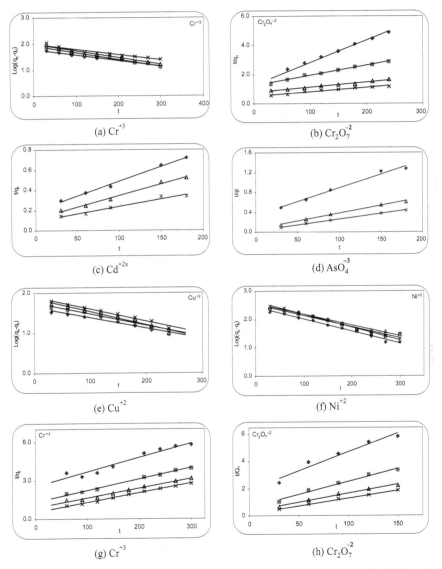

(a) Cr^{+3}

(b) $Cr_2O_7^{-2}$

(c) Cd^{+2x}

(d) AsO_4^{-3}

(e) Cu^{+2}

(f) Ni^{+2}

(g) Cr^{+3}

(h) $Cr_2O_7^{-2}$

Fig. 7.3 Contd.

Fig. 7.3 Contd.

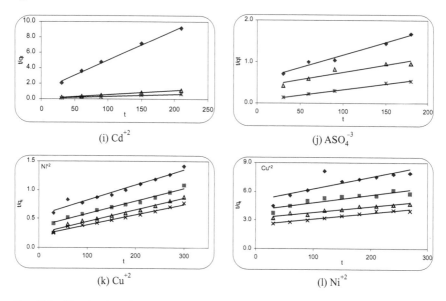

(i) Cd^{+2}

(j) ASO_4^{-3}

(k) Cu^{+2}

(l) Ni^{+2}

FIG. 7.3 Kinetic plots for the sorption of metal ions on live (a to f) and dead (g to l) *Spirulina* sp.

The rate constants for the sorption of metal ions on live *Spirulina* sp. are less compared to the same for the dead *Spirulina* sp., the following explanation may lend support to this observation. Live *Spirulina* sp. cell may be represented as (Koning 1994, Doshi et al. 2007b)

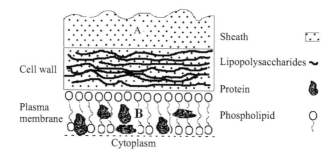

The metal ions will get adsorbed on surface A and some of them will diffuse through the different layers to walk on to surface B. There are membrane bound enzymes that are known to aid in such transport. In other words, diffusion through these layers will, to some extent, control the adsorption process. These will possibly slower the rate of sorption. It is also likely that when this biomass is treated with pollutants, glutamate is secreted due to stress-induced enzymatic reactions (Jiunn-Tzong Wu et al. 1998, Siripornadulsil et al. 2002). This glutamate is the precursor of proline. Consequently, glutamate is transformed into proline in course of time. The glutamate/proline aids in the sorption process. In addition to the layers A and B, other layers also act as an adsorbing surface to take up the metal ions. As a result the uptake by live *Spirulina* sp. will be high. In case of dead *Spirulina* sp. the material is dried up and water is lost and dead *Spirulina* sp. is just like any other chemical compound where membrane bound enzymes are denatured and non-functional. Thus, in dead *Spirulina* sp. there is no enzymatic activities as in live *Spirulina* sp. to have chemicals that aid in sorption process. This possibly explains the greater metal uptake by live *Spirulina* sp. compared to dead *Spirulina* sp.

KINETICS OF ADSORPTION OF CADMIUM

It was observed that the kinetic data follows second order equation better for the sorption of Cd^{+2} by both live and dead *Spirulina* sp. Here the plot t/q_t against t (min) gives a straight line (Fig. 7.3). The calculated values of q_e agree fairly well with the experimental q_e (Table 7.3) and this fact lends support to the validity of second order kinetic for this sorption. It is gratifying that in case of adsorption of Cd^{+2} by live *Spirulina* the rate constant as expected is greater than that by dead *Spirulina*.

Kinetics of AsO_4^{-3} Uptake

Kinetics of arsenate uptake by live and dead *Spirulina* (Doshi et al. 2008) follows second order rate equation (equation-4). The plot of t/q_t vs. t (min) gives straight lines (Fig. 7.3). The parameters obtained from kinetic equation are shown in Table 7.3. The calculated values of q_e agree fairly well (Table 7.3) with experimental q_e and this fact lends support to the validity of second order kinetics for the present sorption process. It has been observed that the rate of adsorption of ions Cu^{+2}, Ni^{+2}, Cr^{+3}, $Cr_2O_7^{-2}$ and Cd^{+2} by live biomass is slower than that by dead *Spirulina* as explained earlier. But the reverse is true for the uptake of arsenate by *Spirulina*. The uptake of AsO_4^{-3} by dead biomass is found to be slower. Further work is required to understand this reverse trend.

INFRARED SPECTRA

IR spectra of biomass and metal treated biomass were obtained on Perkins Elmer FTIR (spectrum GX). The IR spectra of *Spirulina* and metal treated *Spirulina* has been reported to contain functional groups like sulphonic, carboxylic (fatty acid and amino acids), phosphate, amide, hydroxyl (polysaccharide) etc. (Campanella et al. 1998). The very broad

Table 7.3 Kinetics Data for the Uptake by Dead and Live *Spirulina* sp.

Metal	Dead *Spirulina* sp. (DSP)*			Live *Spirulina* sp. (LSP)**		
	q_e		Rate constant $\times 10^3$	q_e		Rate constant $\times 10^3$
	Experimental	Calculated		Experimental	Calculated	
Cr^{+3}	65.0	62.7	5.100	234.7	212.7	0.041
	111.6	97.5	6.000	267.0	256.4	0.054
	134.8	99.0	4.800	285.7	277.7	0.056
$Cr_2O_7^{-2}$	27.0	36.2	0.388	204.6	277.8	0.017
	44.9	52.1	0.595	271.6	333.3	0.018
	68.7	75.2	0.528	353.7	370.4	0.021
Cu^{+2}	50.0	57.2	6.400	96.3	90.9	0.137
	62.0	73.3	7.600	160.0	113.6	0.161
	73.5	80.5	7.100	243.0	204.1	0.055
Ni^{+2}	228.1	269.4	9.900	377.5	357.1	0.024
	306.7	325.5	9.000	533.4	476.2	0.062
	410.3	429.5	10.600	1451.1	1428.6	0.038
Cd^{+2}	35	26	1.34	259	345	0.04
	183	182	0.60	435	435	0.04
	313	333	0.19	601	714	0.02
AsO_4^{-3}	120	156	0.082	176	182	0.090
	213	222	0.062	321	323	0.158
	365	370	0.117	475	455	0.142

*: The uptake of Cr^{+3}, Cu^{+2} and Ni^{+2} by DSP follows first order kinetics and uptake of Cd^{+2}, $Cr_2O_7^{-2}$, AsO_4^{-3} follows second order kinetics.

**: For LSP second order kinetics is obeyed for all the ions.

band at ~3305 cm^{-1} in *Spirulina* sp. is due to hydrogen bonded hydroxyl group. This is also coupled with stretching vibrations of NH_2 moiety. The very strong absorption at 1040–1035 cm^{-1} in *Spirulina* sp. is assigned to stretching of C–O in polysaccharides (Rathinam et al. 2004). This has possibly contribution from SO and PO stretching vibrations (Loukidou et al. 2004). The peak at 868 and 580, 440 cm^{-1} in *Spirulina* sp. may be traced to stretching and bending modes respectively of phosphate moiety (Nakamoto 1963).

Infrared spectra of Cd^{+2} treated biomass

Live *Spirulina* were treated with Cd^{+2} ions for different lengths of time namely 0.5, 1.5, 2.5, 3.5 and for more than 12 hours. The main features of these spectra are quite informative and they offer an insight into the process of metal uptake. The IR spectrum of *Spirulina* treated with Cd^{+2}

for 0.5 hour shows a peak at 1740 cm^{-1} (absent in untreated *Spirulina*), that may be assigned to νC=O of O=C–O moiety. The intensity of this peak decreases as the time allowed for interaction increases. This peak appears as shoulder in *Spirulina* sample treated with Cd^{+2} upto 2.5 hour. When treated with Cd^{+2} ions for 3.5 hour or more this peak almost disappears. A likely explanation for this is provided below.

The peaks at 1538 and 1430 cm^{-1} in *Spirulina* are assigned to νasy COO^{-} and νsym COO^{-} respectively. *Spirulina* treated with Cd^{+2} shows absorption around ~1525 and 1406 (~1390) cm^{-1}. Untreated *Spirulina* has absorption at 1240 cm^{-1} which may be traced to νP=O. The samples treated with Cd^{+2} ions for 0.5 and 1.5 hour show peak around 1150 cm^{-1} in addition to the peak at 1225 cm^{-1}. These absorptions possibly have contribution from νC–O. It is suggested by Doshi et al. (2007b) that Cd^{+2} metal binds with carboxylic moiety in two ways:

type-I   type-II

When type-I is present the C=O stretching possibly appears at 1740 cm^{-1}. With increasing time it is likely that type-I goes over to type-II, thereby decreasing the intensity of C–O and C=O with time. Interestingly when *Spirulina* is treated with Cd^{+2} for 2.5 and 3.5 hour the peak at ~1150 cm^{-1} disappears.

There is also the possibility of another route by which the absorption at ~1740 cm^{-1} might arise. The band around 1650 cm^{-1} in *Spirulina* may be attributed to coupled vibrations of C=O and C=N stretching and NH$_2$ bending, arising from the amino acids/amides present in the biomass. *Spirulina* treated with Cd^{+2} ions for 0.5 hour and 1.5 hour show very strong absorption in the region ~1650 cm^{-1} that splits into two peaks. It is likely that when this biomass is treated with Cd^{+2} ions, glutamate is secreted due to stress induced enzymatic reactions. This glutamate (Jiunn-Tzong Wu et al. 1998, Siripornadulsil et al. 2002) the precursor of proline is transformed into the latter in course of time. So at the beginning the concentration of amino-acids will increase leading to the increase in the intensity of adsorption ~ 1650 cm^{-1}. When the biomass is treated with Cd^{+2} ions for longer period of time the intensity of the said peak decreases for reasons stated above. The formation of glutamate will possibly also contribute to the absorption ~1740 cm^{-1}.

The involvement of phosphate in metal binding is very interesting. From the IR spectra of the different Cd^{+2} treated live *Spirulina* samples, it appears that at the beginning the Cd^{+2} ions interact fairly strongly with phosphate. However, live *Spirulina* treated with Cd^{+2} for 3.5 hour shows practically the same IR peaks for phosphate as seen in untreated

Spirulina. When the biomass is treated with Cd^{+2} ions for a greater length of time (16 hour) the interaction of metal ion with phosphate again becomes very significant, as evident from the disappearance of peak at 1240 cm^{-1}, the shifting of 1048 cm^{-1} to 1090 cm^{-1} and the appearance of new bend at 1170 cm^{-1}. As stated earlier, when live *Spirulina* sp. is treated with Cd^{+2}, due to stress induced enzymatic activities, chemicals will be secreted that bind with metal ion. So, for a certain length of time from the beginning, the interaction of Cd^{+2} with phosphate will slow down. After a long time (~16 hour) the enzymatic activity will stop or become extremely little. Under such conditions the interaction of Cd^{+2} with phosphate will again become pronounced (Doshi et al. 2007b).

It seems that the carboxylic and phosphate groups and to lesser extent the amino group are responsible for the metal uptake in live *Spirulina*. With dead *Spirulina* the said groups are also responsible for metal uptake. Interestingly, Cd^{+2} treated dead *Spirulina* shows a shoulder ~1740 cm^{-1} and the band ~1160 cm^{-1} suggesting the likely formation type-I bonding with carboxylic acid to a small extent.

Infrared Spectra of Ni^{+2}, Cu^{+2}, Cr^{+3} and $Cr_2O_7^{-2}$ Treated Dead *Spirulina* sp.

It has already been stated that a very strong absorption assigned to C-O stretching coupled with stretching of SO and PO appears around 1040-1035 cm^{-1}. In Ni^{+2}, Cr^{+3} and $Cr_2O_7^{-2}$ treated biomass this band registers blue shift but in Cu^{+2} treated three peaks are observed, namely at 1162, 1059 and 987 cm^{-1}. The peak at 868 cm^{-1} in *Spirulina* sp. is attributed to the PO stretching modes of phosphate moiety. This absorption disappears on metal uptake. The bending mode of phosphate at 571 cm^{-1} in *Spirulina* sp. shows an appreciable shift when biomass takes up metal ions. From the preceding discussion, the interaction between hydroxyl/phosphate moieties with the metal ions may be inferred (Doshi et al. 2007a). The phosphates are from the phospholipids backbone and hydroxyls of phytosterols possibly take part in the interactions. *Spirulina* sp. contains amide group and amino acids and band at 1651 cm^{-1} is attributed to coupled vibrations of C=O and C=N stretching with NH_2 bending. In Cu^{+2} treated *Spirulina* sp., this band does not change, but in Ni^{+2}, Cr^{+3} and $Cr_2O_7^{-2}$ treated *Spirulina* sp., this band becomes broad and two peaks at ~1658 cm^{-1} and 1640 cm^{-1} are observed. It may be traced to binding of metal ions with NH_2 group. Asymmetric and symmetric stretching vibrations of carboxylate group in *Spirulina* sp. assigned to absorptions at 1525 cm^{-1} and 1406.5 cm^{-1} respectively show significant change when the biomass takes up metal ions, suggesting the participation of carboxylate moiety (Xue et al. 1988).

Infrared Spectra of Ni^{+2}, Cu^{+2}, Cr^{+3} and $Cr_2O_7^{-2}$ Treated Live *Spirulina* sp.

IR spectra of metal treated live *Spirulina* sp. and dead *Spirulina* sp. are not exactly the same. The important difference is observed at 1650 cm^{-1} on uptake of metal ions by live *Spirulina* sp. This band changes (red shift) appreciably in metal treated live species, which was not the case for dead *Spirulina* sp., indicating significant interaction of metal ions with the amide/amino group in the former. Membrane bound transport enzymes in live *Spirulina* sp. help in intracellular interaction of metal ions with amino acid. The carboxylate moiety of fatty acids also takes part in the sorption process. The interaction with carboxylate is stronger in Ni^{+2} and $Cr_2O_7^{-2}$ treated biomass, while in case of Cu^{+2} and Cr^{+3}, the position of the peaks due to carboxylate change but intensity does not change appreciably (Doshi et al. 2007a). As in dead, in live *Spirulina* sp. too metal ions interact with hydroxyl and phosphate moiety. This interaction is very pronounced with $Cr_2O_7^{-2}$ and Cu^{+2} ions.

In case of metal uptake by dead *Spirulina* sp., primarily phosphate, hydroxyl and carboxylate groups take the lead role. The high metal uptake by live *Spirulina* sp. may be traced to the metal transport through the cell in presence of active enzymes thereby all the functional groups of intercellular biopolymer matrix. Live algae possess intracellular polyphosphates which participate in metal sequestration, as well as algal extracellular polysaccharides that serve to chelate or bind metal ions (Zhang and Majidi 1994, Tropis et al. 1996).

Infrared Spectra of AsO_4^{-3} Treated *Spirulina*

The peaks due to NH_2 and OH stretching do not show any noticeable change in both live and dead *Spirulina* treated with arsenate. The peak around 1650 cm^{-1} assigned to a coupled vibrations of NH_2 bending, CN stretching and CO stretching possibly of the amino acids, also remains practically unchanged in treated *Spirulina*.

The peaks assigned to asy COO$^-$ and sym COO$^-$ in *Spirulina* do not show any major changes, when live and dead *Spirulina* sp. is treated with Na_2HAsO_4 salt solution. Interestingly, a shoulder at 1725 cm^{-1} is observed for As(V) treated live *Spirulina* sp. This may be traced to the formation of glutamate, the precursor of proline, due to stress induced enzymatic activities. The glutamate possibly contains unionized carboxylic acid group (http://en.wikipedia.org/wiki...) which will show C=O stretching.

The very strong absorption at ~1040 cm^{-1} in *Spirulina* sp. is assigned to stretching of C–O in polysaccharides (Rathinam et al. 2004). This

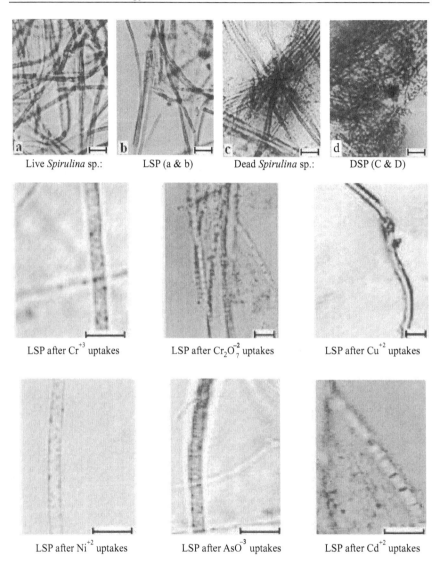

Live *Spirulina* sp.: LSP (a & b) Dead *Spirulina* sp.: DSP (C & D)

LSP after Cr^{+3} uptakes   LSP after $Cr_2O_7^{-2}$ uptakes   LSP after Cu^{+2} uptakes

LSP after Ni^{+2} uptakes   LSP after AsO^{-3} uptakes   LSP after Cd^{+2} uptakes

Plate 7.1 Figure obtained from image analyzer microscope, Carl Zeiss KS 300 (⊢⊣ : 20 μ).

possibly has contribution from SO and PO stretching vibrations. The peak at 868 and 590 cm^{-1} in *Spirulina* sp. may be traced to stretching and bending modes respectively of phosphate moiety (Sabine and Cliff 2001). The IR peak at 1040 cm^{-1} registers very significant change in As (V) treated *Spirulina* sp. and appears at ~1090 cm^{-1}. It is suggested that the

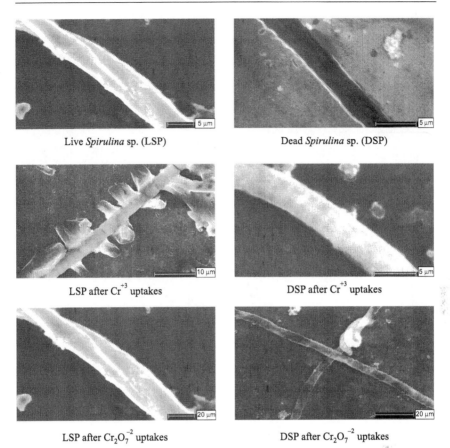

Live *Spirulina* sp. (LSP)   Dead *Spirulina* sp. (DSP)

LSP after Cr^{+3} uptakes   DSP after Cr^{+3} uptakes

LSP after $Cr_2O_7^{-2}$ uptakes   DSP after $Cr_2O_7^{-2}$ uptakes

Plate 7.2(a) SEM micrograph of live and dead *Spirulina* sp. and treated ones.

AsO_4^{-3} ions will replace phosphate (Mohammed et al. 2002). This is primarily responsible for the uptake of arsenate. In live *Spirulina* due to enzymatic activities this transport phenomenon is more pronounced. The peaks of arsenate treated *Spirulina* at 822 cm^{-1} and 970 cm^{-1} are possibly due to arsenate [As (V)]. In arsenate treated live *Spirulina* a peak is observed at 791 cm^{-1} that may be traced to arsenite [As (III)] (Sabine and Cliff 2001). It is likely that As (V) is reduced to As (III) by enzymetic activities.

SCANNING ELECTRON MICROSCOPY

The SEM micrographs are shown in Plates 7.2(a) and 7.2(b). The width of the filament of live *Spirulina* sp. varies from 2.00 to 5.0 μm and as expected the cell shrinks when it is dried. In dry species the width of the filament lies in the range of 1.4 to 3.7 μm. The difference in the surface morphology after the metal uptake by live and dead species is evident from Plate 7.2(a). The surface of the biomass becomes rough after metal uptakes. It has already been stated in the preceding section on IR spectroscopy and kinetic study that live *Spirulina* sp. take up more metal ions than the dead one. A close examination of the SEM micrograph of the

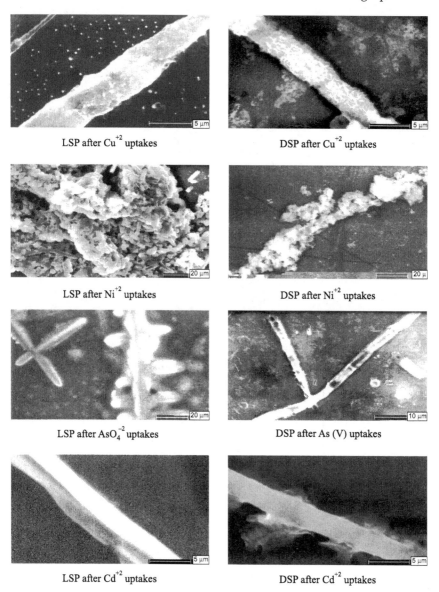

LSP after Cu^{+2} uptakes   DSP after Cu^{+2} uptakes

LSP after Ni^{+2} uptakes   DSP after Ni^{+2} uptakes

LSP after AsO_4^{-2} uptakes   DSP after As (V) uptakes

LSP after Cd^{+2} uptakes   DSP after Cd^{+2} uptakes

Plate 7.2(b) SEM micrograph of live and dead *Spirulina* sp. and treated ones.

treated samples also support this fact, the width of the cell of treated live biomass is more compared to the same for dried species. The average size of the most of the metal treated cells in live *Spirulina* sp. is in the range Ni^{+2} (3.0-17.5 μm), Cu^{+2} (2.9-11.0 μm), Cd^{+2} (3.6-10.6 μm), Cr^{+3} (2.7-5.6 μm), $Cr_2O_7^{-2}$ (4.8-9.3 μm) and AsO_4^{-3} (5.0-10.5 μm) while in dead *Spirulina* sp. the same is found to be in the range Ni^{+2} (3.8-13.5 μm), Cu^{+2} (3.1-10.2 μm), Cd^{+2} (4.1-7.3 μm), Cr^{+3} (2.0-4.5 μm), $Cr_2O_7^{-2}$ (3.9-6.8 μm) and AsO_4^{-3} (2.0-4.6 μm). The cell sizes are not in conformity with the sorption results. This may be traced to the fact that many cells are aggregated when treated with metal ions. Thus cells of different sizes were obtained. When treated with metal ions there is the possibility of transport of these ions through the cell membrane in live biomass that may be responsible for more uptakes in this species. The SEM micrograph of metal treated live *Spirulina* sp. shows metal ions protruding in the strand of the cell, supporting membrane transport process. Images (Plate 7.1) obtained from the image analyzer also support the results of fluorescence and SEM studies.

FLUORESCENCE MICROSCOPY

The fluorescence microscopy images of live *Spirulina* sp. and metal treated biomass are shown in Plate 7.2. Live *Spirulina* sp. when exposed to UV radiation gives red fluorescence (Chen et al. 1996). The images of live *Spirulina* sp. (pure) show the presence of a continuous strand, while the metal treated biomass shows a discontinuous broken strand having different sizes. The amount of metal uptake is responsible for the extent of discontinuity of the strand at different places. The live *Spirulina* sp. emits more fluorescence radiation than the one treated with metal ions. It is clear from the images (Plate 7.2) that the biomass has taken more Ni^{+2} than Cr^{+3} and $Cr_2O_7^{-2}$. The images of Cu^{+2} treated *Spirulina* sp. apparently show more Cu^{+2} uptake than Ni^{+2} which is not the case. We have observed microscopically that to an extent (about one week) Ni^{+2}, Cr^{+3} and $Cr_2O_7^{-2}$ helps the *Spirulina* sp. to survive, but in case of Cu^{+2} the biomass does not survive even for a day. In other words Cu^{+2} is highly toxic for *Spirulina* sp. and possibly due to this the image of Cu^{+2} treated *Spirulina* sp. shows the presence of very less fluorescence emitting species. Live *Spirulina* treated with Cd^{+2} and AsO_4^{-3} also show discontinuous broken strand of different sizes in the fluorescence microscopic images. Although the uptake of AsO_4^{-3} is more than Cd^{+2}, the images of AsO_4^{-3} treated *Spirulina* show more continuity than the Cd^{+2} treated one. It appears that cadmium is more toxic to *Spirulina* than arsenate.

Interestingly, the photographs (Plate 7.1) of metal treated *Spirulina* obtained from image analyzer also support the observation of fluorescence microscopic studies.

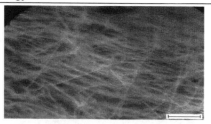

Live *Spirulina* sp. (LSP)

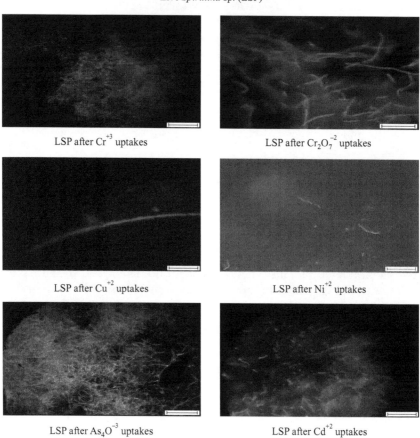

LSP after Cr^{+3} uptakes LSP after $Cr_2O_7^{-2}$ uptakes

LSP after Cu^{+2} uptakes LSP after Ni^{+2} uptakes

LSP after As_4O^{-3} uptakes LSP after Cd^{+2} uptakes

Plate 7.3 Fluorescence microscopic photographs (⊢⊣ : 40 μ) of live Spirulina sp. (LSP) and treated ones.

Color image of this figure appears in the color plate section at the end of the book.

REFERENCES

Allen, M. 1968. Simple conditions for growth of unicellular blue–green algae on plates. J. Phycol., 4: 1-4.

Augusto da Costa, A.C. and F.P. de Franca. 1998. Short communication: Cadmium uptake by *Spirulina maxima*: Toxicity and mechanism. World Journal of Microbiology and Biotechnology 14: 579-581.

Augusto da Costa, A.C. and F.P. de Franca. 2003. Cadmium interaction with microalgal cells. Cyanobacterial cells, and seaweeds, toxicology and biotechnological potential for wastewater treatment. Marine Biotechnology 5: 149-156.

Belay, A. 1997. Mass culture of *Spirulina* outdoors—The Earthrise Farms Experience. *In:* A. Vonshak [ed.]. *Spirulina platensis* (Arthrospira): Physiology, Cell Biology and Biotechnology. Taylor & Francis, London, pp. 131-158.

Campanella, L., G. Crescentini, P. Avino, and A. Moauro. 1998. Determination of macrominerals and trace elements in the alga, *Spirulina platensis*. Analusis, 26: 210-214.

Cetinkaya, D.G., Z. Aksu, A. Ozturk, and T.A. Kutsal. 1999. Comparative study on heavy metal biosorption characteristics of some algae. Process Biochemistry 34: 885-892.

Chen, F., Y.M. Zhang, and S.Y. Guo. 1996. Growth and phycocyanin formation of *Spirulina platensis* in photoheterotrophic culture. Biotechnol Lett., 18: 603-608.

Chojnacka, K., A. Chojnacki, and H. Gorecka. 2005. Biosorption of Cr^{3+}, Cd^{2+} and Cu^{2+} ions by blue–green algae *Spirulina* sp.: Kinetics, equilibrium and mechanism of the process. Chemosphere 59: 75-84.

Ciferri, O. 1983. *Spirulina*, the edible microorganism. Micobiol Review 47: 551-578.

Clemens, S., E.J. Kim, D. Neumann, and J.I. Schroeder. 1999. Tolerance to toxic metals by a gene family of phytochelatin syntheses from plants and yeast. EMBO J. 18: 3325-3333.

David, K. and B. Volesky. 1998. Advances in biosorption of heavy metals. Trends in Biotech. 16: 291-300.

Doke, J., K.V. Raman, and V.S. Ghole. 2005. Bioremediation potential of *Spirulina* sp.: Toxicity and sorption studies of Co and Pb. International Journal on Algae 7: i2.30, 118-128

Dokken, K., G. Gamez, I. Herrera, K.J. Tiemann, N.E. Pingitore, R.R. Chianelli, and J.L. Gardea-Torresdey. 1999. Characterization of Cr(VI) bioreduction & Cr(III) Binding to *Alfalfa* biomass. Proceedings of the Conference on Hazardous Waste Research, 101-113.

Doshi, H., A. Ray, I.L. Kothari, and B. Gami. 2006. Spectroscopic and SEM studies on bioaccumulation of pollutants by algae. Current Microbiology 53: 148-157 and references therein.

Doshi, H., A. Ray, and I.L. Kothari. 2007a. Bioremediation potential of live and dead *Spirulina*: Spectroscopic, kinetics and SEM studies. Biotechnology and Bioengineering 96: 1051-1063.

Doshi, H., A. Ray, and I.L. Kothari. 2007b. Biosorption of Cadmium by Live and Dead *Spirulina*: IR Spectroscopic Kinetics, and SEM Studies. Current Microbiology 54: 213-218.

Doshi, H., A. Ray, and I.L. Kothari. 2009. Live and Dead *Spirulina* sp. to Remove Arsenic (V) in Water. International J. Phytoremediation 11(1): 53-64.

Eccles, H. 1999. Treatment of metal-contaminated wastes: Why select a biological process? Trends in Biotechnology 17: 462-465.

Gardea-Torresdey, J.L., J.L. Arenas, N.M.C. Francisco, K.J. Tiemann, and R. Webb. 1998a. Ability of immobilized cyanobacteria to remove metal ions from solution and demonstration of the presence of metallothionein genes in various strains. Journal of Hazardous Substance Research 1: 2-1 to 2-18.

Gardea-Torresdey, J.L., A. Hernandez, K.J. Tiemann, J. Bibb, and O. Rodriguez. 1998b. Adsorption of Toxic Metal Ions from Solution by Inactivated Cells of *Larrea tridentata* (Creosote Bush). Journal of Hazardous Substance Research 1: 3-1 to 3-16.

Goldberg, S. and C.T. Johnston. 2001. Mechanisms of Arsenic Adsorption on Amorphous Oxides Evaluated Using Macroscopic Measurements, Vibrational Spectroscopy and Surface Complexation Modeling. Journal of Colloid and Interface Science 234: 204-216.

Hadi, B., A. Margaritis, F. Berruti, and M. Bergougnou. 2003. Kinetics and Equilibrium of Cadmium Biosorption by Yeast Cells *S. cerevisiae* and *K. fragilis*. International Journal of Chemical Reactor Engineering, 1(A47): 1-18.

Hansen, H.K., P. Nunez, and R. Grandon. 2006. Electrocoagulation as remediation tool for wastewaters containing arsenic, Minerals Engineering 19(5): 521-524.

Heitzer, A. and G.S. Sayler. 1994. Monitoring the efficacy of bioremediation. Trends in Biotech. 11(8): 334-343.

Hernandez, E. and E.J. Olguín. 2002. Biosorption of heavy metals influenced by the chemical composition of *Spirulina* sp. (Arthrospira) biomass. Environ Technol., 23(12): 1369-1377.

Henrikson, R. 1994. Microalga *Spirulina*, superalimento del futuro. Ronore Enterprises, 2nd ed. Ediciones Urano, Barcelona, Espana.

Ho, Y.S. and G. Mckay. 1998. Sorption of dye from aqueous solution by peat. Chemical Engineering Journal 70: 115-124.

Holan, Z.R., B. Volesky, and I. Prasetyo. 2004. Biosorption of cadmium by biomass of marine algae. Biotechnology and Bioengineering 41(8): 819-825.

Hong, C. and P. Shan-Shan. 2005. Bioremediation potential of *Spirulina*: Toxicity and biosorption studies of lead. J of Zhejiang Univ SCI, 6B: 171-174.

http://en.wikipedia.org/wiki/Monosodium_glutamate

http://www.aseanbiotechnology.info/scripts/count_article.asp? Article_code=22003320

Hussein, H., S.F. Ibrahim, K. Kandeel, and H. Moawad. 2004. Biosorption of Heavy Metals from Waste Water Using *Pseudomonas* sp. Electronic Journal of Biotechnology 7: 38-46.

Jennett, J.C., J.E. Smith, and J.M. Hassett. 1982. Factors Influencing Metal Accumulation by Algae. NTIS PB 83149377, 124.

Jiunn-Tzong Wu, Ming-T. Hsieh, and Lai-Chu Kow. 1998. Role of proline accumulation in response to toxic copper in *Chlorella* sp. cells. J. Phycol., 34: 113-117.

Kadivelu K. and C. Namasivayam. 2000. Agriculture byproducts as metal adsorbent: Sorption of lead (II) from aqueous solution onto coir-pith carbon. Environmental Technology 21(10): 1091-1097.

Koning, R.E. 1994. http://koning.ecsu.ctstateu.edu/plant_biology/cyanophyta.html

Kuppasamy, V., R.J. Joseph, P. Kandasamy, and V. Manickam. 2004. Copper removal from aqueous solution by marine green alga *Ulva reticullata*. Electronic Journal of Biotechnology 7: 61-71.

Loukidou Maria, X., I. Zouboulis Anastasios, D. Karapantsios Thodoris, and A. Matis Kosatas. 2004. Equilibrium and kinetic modeling of Chromium (VI) biosorption by *Aeromonas caviae*. Colloids and Surface A: Physicochem. Eng. Aspects, 242: 93-104.

Mohammed, J.A., F. Jörg, and A.M. Andy. 2002 Uptake Kinetics of Arsenic Species in Rice Plants. Plant Physiol 1: 128; 3: 1120-1128.

Momotaj, H. and A.Z.M. Hussain Iftikhar. 2001. Effect of *Spirulina* on Arsenicosis Patients in Bangladesh: Arsenic in Drinking Water: An International Conference at Columbia University, New York, November, 26-27.

Nakamoto, K. 1963. Infrared spectra of inorganic and co-ordination compounds. John Wiley and Sons, New York, NY. p. 107.

Norberg, A. and S. Rydin. 1984. Development of a Continuous Process for Metal Accumulation by *Zoogloea ramigeru*. Biotechnol. and Bioeng. 26: 265-268.

Parmeggiani, A.C. and J.C. Masini. 2003. Evaluating scatchard and differential equilibrium functions to study the binding properties of Cu (II) to the source of mixed species of lyophilized *Spirulina*. J Braz. Chem. Soc., 14: 416-424.

Rangsayatorn, N., E.S. Upatham, M. Kruatrachue, P. Pokethitiyook, and G.R. Lanza. 2002. Phytoremediation potential of *Spirulina* (*Arthrospira*) *platensis*: Biosorption and toxicity studies of cadmium. Environ Pollut. 119: 45-53.

Rathinam, A., M. Balaraman, R.J. Raghava, N.B. Unni, and R. Thirumalachari. 2004. Bioaccumulation of chromium from tannery waste water: An approach for chrome recovery and reuse. Environ. Sci. Technol., 38: 300-306.

Richard, K. and H.H. Ronald. 1996. The study of *Spirulina* (Effect on the AIDS virus, cancer and the immune system) http://www.*Spirulina*.com/SPLNews96.html

Richmond, A. 1992. Mass culture of cyanobacteria. *In:* N. Mann, and N. Carr [eds.]. Photosynthetic Prokaryotes. 2nd ed. Plenum Press, New York and London. pp. 181-210.

Sabine G. and J.T. Cliff. 2001. Mechanism of Arsenic Adsorption on Amorphous Oxides Evaluated using Macroscopic Measurements, Vibrational Spectroscopy, and Surface Complexation Modeling. Journal of Colloid and Interface Science. 234: 204-216.

Saxena, P.N., M.R. Ahmad, R. Shyan, and D.V. Amla. 1983. Cultivation of *Spirulina* in sewage for poultry feed. Experientia 39: 1077-1083.

Shekharam, K., L. Ventakaraman, and P. Salimath. 1987. Carbohydrate composition and characterization of two unusual sugars from blue–green algae *Spirulina platensis*. Phytochem 26: 2267-2269.

Siripornadulsil, S., S. Traina, D.P.S. Verma, and R.T. Sayre. 2002. Molecular mechanisms of proline-mediated tolerance to toxic heavy metals in transgenic micro algae. Plant Cell 14: 2837-2847.

Siripornadulsil, S., S. Traina, D.P.S. Verma, and R.T. Sayre. 2002. Molecular mechanisms of proline-mediated tolerance to toxic heavy metals in transgenic microalgae. Plant Cell 14: 2837-2847.

Slotton, D.G., C.R. Goldman, and A. Frank. 1989. Commercially grown *Spirulina* found to contain low levels of mercury and lead. Nutrition Reports International 40: 1165-1172.

Switzer, L. 1980. *Spirulina*, the Whole Food Revolution. Proteus Corporation, USA. pp. 1-69.

Tropis, M., F. Bardou, B. Bersch, M. Daffe, and A. Milon. 1996. Composition and phase behaviour of polar lipids isolated from *Spirulina* maxima cells grown in a perdeuterated medium. Biochim Biophys Acta 1284(2): 196-202.

Volesky, B. 1990. Biosorption of Heavy Metal. CRC Press, R Boca Raton., FL, 36.

Volesky, B., H. May, and Z.R. Holan. 2004. Cadmium biosorption by *Saccharomyces cerevisiae*. Biotechnology and Bioengineering 41(8): 826-829.

Wu, Jiunn-Tzong, T. Hsieh Ming, and Kow, Lai-Chu. 1998. Role of proline accumulation in response to toxic copper in *Chlorella* sp. cells. J. Phycol., 34: 113-117.

Xue, H.B., W. Stumm, and L. Sigg. 1988. The binding of heavy metals to algal surfaces. Wat. Res., 22: 917-926.

Zhang, W. and V. Majidi. 1994. Monitoring the cellular response of *Stichococcus bacillaris* to exposure of several different metals using in vivo 31P NMR and other spectroscopic techniques. Environ. Sci. Technol., 28: 1577-1581.

CHAPTER

Extant Analogues of the Microbial Origins of Life

Brendan P. Burns, Falicia Goh, Michelle A. Allen,
Rachael Shi and Brett A. Neilan[*]

INTRODUCTION

Are we alone? What is the origin of life? These are some of the most profound questions of humankind, capturing the imagination of scientists and lay people alike for centuries. The search for life elsewhere is dependent on applying lessons learnt from studying the Earth, and thus it is important to focus on studies of Early Earth analogue communities. A major challenge in science is also to identify modern living systems that present unique opportunities to address fundamental questions in fields ranging from microbial ecology, evolution, chemical biology, functional genomics, and biotechnology. Stromatolites represent such a system. One of the earliest pieces of evidence of planetary life is in fact contained in the microfossils of stromatolites. This chapter will focus on microbial mats and stromatolites as living analogues of Early Earth communities. These extant analogues provide an insight into the nature of ancient microbial systems that dominated early life on Earth (McNamara and Awramik 1992), and may also provide clues as to their resilience over such immense periods of geological time. Emphasis will

School of Biotechnology and Biomolecular Sciences, and Australian Centre for Astrobiology, The University of New South Wales, 2052 Australia
*Corresponding author: E-mail: b.neilan@unsw.edu.au

be placed on recent work carried out on the microbial diversity of the Shark Bay stromatolites, as these systems are one of the most extensively studied to date in terms of their microbial communities. We will also discuss how these early life analogues fit into the emerging and exciting field of astrobiology, a multi-disciplinary field of science that allows us to address fundamental questions on our own origins and existence. Astrobiology is the study of life in the universe — its origins, evolution, distribution, and future. It is a challenging science that depends on strong interdisciplinary collaboration between biologists, geologists, astronomers, chemists and others. Aside from the advancement of human knowledge, research in these and other 'extreme' environments has the potential to impact on areas such as economics, theology, and ethics, as well as other philosophical issues that may ultimately define who we are. Finally in this chapter, the biotechnological potential of some of the stromatolite research as well as stress on the importance of conservation of these fascinating 'living rocks' will be introduced.

MICROBIAL MATS

Microbial mats once dominated the Earth. The very earliest mats were most likely simple single species biofilms, which then diversified into the more functionally complex systems which are seen today (Nisbet and Fowler 1999). They are stratified microbial communities and typically develop at the interface of water and a solid medium (such as intertidal areas), and they form laminated multilayers (Figure 8.1). Petri dish diameters are 11 cm. This stratified layer of microorganisms of different metabolisms and physiologies causes a steep gradient of various nutrients; boundaries are very well defined based on the diffusion rate of various molecules so microorganisms will position themselves accordingly. A rapid internal cycling of carbon, nitrogen, sulphur, and phosphate supports high rates of mat metabolism. This very efficient recycling of key nutrients observed within microbial mats (Des Marais 2003), results in extremely productive systems.

Cyanobacteria or other phototrophic microbes dominate primary production of microbial mats, however these systems are home to an incredibly diverse range of metabolisms. These include oxygenic photosynthesis, aerobic heterotrophy, chemolithotrophy, sulphate reduction, fermentation, and methanogenesis, and indicate that these are very functionally complex ecosystems. The microenvironments of these mats are extremely dynamic and fluctuate on a daily basis, particularly the depth of nutrients such as oxygen, which is depleted at night in the upper layers when photosynthesis ceases (Des Marais 2003). Therefore

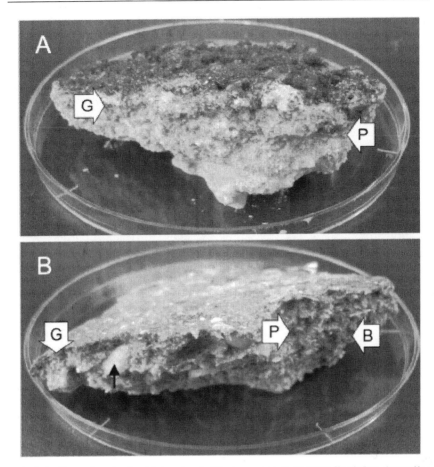

FIG. 8.1 Examples of microbial mats of differing morphology from Shark Bay, Australia. (A) Pustular mat and (B) smooth mat. White arrows indicate zones of green (G), pink (P) and black (B) colouration within the mats. The small black arrow indicates a *Fragum erugatum* shell trapped in the smooth mat.

many organisms within a mat are either motile to move with these changing nutrient boundaries, or have developed broad metabolic diversity and/or relevant protective mechanisms such as UV-screening compounds (Kruschel and Castenholz 1998). The many and varied positive and negative interactions that occur within mats allows tight biological control of the microenvironment and hence greater survival than would be possible for individual species alone (Paerl et al. 2000). This regulation of the microenvironment also directly or indirectly affects trapped or underlying sediments, as well as any mineral precipitation or dissolution. Microbial mats can be found in a diverse range of environments around the world, including the hypersaline setting of Guerrero Negro (Des Marais 1995), geothermal hot springs at Yellowstone National Park USA (Walter et al. 1976), the cold dry valleys of Antarctica (Jungblut et al. 2005), and alkaline sulphidic springs of Russia (Namsaraev et al. 2003). When the cumulative record of microbial mat activities becomes lithified the resulting geological structure is known as a stromatolite.

STROMATOLITES

Stromatolites are abundant in the Earth's rock record, and these structures, along with the microfossils, molecular biomarker molecules and mineral isotopic compositions contained within them, represent our earliest evidence of life on Earth (Schopf 2006). Stromatolites are not however in themselves living organisms, but rather biologically created sedimentary formations (McNamara and Awramik 1992). As alluded to above, stromatolites are basically microbial mats in which a rock-layer of sand or precipitated minerals is present, produced by the trapping and/ or precipitation activity of resident microorganisms. The biogenic origin of fossilized stromatolites has been keenly debated (Schopf et al. 2002, Brasier et al. 2002), following from earlier suggestions that stromatolites could result from purely abiotic processes (Grotzinger and Rothman 1996). However, very recent findings have shown the occurrence of multiple different morphologies within single sedimentary facies, strengthening the argument for a microbial origin for ancient stromatolites (Allwood et al. 2006, Altermann et al. 2006, Schopf 2006). There is as such now widespread and growing acceptance of the biogenicity of even the very oldest stromatolites, such as the 3.43 billion years old forms in the Pilbara Craton, Australia (Allwood et al. 2006, Awramik 2006).

Microfossils preserved within Archaean and Proterozoic stromatolites further increase the significance of these structures for our understanding of the evolution of life on Earth. Possible filamentous cyanobacterial forms are dated at 3.45 billion years old (Schopf, 1993, Hofmann et al. 1999) and indisputable cyanobacterial forms have been dated at 2 billion years old (Golubic and Hofmann 1976). The presence of possible phototrophs so early in Earth's history is intriguing, and a number of proposals regarding the origin of photosynthesis and evolution of photosynthetic bacteria have been made (Awramik 1992, Lazcano and Miller 1994, Olson 2006, Tomitani et al. 2006). Molecular fossils of biological lipids (biomarkers), and their sedimentary carbon isotope ratios can also be analyzed even in the absence of preserved microfossils, and have been used to indicate the presence of various microbial metabolisms in ancient samples (McKirdy 1976, Kakegawa and Nanri 2006). Further, minerals within fossil stromatolites can provide valuable information regarding ancient seawater chemistries, climates and other environmental parameters that influenced their deposition (Playford 1990, Grotzinger and Knoll 1999, van Kranendonk 2006).

The dominance of stromatolites in the early rock record and up until some 600 million years ago indicates that these microbial communities are a highly persistent mode of life and a significant stage in the Earth's evolution. Indeed the oxygenation of the Earth's atmosphere is attributed to the oxygenic photosynthesis and other gas production performed by Archaean stromatolite communities (Gebelein 1976, Hoelher et al. 2001). This major step in the evolution of the geosphere contributed to the further evolution of the biosphere, such as the diversification of eukaryotic organisms observed in the Cambrian explosion. It has even been suggested that some eukaryotes may have originated from a fusion of symbiotic partners in microbial mat and stromatolite ecosystems (Nisbet and Fowler 1999). The accompanying increased competition for habitats and limiting nutrients is thought to have been one of the reasons for the decline in stromatolite numbers and diversity. As higher plants evolved there was significant competition for sunlight and available nutrients, potentially affecting major photosynthetic bacteria such as the cyanobacteria. Furthermore the evolution of higher eukaryotes brought with it grazing pressures as microbial mats and stromatolites were a significant source of energy and carbon (McNamara and Awramik 1992). It is hypothesized that a combination of these effects gradually restricted stromatolite systems to protected or 'extreme' environments (such as elevated conditions of desiccation, temperature, or salinity) from which a proportion of metazoans and/or other predatory organisms were excluded (Awramik 1971, Riding 2006). It is somewhat ironic that although stromatolite systems may have played a crucial role in ultimately giving rise to eukaryotes, this step may have also brought about their decline in dominance.

In contrast to the abundance of fossilized stromatolites, extant stromatolites are rare. The best-studied modern stromatolites are those forming in open marine waters in Exuma Sound, Bahamas (Macintyre et al. 2000), and those from a hypersaline marine environment, that of Shark Bay on the western coast of Australia (Logan 1961). Extant stromatolites have also been discovered in a few other locations, including in a Tongan caldera (Kazmierczak and Kempe 2006), in Green Lake, New York (Eggleston and Dean 1976) and in Lake Clifton, Australia (Moore 1987), however here the concentration will primarily be on the microbiology of the living analogues in the Bahamas and, in particular, Shark Bay.

BAHAMAN STROMATOLITES

From studies on the Bahaman stromatolites and associated microbial mats a great deal about microbial mat lithification and stromatolite formation has been elucidated. Carbon, nitrogen, oxygen and sulphur cycles have been studied (Pinckney et al. 1995), and the cyanobacteria, aerobic heterotrophs, anoxygenic phototrophs, sulphate reducers, sulphide oxidizers and fermenters have been identified as the key metabolic groups within the stromatolite community (Dupraz and Visscher 2005). Certain species of sulphate-reducing bacteria have been identified by 16S rDNA sequencing, and their location within the community pinpointed using Fluorescent in situ Hybridization (Baumgartner et al. 2006).

The surface population of the stromatolites was found to cycle between several community types (Reid et al. 2000). Initially a pioneer community of the filamentous cyanobacterium *Schizothrix* sp. formed, which bound and trapped carbonate sand grains. Next a heterotrophic bacterial biofilm containing abundant extracellular polymeric substances (EPS) developed, during which time cyanobacterial EPS production and bacterial sulphate reduction regulated calcium carbonate precipitation (Visscher et al. 1998, Kawaguchi and Decho 2002, Decho et al. 2005). Finally a climax community developed, dominated by the endolithic cyanobacterium *Solentia* sp. which strengthened the lithified layer by fusing adjacent carbonate grains through microboring (Macintyre et al. 2000). Cycling between these community types lead to the development of a laminated lithified stromatolite structure. In addition to the study of stromatolite communities, the important question of why some mats lithify while others do not, was also addressed. Possible answers include the need for an uncoupling of the metabolism of the key functional groups by spatial and/or temporal separation, and the influence of physicochemical properties such as calcium carbonate saturation indices and iron availability (Visscher et al. 1998, Dupraz and Visscher 2005). From this current information on mat lithification and stromatolite formation processes in the Bahamas, it is possible to infer analogous processes that also may have occurred within Pre-Cambrian stromatolites.

SHARK BAY STROMATOLITES

The stromatolites of Hamelin Pool at Shark Bay (Australia) are well recognized as the best examples of living marine stromatolites (Figure 8.2). Hamelin Pool is the innermost basin of Shark Bay, a shallow hypersaline bay on the western coast of Australia. Across the mouth of

FIG. 8.2 Intertidal stromatolites in Hamelin Pool, Shark Bay, Western Australia.

Hamelin Pool is a sea-grass covered sandbank that restricts water flow into Hamelin Pool. Combined with the high evaporation rates, low rainfall, and the lack of freshwater input from the extremely arid land surrounding the bay, this has resulted in salinity that reaches at least twice that of normal seawater (Arp et al. 2001). As alluded to earlier, this increase in salinity in this area may result in a reduction in the level of grazing by higher eukaryotes.

Extant stromatolites of Hamelin Pool are relatively young; the oldest were radiocarbon dated to be 1000–1250 years old, with very slow growth rates of around 0.4 mm/year (Chivas et al. 1990). Early studies often reported only taxonomy and physiological properties of the dominant type of cyanobacteria found in Shark Bay stromatolites (Logan et al. 1974), however recent reports have begun to reveal an incredible microbial diversity of these formations, allowing researchers to make more specific and informed inferences about stromatolite functional complexity. One of these studies was the first polyphasic examination of the microbial communities of Shark Bay stromatolites, combining culture-dependent and culture-independent nucleic acid based methods (Burns et al. 2004). This study showed the stromatolite community was characterized by organisms of the cyanobacterial genera *Synechococcus, Xenococcus, Microcoleus, Leptolyngbya, Plectonema, Symploca, Cyanothece,*

Pleurocapsa, Prochloron and *Nostoc.* Several cyanobacteria isolated from the extant analogues in Shark Bay were also filamentous (Figure 8.3), a characteristic known to aid sediment trapping in stromatolites (Reid et al. 2000). Scale bar is 10 mm in each image. Extracellular polymeric substances, known to be produced by several cyanobacteria identified also provide stromatolite structure by providing an adhesive matrix to physically bind sediment, as well as providing nucleation sites that promote carbonate precipitation (Arp et al. 2001). Phylotypes related to *Synechococcus* were also observed in stromatolite 16S rDNA libraries (Burns et al. 2004), and the outermost cell surface of *Synechococcus* has been shown to have a role in fine-grain mineral formation (Schultze-Lam et al. 1992). Interestingly, this formation occurs with both live and dead cells, so such a process could be important in stromatolite lithification even after cell death. A Precambrian counterpart to *Synechococcus, Eosynechococcus,* has been described in ancient stromatolites (Hofmann

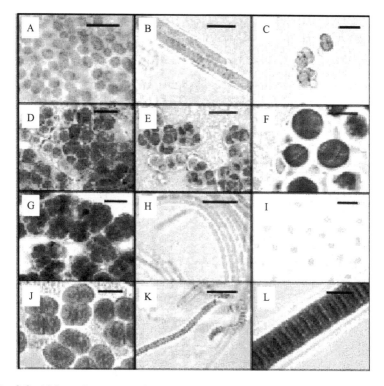

FIG. 8.3 Light microscopy of cyanobacterial isolates from Shark Bay mats/ stromatolites. A) *Euhalothece*, B) *Microcoleus*, C) *Pleurocapsa*, D) *Pleurocapsa*, E) *Stanieria*, F) *Chroococcidiopsis*, G) *Xenoccocus*, H) *Halomicronema*, I) *Halothece*, J) *Chroococcus*, K) *Spirulina*, L) *Lyngbya*.

1976), and it has been suggested that characteristics of Shark Bay microbial mats may allow the preservation of cyanobacterial cells as microfossils (López-Cortés 1999).

In addition to the considerable salinity and desiccation stress stromatolites must tolerate, high temperatures and the relatively thin atmospheric ozone layer contribute to a high ambient UV irradiance in Hamelin Pool (Palmisano et al. 1989). The sheath pigments known to be present in many of the surface-dwelling cyanobacteria identified (Burns et al. 2004) are likely to play a photoprotective role in screening deeper members of the community from physiological damage. Of further interest was the finding that a number of 16S rDNA clones clustered with the genus *Prochloron* (Burns et al. 2004). *Prochloron* is symbiotic with didemnid ascidians and to date there is no report of its existence as a free-living organism (Kühl and Larkum 2002). There are also no reports of ascidians in Shark Bay, and thus the discovery of potentially free-living *Prochloron* associated with stromatolites certainly warrants further study.

As described earlier, non-cyanobacteria microorgamisms are also prominent in these systems, and another recent study on the Shark Bay stromatolites revealed that the most prominent sequences were in fact novel proteobacteria (Papinaeu et al. 2005). It is quite interesting that in contrast to earlier notions, cyanobacteria do not appear to dominate in these stromatolites, though it is quite likely they still have major roles in primary production (Papinaeu et al. 2005). Both of the studies using molecular methods concluded that many of the stromatolite microorganisms were unique with no close relatives in the database (Burns et al. 2004, Papinaeu et al. 2005), and these microorganisms may also possess novel physiologies vital to the survival, integrity, and persistence of stromatolites. An example of this is a novel archaeon that was recently isolated and characterized from the Shark Bay stromatolites is named *Halococcus hamelinensis* (Goh et al. 2006). This organism has an oxidase negative phenotype, whereas all recognized *Halococcus* species are oxidase positive. Characterization of other novel microorganisms identified from extant stromatolites may reveal further unique metabolisms or pathways. Furthermore a plethora of archaea have been identified from living stromatolites for the first time (Burns et al. 2004, Papinaeu et al. 2005), and although their exact roles in stromatolite biology is unknown, they are likely to be important community members involved in nutrient cycling. As some Halobacterial species have been shown to be capable of fixing CO_2 (Javor 1988), it would be intriguing to ascertain whether halobacterial species in stromatolites are involved in the calcification process in addition to cyanobacteria.

These recent studies have focused on microbial community identification, as at the microscale most relevant to bacteria, a vital factor that affects stromatolite formation is the presence of other organisms. An immense diversity of prokaryotic life associated with modern stromatolites has been revealed, and combined with our knowledge on the prevailing environmental conditions reveals an intimate association between biotic and abiotic factors in stromatolite formation. Most studies on both the Shark Bay and Bahaman stromatolites also revealed that eukaroytes were scarce in these extant formations (Reid et al. 2000, Burns et al. 2004, Papinaeu et al. 2005), though one study has documented various flagellates in Shark Bay (Al-Qassab et al. 2002). This supports original notions that modern stromatolites appear to thrive in environments that exclude most higher organisms. In addition, the differences observed between microbial community compositions of extant stromatolites in different locations (Reid et al. 2000, Burns et al. 2004, Papinaeu et al. 2005), suggests that different stromatolite morphotypes will depend on the community present and therefore will be determined by it. For example specific microorganisms may trap sediments differently (Papinaeu et al. 2005), resulting in different stromatolite structures. The numerous novel or unknown microbes identified may play pivotal roles in stromatolite systems that are not yet understand, and may also be key biotechnological resources.

BIOTECHNOLOGICAL POTENTIAL OF STROMATOLITES

In addition to addressing important evolutionary questions that may shed light on how stromatolites have survived for such an immense period of time, living stromatolites have also been considered as a potential biotechnological resource. The search for new groups of microorganisms in this environment is also driven by the philosophy that unique biochemical pathways could be uncovered, leading to the discovery and production of novel secondary metabolites. It has even been suggested that the very first secondary metabolites may have been antibiotics produced over 3.5 billion years ago in microbial mats present on Early Earth (Cavalier-Smith 1992). In addition many organisms are known to produce bioactive secondary metabolites in response to environmental stresses (Baker et al. 1995), and the stromatolitic microorganisms thriving in the hypersaline reaches of Shark Bay may have developed similar defence mechanisms. The substantial biodiversity that has been observed in these stromatolites at the species level (Burns et al. 2004, Papinaeu et al. 2005), suggests there may also be significant molecular diversity in this environment with the potential to discover many more novel products of medical or industrial applications.

The cyanobacteria in particular are a potentially rich source of these compounds (Neilan et al. 1999, Burja et al. 2001). A large number of these bioactive compounds, commonly secondary metabolites, are produced non-ribosomally via large modular enzyme complexes known as non-ribosomal peptide synthetases (NRPS) and polyketide synthases (PKS). Numerous compounds produced by these complexes of medical importance have already been isolated, including those with antibacterial, antiviral, fungicide, immunosuppressive, enzyme inhibiting, and cytotoxic activity (Smith et al. 1994, Panda et al. 1998).

Preliminary studies of novel cyanobacteria isolated from extant stromatolites has revealed for the first time the genetic potential of stromatolite microorganisms to produce biometabolites (Burns et al. 2005). Several isolates revealed the presence of a number of NRPS and PKS genes, including the first report of *Leptolyngyba* sp. containing putative NRPS genes. Sequence analysis indicates that the enzymes encoded by these genes may be responsible for the production of different secondary metabolites, such as those with antibacterial or antifungal effects. Mass spectral analysis also allowed the putative identification of the cyclic peptides cyanopeptolin S and 21-bromo-oscillatoxin A (a protease inhibitor and a putative anticancer agent, respectively), in these stromatolite isolates. Preliminary work has also identified putative anti-proliferative (HeLa) and anti-protozoal activity in several isolates, and the latter activity in particular may also help stromatolite organisms survive potential grazing pressures, thus increasing our understanding of the resilience of these systems through evolutionary time. Future detailed studies of these isolates can involve characterizing the putative NRPS and PKS genes identified and screening isolates for bioactivities of potential novel value.

However, these techniques are not capable of studying the proportion of microorganisms present in living stromatolites that are unculturable. Whilst one of the easiest ways to obtain information of bioactive capability of microbes is through culture, obtaining representative organisms in culture can be extremely difficult, as many culture approaches recover only a small, distorted fraction of total bacteria (Amman et al. 1995, Spiegelman et al. 2005). Employing genetic-based techniques such as metagenomics to stromatolites is desirable to access these microorganisms, in order to identify potentially useful metabolites and further our understanding of the evolutionary significance of stromatolites. Comparative studies between a stromatolite metagenomic library and existing specific gene libraries can both complement and provide additional information to existing data on this environment. In particular, the ability to clone large sequences (~40 kb) results in entire gene clusters that can be studied, in order to enhance the likelihood of isolating gene clusters encoding biosynthetic pathways for secondary metabolites. Preliminary studies applying metagenomics to extant stromatolites are underway (unpublished data), and applying both sequence based and functional screens to these metagenomic libraries holds exciting potential.

Such screens for new drug leads are increasingly important due to recent developments of antibiotic resistant infectious microorganisms and cancerous tumours. Genetic screening methods and bioassays applied to stromatolites will not only identify unique chemistry but will provide a greater understanding of the diversity and ecology of this class of molecules in these ecosystems. Natural products in stromatolites may act as chemical defences, mediating competitive interactions, or alternatively may enhance favourable symbiotic relationships. Stromatolites are a potentially rich source of bioactive natural products, however while it is important to harness the chemical diversity generated, it is equally important to ensure that these ecosystems are conserved.

CONSERVATION

Apart from the potential basic and applied benefits of research into extant analogs of early life, the assessment of molecular biodiversity and functional complexity of stromatolites is intrinsic to the conservation of such fragile and unique ecosystems. Although these complex systems have survived for 80% of Earth's entire history they are still very susceptible to human impact. Understanding microbial biodiversity is fundamental to stromatolite conservation because loss of biodiversity, starting at the primary producer level, can indicate detrimental changes in the environment. Loss of the organisms that help create stromatolites would inevitably result in the loss of stromatolites. Population fluxes of resident microbial communities may act as an indicator of possible threats to living stromatolites. Preliminary studies of samples taken at different time points over several years have already suggested a loss of biodiversity over time (unpublished data), and this information is critical for the management and conservation of these biological resources. There is also already a documented increase in phosphate levels in waters harbouring the living stromatolites in Western Australia due to agricultural run-off (McNamara and Awramik 1992), and this could disrupt the state of equilibrium in the community and have severe consequences on stromatolite survival.

It is important that researchers continue to carefully monitor human activities around stromatolites to improve their management. As the study of stromatolites increases our understanding of long-term stability and change, from a local to a planetary scale, such structures should be protected wherever possible as natural science preserves. By understanding the biodiversity and the processes involved in their formation one may ensure that potential hazards to this rare biological system are readily detected and controlled, thus allowing their continued existence.

CONCLUSIONS AND FUTURE PERSPECTIVES

Stromatolites are excellent natural laboratories for the study of microbial ecosystems that may have shaped the biology of the Early Earth (Des Marais 2003). Although valuable, there is unfortunately limited information one can gain from studying fossil stromatolites, and in contrast, studying extant stromatolites provides one with a glimpse of what life may have been like on the Early Earth: the kinds of complex microbial interactions that occur, and what kind of metabolisms/physiologies may have been important in early microbial communities. Indeed evidence suggests Archaean life forms may have been relatively advanced (Altermann et al. 2006), and thus metabolic pathways observed in present stromatolites may have already been utilized by their ancient counterparts. Although culture-independent molecular analyses alone do not allow one to absolutely determine whether sequences represent active stromatolite organisms, the studies discussed in this chapter on extant stromatolites show that one can take advantage of the phylogenetic affinity with well-studied species to make predictions about the metabolic contributions of organisms identified. Researchers can now build constructively on this platform of microbial community analyses by targeting specific functional genes and enzyme activities, as well as conducting large-scale functional genomics studies, thereby furthering our knowledge on how individual and combined physiologies contribute to stromatolite systems. The close physical association of microorganisms in this setting may also facilitate horizontal gene transfer of adaptive and evolutionarily significant traits such as antibiotic resistance, which has implications both for organism evolution and the field of biotechnology.

Finally, the identification of stable biomarkers in stromatolites that are uniquely produced by microorganisms is an area of increasing interest. Characterizing the breadth of biomarkers (such as lipids, characteristic pigments, etc.) in ancient microbial systems such as microbial mats and stromatolites, has important implications in the field of astrobiology, with the exciting potential for using the knowledge gained in the rational search and detection of possible biosignatures of life outside Earth. Furthermore, communities of microbes in stromatolites are responsible for the production of important trace gases (such as photosynthetic oxygen), and the use of remote sensing to interpret infrared spectra may help one to identify biological signatures arising from life on distant planetary atmospheres (Des Marais 2002). The quest to understand early life on Earth and the prospects for life elsewhere addresses some of the most profound questions of humankind, and one of the extant analogues of early life, stromatolites, may be the key in providing these answers.

ACKNOWLEDGEMENTS

This work was supported by grants from the Australian Research Council.

REFERENCES

Allwood, A.C., M.R. Walter, B.S. Kamber, C.P. Marshall, and I.W. Burch. 2006. Stromatolite reef from the Early Archaean era of Australia. Nature 441: 714-718.

Al-Qassab, S., W.J. Lee, S. Murray, and D.J. Patterson. 2002. Flagellates from stromatolites and surrounding sediments in Shark Bay, Western Australia. Acta Protozol. 41: 91-144.

Altermann, W., A. Kazmierczak, A. Oren, and D.T. Wright. 2006. Cyanobacterial calcification and its rock-building potential during 3.5 billion years of Earth history. Geobiology 4: 147-166.

Amann, R.I., W. Ludwig, and K.H. Schleifer. 1995. Phylogenetic identification and in situ detection of individual microbial cells without cultivation. Microbiol. Rev. 59: 143-169.

Arp, G., A. Reimer, and J. Reitner. 2001. Photosynthesis-induced biofilm calcification and calcium concentrations in Phanerozoic oceans. Science 292: 1701-1704.

Awramik, S.M. 1971. Precambrian columnar stromatolite diversity: Reflection of metazoan appearance. Science 174: 825-827.

Awramik, S.M. 1992. The oldest records of photosynthesis. Photosynth. Res. 33: 75-89.

Awramik, S.M. 2006. Respect for stromatolites. Nature 441: 700-701.

Baker, J.T., R.P. Borris, and B. Carte. 1995. Natural product drug discovery and development: New perspectives on international collaboration. J. Nat. Prod. 58: 1325-1357.

Baumgartner, L.K., R.P. Reid, C. Dupraz, A.W. Decho, D.H. Buckley, J.R. Spear, K.M. Przekop, and P.T. Visscher. 2006. Sulfate reducing bacteria in microbial mats: Changing paradigms, new discoveries. Sediment. Geol. 185: 131-145.

Brasier, D.M., O.R. Green, and A.P. Jephcoat. 2002. Questioning the evidence of Earth's oldest fossils. Nature 416: 76-81.

Burja, A.M., B. Banaigs, and E. Abou-Mansour. 2001. Marine cyanobacteria—a prolific source of natural products. Tetrahedron 57: 9347-9377.

Burns, B.P., F. Goh, M. Allen, and B.A. Neilan. 2004. Microbial diversity of extant stromatolites in the hypersaline marine environment of Shark Bay, Australia. Environ. Microbiol. 6: 1096-1101.

Burns, B.P., A. Seifert, F. Goh, F. Pomati, and B.A. Neilan. 2005. Genetic potential for secondary metabolite production in stromatolite communities. FEMS Microbiol. Lett. 243: 293-301.

Cavalier-Smith, T. 1992. Origins of secondary metabolism. *In:* D.J. Chadwick and J. Whelan [eds.]. Secondary Metabolites: Their Function and Evolution. John Wiley, Chichester. pp. 67-87.

Chivas, A.R., T. Torgersen, and H.A. Polach. 1990. Growth rates and Holocene development of stromatolites from Shark Bay, Western Australia. Aust. J. Earth Sci. 37: 113-121.

Decho, A.W., P.T. Visscher, and R.P. Reid. 2005. Production and cycling of natural microbial exopolymers (EPS) within a marine stromatolite. Palaeogeogr. Palaecol. 219: 71-86.

Des Marais, D.J. 1995. The biogeochemistry of hypersaline microbial mats. Adv. Microb. Ecol. 14: 251-274.

Des Marais, D. 2003. Biogeochemistry of hypersaline microbial mats illustrates the dynamics of modern microbial ecosystems and the early evolution of the biospehere. Biol. Bull. 204: 160-167.

Des Marais, D.J, M.O. Harwit, and K.W. Jucks. 2002. Remote sensing of planetary properties and biosignatures of extrasolar terrestrial planets. Astrobiology 2: 153-181.

Dupraz, C. and P.T. Visscher. 2005. Microbial lithification in marine stromatolites and hypersaline mats. Trends. Microbiol. 13: 429-438.

Eggleston, J.R. and W.E. Dean. 1976. Freshwater stromatolitic bioherms in Green Lake, New York. *In:* M.R. Walter [ed.]. Stromatolites. Elsevier Scientific Publishing Company, Amsterdam. The Netherlands. pp. 479-488.

Gebelein, C.D. 1976. The effects of the physical, chemical and biological evolution of the Earth. *In:* M.R. Walter [ed.]. Stromatolites. Elsevier Scientific Publishing Company, Amsterdam, The Netherlands. pp. 499-515.

Goh, F., S. Leuko, M.A. Allen, B.A Neilan, and B.P. Burns. 2006. *Halococcus hamelinensis* sp. nov., a novel halophilic archaeon isolated from stromatolites in Shark Bay, Australia. Int. J. Syst. Evol. Microbiol. 56: 1323-1329.

Golubic, S. and H.J. Hofmann. 1976. Comparison of Holocene and mid-Precambrian Entophysalidaceae (Cyanophyta) in stromatolitic algal mats: Cell division and degradation. J. Paleontol. 50: 1074-1082.

Grotzinger, J.P. and D.H. Rothman. 1996. An abiotic model for stromatolite morphogenesis. Nature 383: 423-425.

Grotzinger, J.P. and A.H. Knoll. 1999. Stromatolites in Precambrian carbonates: Evolutionary mileposts or environmental dipsticks? Annu. Rev. Earth Planet. Sci. 27: 313-358.

Hoehler, T.M., B.M. Bebout, and D.J. Des Marais. 2001. The role of microbial mats in the production of reduced gases on the early Earth. Nature 412: 324-327.

Hofmann, H.J. 1976. Precambrian microflora, Belcher Island, Canada: Significance and systematics. J. Paleontol. 50: 1040-1073.

Hofmann, H.J., K. Grey, and A.H. Hickman. 1999. Origin of 3.45 Ga coniform stromatolites in Warrawoona Group, Western Australia. Geol. Soc. Am. Bull. 111: 1256-1262.

Javor, B.J. 1988. CO_2 fixation in halobacteria. Arch. Microbiol. 149: 433-440.

Jungblut, A-D., I. Hawes, D. Mountfort, B.P. Burns, and B.A. Neilan. 2005. Diversity within cyanobacterial mat communities in variable salinity meltwater ponds of McMurdo Ice Shelf, Antarctica. Environ. Microbiol. 7: 519-529.

Kakegawa, T. and H. Nanri. 2006. Sulfur and carbon isotope analyses of 2.7 Ga stromatolites, cherts and sandstones in the Jeerinah Formation, Western Australia. Precambrian Res. 148: 115-124.

Kawaguchi, T. and A.W. Decho. 2002. A laboratory investigation of cyanobacterial extracellular polymeric secretions (EPS) in influencing $CaCO_3$ polymorphism. J. Cryst. Growth 240: 230-235.

Kazmierczak, J. and S. Kempe. 2006. Genuine modern analogues of Precambrian stromatolites from Caldera lakes of Niuafo'ou Island, Tonga. Naturwissenschaften 93: 119-126.

Kruschel, C. and R. Castenholz. 1998. The effect of solar UV and visible irradiance on the vertical movements of cyanobacteria in microbial mats of hypersaline waters. FEMS Microbiol. Ecol. 27: 53-72.

Kühl, M. and A.W.D. Larkum. 2002. *In:* J. Seckbach [ed.]. Cellular Origin and Life in Extreme Habitats. Vol. 3. Kluwer, Dordrecht. pp. 273-290.

Lazcano, A. and S.L. Miller. 1994. How long did it take for life to begin and evolve to Cyanobacteria? J. Mol. Evol. 39: 546-554.

Logan, B.W. 1961. Cryptozoon and associate stromatolites from the Recent, Shark Bay, Western Australia. J. Geol. 69: 517-533.

Logan, B.W., P. Hoffman, and CD. Gebelein. 1974. Algal mats, cryptalgal fabrics, and structures, Hamelin Pool, Western Australia. American Association of Petroleum Geologists Memoir 22: 140-194.

López-Cortés, A. 1999. Paleobiological significance of hydrophobicity and adhesion of phototrophic bacteria from microbial mats. Precambrian Res. 96: 25-39.

Macintyre, I.G., L. Prufert-Bebout, and R.P Reid. 2000. The role of endolithic cyanobacteria in the formation of lithified laminae in Bahamian stromatolites. Sedimentology 47: 915-921.

McKirdy, D.M. 1976. Biochemical markers in stromatolites. *In:* M.R. Walter [ed.]. Stromatolites. Elsevier Scientific Publishing Company, Amsterdam, The Netherlands. pp. 163-191.

McNamara, K.J. and S.M. Awramik. 1992. Stromatolites: A key to understanding the early evolution of life. Sci. Progress Oxford 76: 345-364.

Moore, L.S. 1987. Water chemistry of the coastal saline lakes of the Clifton-Preston Lakeland System, South-western Australia, and its influence on stromatolite formation. Aust. J. Mar. Fresh. Res. 38: 647-660.

Namsaraev, Z.B., V.M. Gorlenko, and B.B. Namsaraev. 2003. The structure and biogeochemical activity of the phototrophic communities from the Bol'sherechenskii alkaline hot spring. Microbiol 72: 193-202. Translated from Mikrobiologiya 72: 228-238.

Neilan, B.A., E. Dittmann, L. Rouhiainen, and T. Boerner. 1999. Nonribosomal peptide synthesis and toxigenicity of cyanobacteria. J. Bacteriol. 181: 4089-4097.

Nisbet, E.G. and C.M.R. Fowler. 1999. Archaean metabolic evolution of microbial mats. Proc. R. Soc. Lond. 266: 2375-2382.

Olson, J.M. 2006. Photosynthesis in the Archean Era. Photosynth. Res. 88: 109-117.

Paerl, H.W., J.L. Pinckney, and T.F. Steppe. 2000. Cyanobacterial-bacterial mat consortia: Examining the functional unit of microbial survival and growth in extreme environments. Environ. Microbiol. 2: 11-26.

Palmisano, A.C., R.E. Summons, and S.E. Cronin. 1989. Lipophilic pigments from cyanobacterial (blue-green algal) and diatom mats in Hamelin Pool, Shark Bay, Western Australia. J. Phycol. 25: 655-661.

Panda, D., K. DeLuca, and D. Williams. 1998. Antiproliferative mechanism of action of cryptophycin-52: Kinetic stabilization of microtubule dynamics by high-affinity binding to microtubule ends. Proc. Natl. Acad. Sci. USA. 95: 9313-9318.

Papineau, D., J.J. Walker, S.J. Mojzsis, and N. Pace. 2005. Composition and structure of microbial communities from stromatolites of Hamelin Pool in Shark Bay, Western Australia. Appl. Environ. Microbiol. 71: 4822-4832.

Pinckney, J., H.W. Paerl, and R.P. Reid. 1995. Ecophysiology of stromatolitic microbial mats, Stocking Island, Exuma Cays, Bahamas. Microb. Ecol. 29: 19-37.

Playford, P.E. 1990. Geology of the Shark Bay area, Western Australia. *In:* P.F. Berry, S.D. Bradshaw, and B.R. Wilson [eds.]. Research in Shark Bay. Report of the France-Australe Bicentenary Expedition Committee. Western Australian Museum. pp.13-31.

Reid, R.P., P.T. Visscher, and A.W. Decho. 2000. The role of microbes in accretion, lamination and early lithification of modern marine stromatolites. Nature 406: 989-992.

Riding, R. 2006. Microbial carbonate abundance compared with fluctuations in metazoan diversity over geological time. Sediment. Geol. 185: 229-238.

Schopf, J.W. 1993. Microfossils of the Early Archean Apex Chert: New evidence of the antiquity of life. Science 260: 640-646.

Schopf, J.W. 2006. Fossil evidence of Archaean life. Philos. T. Roy. Soc. B. 361: 869-885.

Schopf, J.W., A.B. Kuryavtsev, and D.G. Agresti. 2002. Laser-Raman imagery of Earth's earliest fossils. Nature 416: 73-76.

Schultze-Lam S., G. Harauz, and T.J. Beveridge. 1992. Participation of a cyanobacterial S layer in fine-grain mineral formation. J. Bacteriol. 174: 7971-7981.

Skyring, G.W. and J. Bauld. 1990. Microbial mats in Australian coastal environments. *In:* K.C. Marshall [ed.]. Advances in Microbial Ecology 11. Plenum, New York. pp. 461-498.

Smith, C.D., X. Zhang, and S.L. Mooberry. 1994. Cryptophycin: A new antimicrotubule agent active against drug-resistant cells. Cancer Res. 54: 3779-3784.

Spiegelman, D., G. Whissell, and C.W. Greer. 2005. A survey of the methods for the characterization of microbial consortia and communities. Can. J. Microbiol. 51: 355-386.

Tomitani, A., A.H. Knoll, and C.M. Cavanaugh. 2006. The evolutionary diversification of cyanobacteria: Molecular-phylogenetic and paleontological perspectives. Proc. Natl. Acad. Sci. USA. 103: 5442-5447.

van Kranendonk, M.J. 2006. Volcanic degassing, hydrothermal circulation and the flourishing of early life on Earth: A review of the evidence from c. 3490-3240 Ma rocks of the Pilbara Supergroup, Pilbara Craton, Western Australia. Earth-Sci. Rev. 74: 197-240.

Visscher, P.T., R.P. Reid, and B.M. Bebout. 1998. Formation of lithified micritic laminae in modern marine stromatolites (Bahamas): The role of sulfur cycling. Am. Mineral. 83: 1482-1493.

Walter, M.R., J. Bauld, and T.D. Brock. 1976. Microbiology and morphogenesis of columnar stromatolites (*Conophyton, Vacerrilla*) from hot springs in Yellowstone National Park. *In:* M.R. Walter [ed.]. Stromatolites. Elsevier Scientific Publishing Company, Amsterdam, The Netherlands. pp. 273-310.

CHAPTER 9

Biodeterioration of Archaeological Monuments and Approach for Restoration

Preeti Bhatnagar[1*], Abdul Arif Khan[2], Sudhir K. Jain[3] and M.K. Rai[4]

INTRODUCTION

Cultural heritage objects are damaged by various agents like atmospheric agents, condensation or capillary humidity, temperature range, human action and microorganisms. A wide variety of organisms like Bacteria, Fungi, Algae and Plants etc. have been reported in the degradation of monuments (Realini et al. 1985). Microbes are more important than other agents. The intensity of damage is determined by the type and dimension of the microorganism involved, the kind of material used and level of pollution. Deteriorating agents can modify the composition and structure of stone, wood and metal used in artifact. In situ deterioration process is influenced by vital activity of living organisms. Although biodeterioration is an essential process in environmental cycling of matter but this may lead to loss of valuable cultural property. The monument alteration by living organisms is usually indicative of an advanced state of deterioration predetermined by chemical and physical

[1]Department of Microbiology, College of Life Sciences, Cancer Hospital & Research Institute, Gwalior (M.P.), INDIA, E-mail: bhatnagarpreeti01@gmail.com
[2]College of Pharmacy, King Saud University, Riyadh, Saudi Arabia, E-mail: abdularifkhan@gmail.com
[3]School of Studies in Microbiology, Vikram University, Ujjain (M.P.), INDIA, E-mail: sudhirkjain1@rediffmail.com
[4]Department of Biotechnology, SGB Amravati University, Amravati, Maharashtra, INDIA, E-mail: mkrai123@rediffmail.com
*Corresponding author

processes, which lead to increased surface area by formation of fissures that provide area for colonization of living organism. The growth process and vegetative development of organisms have a direct consequence on the growth of the microorganism.

To control biodeterioration problems, three factors must be considered i.e., the organism, environment and the surface to be treated. The alteration of any one of these can impact the growth of biodeteriogens and thereby biodeterioration. The knowledge of type species and their activity is valuable for adequate establishment of conservation strategy. The degree with which these organisms are involved in the stone decay process must be clarified. Conservative interventions can be used to control the biodeterioration process. The methodologies and products are to be selected according to substrate conditions and species to be treated; it will not cause negative interference with materials and with low environmental effect. In order to eradicate the biological growth, strategies are needed to evaluate the treatment of vital importance so that damaged monuments will be effectively and economically conserved.

Types of Substrate Subjected to Biodeterioration

Inorganic substrata

Inorganic materials such as glass, metals and stones are present in monument sites and are subjected to biodeterioration. Though metals and glass are present as an integral part of monuments and archaeological sites, they are less prone to biological attack and usually remain unaffected by microorganisms. A case is reported in which the ancient glass windows of a medieval church were covered by algae, while some corrosion pattern of buried metals can be attributed to an indirect action due to the surrounding microbial activity (Schaffer 1967).

Stones

Stone is damaged by the interaction between stone materials and natural or anthropogenic weathering factors. The decay of rocks and bare monumental stones is a complex process in which physical, chemical and biological mechanisms act together. Various kinds of organisms can grow using the mineral components of a stone and its superficial deposits. Besides that pollutants and their secondary reaction products also have an effect on surfaces; they form deposits of particles, black encrustation and leave secondary reaction products on stone surfaces. These collective actions lead to loss of cohesion with dwindling and scaling of stone and general weakening of the surface structural strength (Figure 9.1).

FIG. 9.1 Building of cultural heritage showing microbial biofilms

The resistance of minerals rocks used as building material to dissolution and destruction is in following order:

Quartz > Mica > K-plagioclase > Na-plagioclase > Ampholite > Oxalates > Dolomites > Pyroxene > Calcite > Feldspatoids > Olivine

Though some exceptions may also occur such as calcium oxalate which is a very insoluble material; it is formed in bulk, during patina layers on rocks generated by biofilm. Despite its relatively low solubility, it is rapidly transformed into calcite and even gypsum during biodeterioration and biotransformation of rock surfaces.

DAMAGE DIAGNOSIS

Integrated methodologies for investigations in laboratory and in the field are intended for safeguarding the architectural and archaeological patrimony. Precise diagnosis is necessary for characterization, interpretation, rating and prediction of weathering damages at monuments and the intervention for the preservation, restoration and the delivery of the archaeological and monumental heritage. Quantitative rating of damages is important for trustworthy damage diagnosis at stone monuments. Damage indices are used as tools for quantification and rating of stone damages and suitable for assessment and certification of preservation measures and for long-term survey and maintenance of stone monuments. For monument preservation, diagnosis is a systematic approach that includes 'anamnesis–diagnosis–therapeutical steps' (Figure 9.2). By anamnesis information, data and documents are obtained, compiled and evaluated in order to explain the characteristics of monument and history as prior interventions, preservation treatments; natural or anthropogenic impacts like earthquakes, fire, war, etc.; history of environmental conditions including air pollution. Damage indices for monuments are determined by weathering forms and damage categories. The damage index gives a reliable and reproducible quantification and rating of stone damages and contributes to monument preservation. Linear and progressive damage index have been used. The damage indices range from 0 to 5.0. According to the defined calculation modes, the linear damage index corresponds to average damage category, whereas the progressive damage index emphasizes proportion of higher damage categories. Differences between linear and progressive damage index increases as proportion of higher damage categories increases (Fitzner et al. 2002).

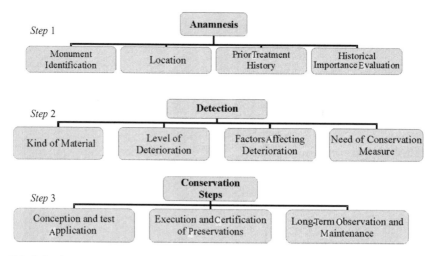

FIG. 9.2 Steps used for anamnesis, detection of damage and conservation steps

Linear damage index (DI_{lin})

$$= \frac{(A.0) + (B.1) + (C.2) + (D.3) + (E.4) + (F.5)}{100}$$

$$DI_{lin} = \frac{B + (C.2) + (D.3) + (E.4) + (F.5)}{100}$$

Progressive Damage Index (DI_{prog})

$$= \sqrt{\frac{(A.0) + (B.1) + (C.2) + (D.3) + (E.4) + (F.5)}{100}}$$

$$DI_{lin} = \sqrt{\frac{B + (C.2) + (D.3) + (E.4) + (F.5)}{100}}$$

A = Area (%) — damage category 0 B = Area (%) — damage category 1
C = Area (%) — damage category 2 D = Area (%) — damage category 3
E = Area (%) — damage category 4 F = Area (%) — damage category 5

DEFINITION OF DAMAGE CATEGORIES

Damage category 0 — no visible damages.
Damage category 1 — very slight damages.
Damage category 2 — slight damages.
Damage category 3 — moderate damages.
Damage category 4 — severe damages.
Damage category 5 — very severe damages.

Plaster, Mortars and Frescoes

These substrata are often colonized by the same organisms that can develop over stone substrata. However, the degree of extension will depend upon the site where these materials are placed in and whether subjected to outdoor or indoor exposure. As in case of frescoes, which are made up of inorganic material but are subjected to deterioration by organisms developing on organic substrata. Probably the restoration measures add organic materials to them and consequently subject them to attack by microorganisms developing on organic substrata (Clapp and Kenk 1963).

Organic substrata

Various artistic objects such as easel and panel paintings, wooden sculptures, library materials, prints and textiles are made up of natural organic materials. Due to their biodegradable nature all of them are usually confined to environments. The biological attack of these objects occurs under poor conservation conditions: high humidity level, soil contact, poor ventilation and rare maintenance operations.

Wood

Wood's moisture content usually above 20% makes it susceptible to microbial attack. Principally micro fungi are involved in wood decay while bacteria and actinomycetes are less important because of their high moisture requirement. They can develop over surface or within internal structures, through the production of exoenzymes leads to change in cell integrity. The activity of fungal strains mainly destroys the structural biological polymers. Insects use wood as a nutrient source, for shelter and egg deposit and are one of the most serious sources of damage for wooden objects kept in museums or indoor environments. Amongst *Coleoptera* or beetles, many species have larvae that destroy wood by living inside it and feeding on it; their infestations are detected by the small flight holes or by the formation of little bore dust deposits. Anobiidae insects are most frequently found in deterioration of furniture, sculpture and other wooden objects. High RH and moderate temperature favor their development. Unlike the Anobiidae, the powder post beetles (Lyctidae) even thrive in dry conditions and attack mainly sapwood. Longhorn beetles (Cerambycidae) can cause serious damage, mainly to structural timber such as roofs or floors (Kirk and Shimada 1985, Allsopp and Seal 1986, Bravery et al. 1987).

   Termites are also common in the tropics and subtropics and often hollow out the wood completely, leaving a thin outer shell of undamaged wood. Termite attack often remains undetected until the whole structure collapses. The range of omnivorous insects may cause incidental damage while scavenging for other food such as cockroaches (Dictyoptera). Waterlogged wood is more often attacked by microaerophilic and anaerobic heterotrophic microorganisms and, in sea sites, marine organisms. Bacterial degradation of wooden materials in sea water can occur in water or in sediment up to a maximum of 60 cm deep while fungal degradation take place in aerobic conditions in the sediment–water interface. Amongst bacteria, commonly cellulolytic aerobic (*Cytophaga, Cellvibrio, Cellfalcicula*) or anaerobic species (*Plectridium, Clostridium*) and common microorganisms such as strains of *Bacillus, Pseudomonas, Arthrobacter, Flavobacterium,* and *Spirillum* are involved (Florian 1988).

In sea water and fresh water environments, some strains of microalgae can also cause chemicophysical damage to immersed woods. Though alterations caused by these microorganisms are not too serious. In sea water, marine borer animals destruct waterlogged wood. They are widely distributed in salt water, prevalent in warm climates, and include different molluscan and crustacean species (Abbate 1967, Abbate et al. 1989).

Paper

Paper is chiefly composed of cellulose and other substances as lignin, hemicelluloses, pectins, waxes, tannins, proteins and mineral constituents. It is a good source of nourishment for heterotrophic microorganisms and organisms and thus more vulnerable to microbial attack (Feilden 1982). The high water content and the hygroscopicity of paper make it more sensitive to the biodegradation. Bacteria, microfungi and actinomycetes, either cellulolytic strains or non-cellulolytic ones with non-specific action can damage paper but, fungi are the most common. Fungi cause alterations on paper producing various kinds of stains, either round or irregular in shape and are colored red, violet, yellow, brown, black, etc., the color of pigments depend on the conditions of growth and the properties of paper (e.g., pH, presence of starch or gelatin sizing, presence of metal, etc.). Fungi can also lead to the discoloration of inks due to tannase, an enzyme that catalyzes the hydrolysis of gallotannate, produced by some strains of *Aspergillus* and *Penicillium* (Kowalik 1980, Strzelczyk 1981, Dhawan and Agrawal 1986). Bacteria attack papers less frequently, require RH higher than 85%. Different metabolic products of all microorganisms produce various organic acids (oxalic, fumaric, succinic, citric, etc.), conditioning the dynamics of bacterial and fungal growth in secondary attacks.

Foxing is a common chromatic alteration of paper which appears as rust colored marks of different shapes. Fungi can be the possible cause of foxing, although presence of heavy metals, like iron, might also favor this kind of alteration (Dhawan 1986, Arai et al. 1988).

Insects may also attack cellulose or can cause damage by utilizing fillers, glues, boards, textile fibers, leather or other constituent elements of paper materials (Kowalik 1980, Strzelczyk 1981).

PARCHMENT AND LEATHER

Parchment is used as writing support and is composed of collagen, keratin and elastin, and a minimal amount of albumin and globulin. The susceptibility of parchments depends on the raw material, its method of production and its conditions of preservation. Collagen can be hydrolyzed by *collagenases* produced by anaerobic bacteria of the genus *Clostridium*. Non-specific proteolytic enzymes produced by many bacteria (e.g. *Bacillus mesenthericus, Pseudomonas, Bacteroides* and *Sarcina* sp.*)* and some fungi of the genera *Aspergillus, Cladosporium, Fusarium, Ophiostoma, Penicillium, Scopulariopsis, Trichoderma* etc. can attack partially decomposed collagen in aerobic conditions. Biodeterioration is influenced by factors such as temperature, humidity, pH and UV. The microbial attack leads to variegated spots, white films and texts fading (Kowalik 1980, Feilden 1982).

Leather's chemical composition is similar to parchment and its susceptibility to biodeterioration and the involved species are almost the same. Under high humidity, bacteria attack untanned leather but not tanned leathers. Tanned leathers are subjected to attack by fungi usually, after tanning; leather has a pH of about 3 to 5, which is more suitable for fungal growth. Fungi that attack tanned leather often belong to lipolytic species and utilize the fats present in leather as a source of carbon. Proteinaceous materials like parchment and leather are also susceptible to attack by insects. Insect damage appears as superficial erosion, deep erosion, holes or loss of material (Varonina et al. 1981, Von Endt and Jessup 1986).

Composite material

Works of art are often made of a combination of different organic and inorganic materials, rather than just one. In general, the risk of biological attack is linked to the most susceptible component. The biodeterioration phenomena on composite materials are similar in their character and effect to those of the single component.

Paintings

Paintings are complex objects made up of a support (canvas, wood, paper or parchment), a preparation layer and a paint layer, the chemical composition of which varies according to the mode of painting, the kind of paints used (oil paints, distemper or watercolor), and the historical period, as in canvas paintings, the preparation is usually made with lime or gypsum with addition of animal or vegetal glue. On smooth surface, several layers of color are present, which consist of pigments mixed with binders of oil or distemper (egg or glue). In paintings on wooden supports, a similar multilayer structure is observed. Paintings on paper can be water colors, gouaches or pastels. The organic components in paintings provide a good source of nutrition to heterotrophic microorganisms and organisms. Biological attack occurs under favorable environmental conditions, and such conditions are often found in museum rooms, old churches or in deposits without any control of the humidity and temperature.

Usually, in paintings on canvas, the microbial attack generally starts from the reverse side, because the glue sizing increases the natural susceptibility of textiles. Subsequently, biodeteriogens go into the canvas reaching the back side of the paint layer, cause cracks and detachment, while the cellulose hydrolysis leads to differences of adhesion between the paint layer and the canvas itself. The presence of substances, such as sizing glue or lining paste used for different treatments of the support, can increase susceptibility (Makies 1981, 1984).

Biological attack on the paint layer is less frequent than on the support and depends on the nature of pigments. The casein and egg distemper, emulsion distemper and linseed oil are more susceptible to biological attack. In contrast, the presence of heavy metals in some pigments, such as lead, zinc or chromium can increase the resistance of the paint layer. Watercolors contain only a small amount of organic binder and are, therefore, as susceptible to microbial deterioration as pastels (Dhawan and Agrawal 1986). Genera of fungi frequently involved in deterioration of the paint layer include species of *Penicillium, Aspergillus* and *Phoma pigmentivora* which disintegrate distemper and oil binders, *Aureobasidium* decompose oil binders, *Geotrichum* develop on casein binders, *Mucor* and *Rhizopus* attack glue (Gallo 1985, Dhawan and Agrawal 1986).

Agents Involved in Deterioration and Destruction

In deterioration, several different agents act together and are interdependent. Possible causes have a very wide range and include meteorological events such as earthquakes, fires, floods as well as terrorism, vandalism, neglect, tourism, previous treatments, wind, rain, frost, temperature fluctuations, chemical attack, salt growth, pollution, biodeterioration, etc.

PHYSICAL AGENTS

Oxidation

Oxidation is a natural decay in air which affects all organic materials; presence of oxidizing pollutants, such as ozone, sulfur dioxide, nitrogen dioxide and cleaning fluids accelerate the process. Changes in temperature or humidity, especially rapid or repeated changes (cycling) of conditions between day and night or from season to season, lead to dimensional instability. Acid deterioration is a more common cause of decay in archival documents and results in a slight discoloration, which progresses in time from yellow to brown. Acidity, heat, moisture or light also accelerate the rate of decay. Temperature also affects the rate of decay. High temperatures increase the rate of chemical reactions; when linked to low relative humidity lead to embitterment of paper and the drying out of adhesives; when allied to high relative humidity encourages fungal growth, the migration of adhesives, oxidation and hydrolysis (Korpi et al. 1998).

Salts

Salt damage is principally attributable to two mechanisms: the crystallization of salts from solution; and the hydration of salts. A third way in which salts could cause damage is through thermal expansion. Any salt is capable of causing crystallization damage, while hydration damage can only be caused by them which exist in more than one hydration state. As sodium chloride, for example, is capable only of crystallization damage, whereas sodium sulfate causes both crystallization and hydration damage because it exists as anhydrous salt thenardite (Na_2SO_4) as well as in the form of the decahydrate (mirabilite) ($Na_2SO_4-10H_2O$). Salt damage does not occur only in an outdoor environment, where the stone is subjected to cycles of rainfall and subsequent drying. It can also ensue indoors, through the hygroscopic action of the salts (Price 1996).

Light

Light is another harmful factor to many artifacts (Brill 1980). Light can interact with chemicals and change their composition and structure. High levels of natural or artificial light, especially from the blue end of the spectrum (ultra-violet light), cause fading in inks and pigments (including dyes used in color photography) and through photo-oxidation may speed up the degradation of the organic materials used as supports for those inks and pigments e.g. the breakdown of lignin in paper. Light weakens the natural cellulose fibers in paper and textiles and causes them to become brittle (McGlinchy 1994). Even low level of light if maintained over a long time can generate problems.

The consequence of light on artifacts is cumulative; it also affects development of biodeteriogens on monument sites (Brill 1980). Light provides the energy for photosynthesis of autotrophic organisms or it may be required for or may stimulate sporulation in fungi and it can have deleterious or lethal effects (Nasim and James 1978). Light also influences pigment production in many fungi. Some normally pigmented fungi and mycobacterium fail to produce pigment or may affect the quantitative production of some pigments if grown in dark (McGlinchy 1994).

Atmospheric Pollution

Atmospheric pollution, especially sulfur dioxide and nitrous oxides, and many industrial processes may react with the materials of which documents are composed, e.g. to accelerate acid deterioration in paper or changes to photographic chemicals. With air pollution, soluble salts represent one of the most important causes of stone decay. The growth of salt crystals within the pores of a stone can generate stresses that are sufficient to overcome the stone's tensile strength and turn the stone to a powder. The deterioration of many of the world's greatest monuments can be attributed to salts. There are many ways in which stonework can become contaminated with salts (Werner 1981, Charola 1984).

Dust and Dirt

Dust and dirt may lead to disfigurement of documents or even lead to abrasion of the materials of which they are composed. In addition they may carry with them chemical pollutants or fungal spores which can attack the documents. Concrete dust may be highly alkaline, which is as damaging as high acidity.

Humidity

Moisture could affect the development of species and rate of deterioration. Moisture found in building envelope could originate from indoor, outdoor or from the materials and construction. In a mild and humid climate, rain is a more critical factor than moisture transfer from exfiltration. In the cold northern region, avoiding moisture accumulation from indoor moisture sources is of more importance (Desmarais 2000). Sporulation and survival of biological agents and growth is greatly affected by moisture, air humidity and microclimate. However, in buildings, water transport is strongly influenced by extensive mycelial spread thereby allowing the water movement in deeper surfaces. Besides that, moisture content of the substrate, temperature and time/exposure may often influence the effectiveness of applied conservation methods (Koestler and Salvadori 1996).

TEMPERATURE

Temperature affects the rate of all the processes occurring in microorganisms and it may determine the type of reproduction, the morphology of the organisms and also their nutritional requirements. The colonization of monument substrata is strongly influenced by temperature.

Conservation Measures

Applied conservation measures may also affect biological colonization. There is equally a need to ensure that there are no unforeseen consequences of multiple applications of maintenance coatings.

Biological agents

A biodeteriogen is an organism that causes biodeterioration. A wide variety of biodeteriogens contribute to the deterioration of objects of archaeological significance. The deterioration of stone in buildings and monuments through the action of biological organisms has been known over an extended period. These organisms can cause direct or indirect damage to archaeological sites. Living organisms contribute to the decay of stone and similar materials and, although their action is, generally, of somewhat less importance than certain other deleterious agencies which have been considered, their study presents numerous features of interest (Schaffer 1972). Moreover, the degree of overall damage that can be accredited specifically to biodeteriogens is a matter of controversy since most of biodeterioration occurs in concurrence with other physical and chemical decay processes (Table 9.1). Biodeterioration of monuments possibly will be classified broadly into three categories: biophysical, biochemical and aesthetic deterioration (Caneva et al. 1991, Jain et al. 1993). Fungi or bacterial cells also produce a wide range of microbial produced volatile organic compounds (MVOCs) as 3-methyl-1-butanol, 1-hexanol, 1-octen-3-ol, 2-heptanone, and 3-octanone that can permeate building materials and diffuse into the surrounding air (Pasanen et al. 1996). Both the microbial species and the growth substrate affect the MVOC profile. These compounds are indicators of biological growth in monument buildings.

TABLE 9.1 Factors influencing development of biological agents over monument sites

S.No.	Factors	Types
1	Climatic and environmental factors	Sun, shadow, rain, temperature
2	Inorganic and organic pollutants	C, N, S source
3	Conservation, treatments	Biocides/Surfactants
4	Surface bioreceptivity	Nature of material

BIORECEPTIVITY OF MATERIALS

Bioreceptivity is (Guillitte 1995) the capability of a material to be colonized by one or several groups of living organisms. Otherwise, the sheer occurrence of organisms on surfaces does not automatically lead to destructive action but in some cases could be considered: as only aesthetic detrimental appearance; seeming to be aesthetically pleasing or having a protective role against weather induced belligerence. Bioreceptivity depends upon the property of the material that allows the spread and establishment of flora and/or fauna. In case of stone surfaces it depends primarily on surface roughness, moisture content and chemical composition of the rocks.

The 'primary bioreceptivity' is initial potential of colonization of an intact stone. Then, over a period of time colonizing organisms and environmental factors, encourage 'secondary bioreceptivity'. The applied conservative treatments on stone influence a 'tertiary bioreceptivity'. In 'extrinsic bioreceptivity' the stone colonization is essentially due to the presence of settled matter not associated with the stone beneath (Warscheid et al. 1988, Guillitte and Dreesen 1995). Bioreceptivity index of a particular material can be used to determine the susceptibility of the material to biodeterioration. Suitable specifications are to be used to quantify microbial biomass and to characterize stony material.

TYPES OF CHANGES

The metabolic activities of stone colonizing organisms results in production of various organic, inorganic acids, chelating compounds extra cellular polymers; colored pigments etc. as their growing structures exert the mechanical pressure on stone. These physical and chemical changes influence different types of damage of stone surfaces as described below.

1. Physical (Abrasion, mechanical stress)
2. Chemical (Solubilization, new-reaction products)
3. Aesthetic (Colored patches or patinas and crusts)

a. Physical Changes

Physical damage is influenced by growth of organisms. It leads to the detachment of grains and particles, exfoliation, chipping and increased porosity of surface. The reactive surface area of substrate also increases. Due to the mechanical forces fractures of variable dimensions are developed.

b. Chemical Changes

These changes are developed due to pollutants as well as the microbial growth products. Lichens, bacteria and fungi produce various organic acids like oxalic acid and others secretary metabolites. These agents influence chemical etching around the penetration surface. Chemicals produced by algae, fungi and cyanobacteria may also lead to alkalinization of rock surface and dissolution and fragmentation of smaller grains (Figure 9.3).

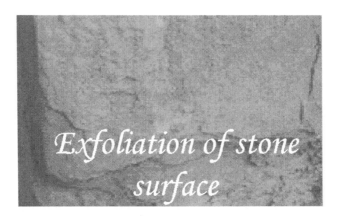

FIG. 9.3 Biochemical deterioration of stone surface

c. Aesthetic Changes

Fluorescence

These are white spots which are produced by dissolved salts when the water on the surface of the porous rocks migrates and evaporates. These salts can be derived from diverse sources such as floors, underground water, birds excreta or older treatments. Sulfate, chlorides, carbonate, nitrates are commonly present in the rocks. If these salts develop beneath the surface of rock monuments they are referred as subsurface fluorescence and if they are in the interior are known as crypto fluorescence.

Patina and Pinta

In Finnish language, pinta apparently means surface while patina means surface change are colored films formed over surfaces of works of art. 'Patina' was at first defined by Filippo Baldinucci, in his 1681 dictionary of art, as the time-dependent darkening of frescoes and oil paintings (Figure 9.4). In the 18th century, the term was commonly used for color changes that occurred on bronze and copper, caused by oxidation of the metals. Patina formation is a complex process, due to exchange of matter and energy between two open systems: the usually solid substrate and the surrounding environment. The natural restrictions of patina formation are the local extension of gradients and the mutual penetration depth of the components. Patina is a static or stabilizing component of the ageing process. Patina formation may lead to considerable increase of the substrate mass in the form of deposits, subaerial biofilms, microbial mats, microstromatolites, sinter, silica skins, crusts, black crusts, internal consolidation, cementation, etc. The mechanical and chemical effect of the mass boost may generate fissures and cracking, exfoliation, desquamation and other alteration processes (Sterflinger et al. 1996, Gorbushina et al. 1998). If the substrate is very soft and the atmospheric and biological deterioration processes can act rapidly a patina cannot be formed (Krumbein and Gorbushina 1996). These may be damaging for the aesthetic view caused by dreadfully dark biofilms as well as by physical and chemical penetration through biodeteriorative agents. In special cases massive biofilms may defend the rock surface better than no

FIG. 9.4 Patina over stone surface

growth and this factor may be increased occasionally through increased organic pollution (Dornieden et al. 2000). However, in recent times microbial pigments, newly deposited stained biogenic minerals and metal ion oxidation and reduction via microbes have been identified as the most obvious source of problems and a stained patina layer on rock surfaces (Sterflinger et al. 1999, Gorbushina et al. 2003, Krumbein, 2003).

Crusts

These are the hard outer layers which are formed due to superficial transformation of strengthened stone substrate. These crusts have determined morphology, strength and color. Physico-chemical nature of crusts is independent of nature of substrate. Black crusts are usually composed of crystals of gypsum (including re-crystallized calcite) mixed with atmospheric particles (spores, pollen, dust and different particulate matter, heavy hydrocarbons) entrapped in the mineral matrix. They originate from wet and dry deposition processes in which sulfuric acid, a sulfur dioxide oxidation product, attacks carbonic rocks, resulting in gypsum formation (Saiz–Jimenez 1993).

Mechanism of Biodeterioration

Deterioration of the archaeological site is influenced by physical, chemical and biological weathering, but most of the mechanisms are interdependent and involve multiple factors. Various bacteria, cyanobacteria, algae, fungi have been found to be involved in degradation. Humidity and sufficient nutrient promotes the colonization of organisms. Principally two underlying mechanisms are responsible for deterioration of archaeological sites. These are grouped as biogeochemical and biophysical.

The growth or movement of an organism or its parts exerts pressure on the surrounding surface material which leads to biophysical deterioration. Hyphae and extensive root systems penetrate deeply into the stone from pre-existing cracks or crevices, causing stresses that lead to physical damage of surrounding stone material. Periodic loosening of attachment devices during repeated wet and dry cycles may also lead to fragmentation. Biophysical processes make surfaces more prone to other deterioration factors, particularly biochemical.

Different types of chemicals produced by organisms lead to biogeochemical deterioration. Inorganic and organic acids produced by organisms have detrimental effect on monuments. These acids decompose stone minerals by producing salts and chelates. An increased volume of soluble salts and chelates may also result in formation of cracks. Insoluble salts and chelates may precipitate on the surface as crusts (Jain, Mishra and Singh, 1993). Carbon dioxide produced by aerobic organisms convert into carbonic acid in an aqueous environment, this acid can dissolve carbonates as marbles, limestone (Caneva, Nugari, Salvadori 1991). Biogenic acids can also affect silica with the glass resulting in a detrimental effect.

$$2SiO_2 + Na_2CO_3 + CaCO_3 \rightarrow Na_2SiO_3 + CaSiO_3 + 2CO_2$$

MICROBES INVOLVED IN DETERIORATION

Various microorganisms including bacteria, fungi, lichens have been reported to cause biodeterioration of stone monuments. The effect of certain organisms, such as bacteria, is still a matter of controversy, but the effect of others, such as the growth of ivy, is generally considered to be detrimental.

Bacteria

Owing to their simple and exacting nutritional requirements, bacteria can develop easily on different surfaces of architectural importance, especially where the surface has a high water content. Several species of bacteria have been detected and secluded from a monument either from its surface or at a depth down to several millimeters, using microbiological techniques (Table 9.2). They can be categorized in two main nutritional groups

- *a)* Chemolithoautotrophs
- *b)* Chemoorganotrophs
- *a)* Chemolithoautotrophs containing bacteria belong to the sulfur and nitrogen cycle.

Thiobacillus sp. (sulfur-oxidizing bacteria) is known to produce high levels of slightly soluble calcium sulfate normally detected on decayed stones (Voute 1969, May et al. 1993). These sulfates are possibly dissolved by rainwater or may be precipitated within the pores of the stone, where, on recrystallization, exert pressure on the pore walls due to an increase in volume, which is manifested as exfoliation. Calcareous stones are especially affected (Lepidi and Schippa 1973). The nitrifying bacteria oxidize reduced nitrogen compounds especially ammonia substrata (by *Nitrosomonas* sp.) derived from the atmosphere and further produce high levels of nitric acid from nitrites and has a direct action on the binding material of a concrete block. It may lead to stone dissolution, powdering, and formation of soluble nitrate salts that appear as efflorescence on the stone surface. Besides this nitrite also catalyzes oxidation of SO_2 to sulfite. However the high level of NOx existing in the polluted atmosphere could

TABLE 9.2 Bacteria found in monuments and objects of archaeological importance

Genera	Surface
Photoautotrophs	
Cyanobacteria	
Anabaena anomala	Andesite
Aphanothece pallida	Andesite
Aphanothece castagnei	Andesite
Aulosira fertilissima	Andesite
Chlorogloea sp.	Sandstone
Chroococcus sp.	Marble, sandstone
Entophysalis sp.	Andesite
Gloeocapsa magma	Andesite
Gloeocapsa pundata	Andesite
Gloeocapsa sp.	Sandstone
Gloeocapsa livida	Marble, sandstone
Gomphosphaeria ponina	Marble, sandstone
Heterohormogonium sp.	Andesite
Lyngbya cinerescens	Marble, sandstone, paintings
Nostoc sp.	Andesite
Oscillatoria annae	Marble, sandstone
Oscillatoria sp.	Limestone
Phormidium angustissimus	Marble, sandstone
Phormidium sp.	Limestone
Porphyrosiphon sp.	Sandstone
Schizothrix sp.	n.s.
Scytonema sp.	Andesite, limestone, sandstone
Stigonema sp.	Andesite
Synechocystis aquatilis	Andesite, marble, sandstone
Tolypothrix sp.	Andesite, sandstone
Chemoautotrophs	
Nitrifying bacteria	
Nitrosomonas sp.	Quartzite, soapstone
Nitrosococcus sp.	n.s.
Nitrobacter sp.	Quartzite, soapstone

Table 9.2 Contd.

Table 9.2 Contd.

Sulfur-oxidizing bacteria	
Thiobacillus sp.	Sandstone, andesite
Chemoheterotrophs	
Arthrobacter	Paintings, wood
Bacillus sp.	Limestone, wood
Desulfovibrio sp.	Sandstone
Pseudomonas sp.	Stone, paintings, wood
Micrococcus sp.	Limestone
Staphylococcus sp.	Limestone
Sarcina sp.	Paintings
Actinomycetes	
Nocardia sp.	n.s.
Micropolyspora sp.	n.s.
Micromonospora sp.	n.s.
Microellobosporium sp.	n.s.
Streptomyces sp.	Sandstone, paintings

n.s.= not specified

have a prevalent role in stone decay processes. Autotrophic nitrifying bacteria oxidize ammonia to nitrite and nitrate ions, which may result in other autotrophic bacteria contributing to dissolution of cations from the stone and surface staining (Wolters et al. 1988). The chemoorganotrophs bacteria, also has the capacity to produce several organic acids that can solubilize the mineral components of stones. Among the wide range of heterotrophs isolated, most could be assigned to the genera *Flavobacterium* and *Pseudomonas*. Heterotrophic bacteria deteriorate by the evolution of biogenic acids (some of which have chelating abilities) that may cause stone dissolution through mobilization of cations such as Ca^{2+}, Fe^{3+}, Mn^{+2}, Al^{+3}, and Si^4 (Wee and Lee 1980). Some heterotrophic bacteria have also been associated with the discoloration of stone surfaces.

Cyanobacteria causes aesthetic damage to monuments by developing various colored microbial films on their surfaces. These microbial films might have significant amounts of adsorbed inorganic materials ensuing from the substrata and detritus. They enhance the local water retention capacity of the stone by dissolving the nearby material, which promotes further increase in their population. Their slimy surfaces facilitate adherence of airborne particles of dust, pollen, oil, and coal ash, thereby giving rise to hard crusts and patinas that are difficult to remove

(Tiano 1987). Their respiration and photosynthesis produce acids as by-products which cause biochemical deterioration of the stone by the etching of mineral components and dissolution of binding minerals, especially in carbonates. Crystal coherence decreases in areas dominated by cyanobacteria, and rainwater detaches the crystals, leading to pitting (Caneva et al. 1994). Biophysical deterioration of the stone may also occur. This may also favor the growth of heterotrophic organisms such as fungi or other bacteria with higher destructive potential (Andreoli et al. 1988).

Fungi

They are present in the atmosphere in an inert state, in conditions such as high temperature and humidity. They can easily develop on outdoor objects and monuments. There are around 50,000 fungal species including molds that are found on decaying vegetation as well as larger fungi such as mushrooms, mildews, smuts, rusts and many other plant diseases causing fungi are also present. Three different decay fungi types are classified and also found in buildings — brown rot, soft rot and white rot. Most of the brown rot and white rot fungi belong to *Basidiomycetes*, but some belong to *Ascomycetes*. Most of the soft rot fungi belong to *Ascomycetes* and *Fungi imperfecti*, e.g. *Chaetomium globosum* and *Phialophora hoffmannii*. Brown rot is the most common decay in buildings having excessive moisture load. Dry rot fungi *Serpula lacrymans* causes the most serious damage in buildings in temperate climates. It is able to transport water through mycelial strands from the source of moisture into dry wood and has been reported to cause destructive damage in buildings and in different materials in temperate climate regions (Viitanen et al. 2000) (Table 9.3).

They damage, by feeding on paper and other organic components of materials, weakening them and often leaving disfiguring stains, which may destroy information. The damage caused by fungi in buildings varies depending on the type of fungal attack. Fungi are chemoheterotrophic organisms, thus, lack the ability to manufacture their own food by using the energy of sunlight (Caneva et al. 1991). Hence, they cannot live on stone, even if it is permanently wet, unless some organic food is present. The residuals of algae and bacteria or their dead cells, decaying leaves, and bird droppings provide such food sources (Sharma et al. 1985). Fungi can degrade stone chemically (Griffin et al. 1991) as well as mechanically (Bassi et al. 1984). Protruding fungal hyphae into decayed stone leads to the biophysical deterioration, by burrowing into stone. While some endolithic fungi produce pitting through a chemical action. Mold fungi do not significantly affect the

TABLE 9.3 Fungi found in monuments and objects of archaeological importance

Genera	Surface
Acremonium sp.	Stone, wood, marble, sandstone, paintings
Aspergillus elegans	Andesite, paintings
Aspergillus flavus	Andesite, limestone, marble, sandstone
Aspergillus nidulans	Marble, sandstone
Aspergillus niger	Limestone, marble, sandstone
Aspergillus versicolor	Marble, limestone
Aureobasidium sp.	Marble, limestone
Blastomyces dermatitis	not specified
Beauveria alba	Paintings
Candida albicans	Basalt
Cephalosporium sp.	Limestone
Cladosporium sp.	Soapstone, quartzite, sandstone, paintings, Andesite
Cunninghamella echinulata	Andesite
Curvularia sp.	Andesite, marble, sandstone
Chaetomium sp.	Paintings
Coniophora puteana	Stone, wood
Fusarium roseum	Andesite, limestone, marble, wood, sandstone
Fibroporia spp.	Stone, wood
Gloeophyllum trabeum	Stone, wood
Gliodadium virens	Andesite
Humicola grisea	Marble, sandstone
Lipomyces neoformans	Stone, sandstone
Oidiodendron	Stone
Macrophoma sp.	Marble, sandstone
Monilia sp.	Limestone
Penicillium multicolor	Andesite, paintings
Penicillium crustosum	Soapstone, quartzite, paintings
Penicillium frequentans	Marble, limestone
Penicillium glabrum	Soapstone, quartzite
Penicillium liladnum	Andesite
Penicillium notatum	Marble, sandstone
Phoma sp.	Stone, wood
Rhizopus arrhizus	Andesite
Stachybotrus sp.	Stone, wood
Serpula lacrymans	Stone, wood
Trichoderma sp.	Stone, wood

strength of materials. The biochemical action of fungi is more important than mechanical degradation. The filamentous fungi deteriorate marble, limestone, granite, and basalt through the action of excreted acids as oxalic and citric acids (May et al. 1993). The acids produced by fungi function as chelating agents that leach metallic cations, such as calcium, iron, or magnesium from the stone surface (Caneva and Salvadori 1988). Oxalic acid also causes extensive corrosion of primary minerals and the complete dissolution of ferruginous minerals through the formation of iron oxalates and silica gels. It is reported that basic rocks are more susceptible to fungal attack than acidic rocks. It has also been shown in the laboratory that fungi such as *Aspergillus niger* were able to solubilize powdered stone and chelate various minerals in a rich glucose medium because they produce organic acids such as gluconic, citric, and oxalic acids (Eckhardt 1985). Fungi also oxidize manganese, causing staining in stone. There have also been reports about similar effects of fungi on stone monuments. Fungi may also cause powdering of stone surfaces. Some black fungi also produce small spherical particles that resemble fly ash (Diakumaku et al. 1996). Microflora may also be capable of producing sulfates.

Mold fungi cause several problems in buildings: stains, discoloration, odor and health problems. Also in the sound buildings mold growth can be found on facades, attics and crawlspaces, which are typical structures exposed to high humidity conditions and attack by mold fungi. These conditions have been verified in many other studies (Burford et al. 2003, Bhatnagar 2008) (Figure 9.5).

Algae

The algal growth on stone surfaces is influenced by dampness, warmth, light and inorganic nutrients, particularly calcium and magnesium (Riederer 1981, Jain et al. 1993). Many algae show a marked sensitivity to pH on the surface, preferring acidic surfaces, but for some this value is not growth limiting. Algae can cause disfigurement and damage to stone monuments (Hyvert 1972, Wee and Lee 1980). Algae can cause direct as well as indirect damage by supporting growth of more corrosive biodeteriogens (Garg et al. 1988). Their growth on surface leads to aesthetic damage which can be seen in the form of various patina and stains of different colors. Depending on the environment, the color of patina may vary as in well-lit and relatively dry environments, patinas on stone surfaces are thin, tough, sometimes green and very often gray or black, while in poorly lit and damp places (interiors of monuments, walls of caves), thick, gelatinous patinas are formed of various colors such as green, yellow, orange, violet and red. Algae may also cause biochemical

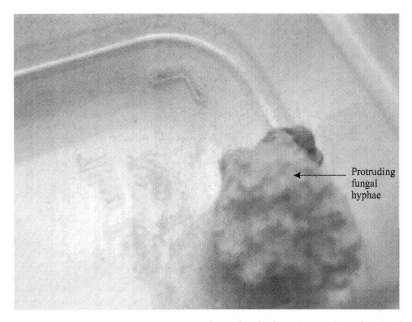

FIG. 9.5 Fungal hyphae growing on the surface of re-hydrated stone in moist chamber method

deterioration (Griffin et al. 1991). These acids either dissolve stone constituents or increase their solubility in water and stimulate migration of salts in stone, causing powdering of its surface. The change in solubility of stone constituents altered properties of the stone, as coefficient of thermal expansion, can add to the sensitivity of stone to physical processes of deterioration. Two major groups of algae— the chlorophytes and bacillariophytes or diatoms—have been isolated from stone monuments in tropical regions (Table 9.4). Among the green algae and cyanobacteria, 'sticky' biotypes were predominant (Caneva et al. 1992, Urzì and de Leo 2001).

Lichens

Together with cyanobacteria they play vital role as pioneer organisms in colonizing rocks. Three types of lichens—foliose, fruticose and crustose—have been isolated in tropical regions (Hale 1980). They can be epilithic (living over stone) or endolithic (living entirely beneath a stone surface) (Table 9.5). They lead to chemical and biophysical damage (Garg et al. 1988, Seaward et al. 1989, Monte 1991, Singh and Sinha 1993). Lichens produce carbon dioxide during respiration which is further transformed within their thallus into carbonic acid. Owing to their slow

TABLE 9.4 Algae on monuments and objects of archaeological importance

Algae	Surface
Achnautes lanceolata	Andesite
Caloneis bacillum	Andesite
Chlorella sp.	Paintings
Chlorococcum humicola	Andesite
Chlosarcinopsis sp.	Stone
Chroococcidiopsis sp.	Stone
Cosmarium sp.	Stone
Cymbella ventricosa	Andesite
Dermococcus sp.	Limestone
Gongrosira sp.	Andesite
Navicula sp.	Andesite, limestone
Nitzschia frustulum	Andesite
Oocystis sp.	Andesite
Pinnularia sp.	Andesite
Pleurastrum sp.	Stone
Pleurococcus sp.	Sandstone, marble, limestone
Pseudopleurococcus sp.	Paintings, stones
Pseudococcomyxa sp.	Paintings
Stichococcus sp.	Stone
Tabellaria sp.	Stone
Trentopohlia sp.	Limestone, andesite
Ulothrix sp.	Stone

metabolic activity, damage by carbonic acid present within their thallus may become significant only after a considerable period of time. The organic acids produced by lichens attack on minerals as calcium, magnesium and iron, as well as silicate minerals such as mica and orthoclase which makes the stone surface appear to be honeycombed with etch pits (Jones and Wilson 1985). Biophysical stone degradation by lichens is primarily due to penetration of the attachment devices of the thallus into the pores, pre-existing cracks and fissures in the stone (Hueck-van der Plas 1968, Riederer 1984). They can cause biopitting as well as weathering (Del Monte 1991, Saiz-Jimenez 1994). Effects of lichen on stone vary with the stone's chemical composition and crystallographic structure. Usually carbonate and ferromagnesium silicate minerals weather easily and show characteristic dissolution features. Olivine grains show deeply penetrating etch pits, while the surface of augite

TABLE 9.5 Lichens found on monuments and objects of archaeological importance

Lichen	Surface
Fruticose	
Roccella fuciformis	Andesite
Roccella montagnie	Granite
Foliose	
Candellaria sp.	Sandstone, limestone
Coccocarpia cronia	Limestone
Collema sp.	Limestone
Dirinaria sp.	Sandstone, limestone
Heterodermia sp.	Sandstone, limestone
Laboria pulmonia	Andesite
Leptogium sp.	Limestone
Parmelia sp.	Sandstone, andesite, granite
Parmeliella pannosa	Limestone
Parmelina minarum	Limestone
Parmotrema crinitum	Limestone
Parmotrema sp.	Limestone
Peltula sp.	Sandstone, lime plaster
Phylliscum sp.	Lime plaster
Physcia sorediosa	Limestone
Pyxine sp.	Sandstone
Sticta weigelii	Limestone
Usnea rubicunda	Limestone
Xanthoparmelia subramigera	Limestone
Crustose	
Caloplaca sp.	Sedimentary rock, limestone
Candelariella sp.	Limestone
Chiodectron sp.	Limestone
Diploschistes sp.	Sandstone
Endocarpon fusitum	Andesite, limestone, lime plaster
Ephebe pubescens	Andesite
Lecanora sp.	Limestone
Leptotrema santense	Limestone
Peltigera malacea	Andesite
Phyllopsora corallina	Limestone
Placynthium nigrum	Andesite
Porina fafinea	Andesite
Septrotrema pseudoferenula	Andesite
Thermucis velutina	Andesite
Verrucaria rupestris	Andesite

grains show deep, cross-cutting, trenchlike features. Feldspars are typically more resistant to lichen activity, though calcium-rich feldspar might be easily reduced. Quartz does not seem to have been affected much by lichens, but it is not completely invulnerable. (Hallbaur and Jahns 1977, Jones et al. 1981).

The fungal part of the lichen produces oxalic acid in addition to other organic acids with the age of the lichen acid accumulation increasing and more prevalent in calcium-loving species. Calcium carbonate of the stone is slightly soluble in water and is attacked by oxalic acid. As a result many lichens penetrate into the stone material and become endolithic (Syers and Iskandar 1973, Singh and Sinha 1993). Crystals of calcium oxalate monohydrate, calcium oxalate dihydrate, magnesium oxalate dihydrate, and manganese oxalate dihydrate have been identified at the stone-lichen interface (Salvadori and Zitelli 1981, Purvis 1984, Jones and Pemberton 1987).

MOSSES AND LIVERWORTS

Mosses and liverworts chiefly deteriorate stone aesthetically (Table 9.6). Pre-existing humus deposits are required for their development, which may result in accumulation of dead algae. As mosses die, the humus deposits extend, and thereby supporting the growth of more destructive higher plants. Mosses and liverworts may also cause some degree of biochemical disintegration of stone surfaces (Saiz-Jimenez 1994) owing to higher acidity of their rhizoids and have an elevated ability of extracting mineral cations from the stones (Bech-Anderson 1986, Saiz-Jimenez 1994). They also produce carbonic acids as a result of cellular respiration processes that are harmful to stone over an extended period of time. Though, mechanical action is less threatening (Shah and Shah 1992-93, Jain et al. 1993) because these organisms possess rhizoids rather than real roots. The presence of clay in the stone favors their growth (Hyvert 1972). Although, rhizoids have capability to penetrate stone.

HIGHER PLANTS

Plants are photoautotrophic organisms which grow in pre-existing fissures or cracks on stone surfaces, thus increasing the fissure size and cracks and decrease the cohesion between stones (Figure 9.6, Table 9.7). The damage of monuments by higher plants is rather complex and occur by both physical and chemical processes. The biophysical decay is mostly due to the growth and radial thickening of the roots of plants in the stone, which results in an increasing pressure on surrounding areas of the

TABLE 9.6 Mosses and liverworts found on monuments and objects of archaeological importance

Mosses and liverworts	Surface
Aongstroemia orientalis	Andesite
Barbula sp.	Andesite, limestone
Bryum coronatum	Stone, limestone
Calymperes sp.	Limestone
Ectropothecium monumentum	Andesite
Eusomolejeunea clausa	Limestone
Frullania sp.	Limestone
Groutiella schlumbergeri	Limestone
Haplozia javanica	Andesite
Hyalophila involuta	Limestone
Lejeunea sp.	Limestone
Marchantia chenopoda	Limestone
Mastigolejeunea auriculata	Limestone
Neckeropsis undulata	Limestone
Neohyophila sprengelii	Limestone
Octoblepharum albidum	Limestone
Papillaria nigrescens	Limestone
Plagiochila distinctifolia	Limestone
Rhacomitrium tomentosum	Limestone
Sematophyllum caespitosum	Limestone
Stereophyllum cultilliforme	Limestone
Weisia jamaicensis	Limestone

masonry. Woody species and trees are more harmful than herbaceous plants (Riederer 1981, Shah and Shah 1992-93, Mishra et al. 1995). This could result in collapse, detachment and damage of stone monuments (Gill and Bolt 1955, Winkler 1975, Mishra et al. 1995). Some fast-growing species of trees may lower the average moisture content of surrounding clay soil and thus cause sufficient shrinkage to damage the foundation of nearby structures. Though, damage in this way does not significantly affect the stability of monuments (Mishra et al. 1995).

The root exudates as carbohydrates, amino acids, amides, tartaric, oxalic, and citric acids are chiefly responsible for biochemical deterioration (Caneva and Altieri 1988). Biodecay of stone is also influenced by the carbonic acid produced during cellular respiration processes and can attack mineral particles (Honeyborne 1990, Mishra et al. 1995).

FIG. 9.6 Plant growth on a monument site

TABLE 9.7 Plants found on monuments and objects of archaeological importance

Plant	Surface
Acacia arabica	Stone
Adiantum sp.	Andesite
Agropyron repens	Stone
Albizia lebbeck	Stone
Argemone mexicana	Stone
Azadirachta indica	Stone
Boerhavia diffusa	Stone
Calotropis procera	Stone
Canscora sp.	Stone
Capparis sp.	Sandstone
Cassia occidentalis	Stone
Catharanthus roseus	Stone
Chloris barbata	Stone
Coccinia indica	Stone
Commelina bengalensis	Stone
Convolvulus sp.	Stone

Table 9.7 Contd.

Table 9.7 Contd.

Croton bonplandianum	Stone
Cynodon dactylon	Stone
Cyperus sp.	Stone
Dactylodenium aegyptiacum	Stone
Dalbergia sisso	Stone
Datura sp.	Stone
Digitaria adscendens	Stone
Dryopteris sp.	Stone
Eclipta alba	Stone
Euphorbia sp.	Andesite, stone
Ficus sp.	Stone
Fleurya interrupta	Stone
Heliotropium indicum	Stone
Holoptelea integrifolia	Stone
Imperata cylindrica	Stone
Kickxia incana	Stone
Leucas biflora	Stone
Lidenbergia indica	Stone
Lycopodium sp.	Andesite
Mimosa pudica	Stone
Nepeta hindostana	Stone
Nephrolepsis sp.	Andesite
Notochlaena sp.	Andesite
Ophioglossum sp.	Andesite
Oxalis sp.	Stone
Physalis minima	Stone
Piperomia sp.	Andesite
Pityrogramma sp.	Andesite
Saccharum munja	Stone
Sida cardifolia	Stone
Solanum nigrum	Stone
Trachyspermum stidocarpum	Stone
Trichodesma amplexicaule	Stone
Tridax prostala	Stone
Woodfordia fruitcoae	Stone
Zizyphus jujuba	Stone

ANIMALS

Animals are also involved in damage to objects of archaeological importance. Insects feed upon the organic components of documentary materials: as paper or on products as certain adhesives, parchments, leathers. Other animal pests such as rodents may also damage documents by feeding upon them, using them as nesting materials or merely fouling them. Birds also may foul documents. Animal and bird excrement is not only unpleasant and corrosive, but provides food for fungi, microorganisms and insects which may also attack the documents. Bats can also infest monument surfaces, leading to their unpleasant appearance and also increasing organic substrate (Figure 9.7). Arthropod communities on stone build support for algal and lichenic growth.

TABLE 9.8 Animals found on monuments and objects of archaeological importance

Animal	Surface
Balaustium murorum	Stone
Cerobasis lucorum	Stone
Mason bee (Hymenoptera)	Stone, wood
Diptera	Marble
Cerambycidae (Coleoptera)	Wood, stone
Termites	Wood, stone, paper
Cockroaches	Paper
Furniture beetles	Paper, wood
Silverfish	Paper
Molluscs	Submerged monuments
Polychaetes	Submerged monuments
Pigeons	Stone
Crows	Stone
Bats	Painted surfaces, stone monuments

FIG. 9.7 (A) Accumulated bird excreta over stone surface at Gwalior Fort (India) (B) Bats infestation over stone surface at Gwalior Fort (India)

Human Factors

Human factors also influence deterioration; records may suffer from neglect, improper handling and careless handling that can lead to or accelerate damage and deterioration. Vandalism involves tearing or cutting, writing on or marking individual documents; disordering assemblies, or even arson. Besides that tourist affect the integrity by damaging sites (Figure 9.8).

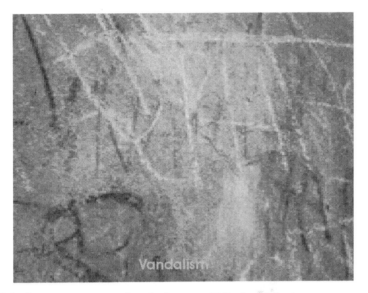

FIG. 9.8 Damage to stone surface by human activity (vandalism)

CONSERVATION METHODS

Once an artifact is created, chemical reactions with environmental factors commence. With time each will affect and threaten the existence of the artifact. The conservation of monuments with respect to biological causes of change and decay will have to be carried out. The conservation process requires the expertize of the art historian and scientist, as well as the experience of the conservator. Systematic analysis based on scientific results and the knowledgeable judgments of the people involved provide an efficient way to carry out the restoration process. Although modern conservation is executed by skilled artists, it combines art with science which emphasizes the necessity of keeping the original intention of the artifact intact and goes a step further than restoration by applying preventive measures to stop potential decay and preserving the integrity of the artifact (Ember 1984, Russell 1985). In some cases, conservation efforts cannot stop the deterioration process or reverse the ensuing damage, but they can help slow the rate of destruction to ensure a longer existence for the artifact. For solving biodeterioration problems, three factors must be considered: the organism, environment and surface. The alteration of any one of these can impact the growth of biodeteriogens and thereby biodeterioration. The basic principles to be followed in conservation (Batchelor et al. 1984, Ember 1984, Bell 1997) are:

- Minimum necessary intervention for the site's continued existence should be performed; new materials must be scientifically tested prior to application.
- Only a minimum loss of the existing material is acceptable; the 'six-inch six-foot' rule should be followed. Any damage repair should be noticeable to an observer six inches away but not six feet way. The restorer should try to blend or tone with the original. Any intervention should be reversible.
- New work should be evidently differentiated from the previous one. The identity, the original intent, and the style of the original composition should be respected. Restorations on artifacts must be documented.

The regular maintenance of cultural heritage objects, both recently restored and non-restored, is necessary in the preservation and to enhance durability of objects. Regular survey and assessment of the objects is the most important activity for safeguarding the cultural heritage. However even when following the best possible maintenance procedures, deterioration of the exposed building materials, including the newly used conservatives, will still unavoidably take place due to the natural corrosive factors such as rain, wind, freezing/thawing cycles and vandalism. The efficacy of the treatments depend on the methods and products chosen, but new growth invariably reoccurs if environmental conditions promoting biological growth are not modified. Therefore the use of conservative materials and intervention into objects should be considered only as minimum necessity for the object's survival and not as an action for providing long-term self-maintenance of the object.

Preventive as well as remedial methods can be used for control and eradication of microorganisms on stone monuments (Ashurst and Ashurst 1988, Ortega-Calvo et al. 1993). Remedial methods are intended at the direct elimination and control of all biodeteriogens. Chemical treatments, mechanical removal, steam cleaning, and low-pressure water washing are the direct means available to eliminate and control the growth of biodeteriogens. Any treatment should be recommended only after accurate diagnosis to detect the colonizing macro- and microflora and evaluate their role in the deterioration process. When organisms and/or microorganisms are not responsible for serious damage, or the effectiveness of the treatment is uncertain, no treatment should be carried out.

PREVENTIVE METHODS

Preventive methods, also called indirect methods, are those that inhibit biological attack on stone monuments by modifying, where possible, the environmental surroundings and physicochemical parameters of a stone surface thereby becoming unfavorable for biological growth. This approach is effective for eliminating undesirable growth because of the strong dependence of the viability of the biodeteriogen on the environment. Though environmental parameters such as humidity, temperature, and light cannot be modified in the outdoor environment, these factors can be controlled in the indoor environment. On the other hand, nutritive factors—such as deposits of organic debris, dirt, pigeon droppings, and the like—can be reduced. It is also possible to alter the physicochemical parameters of the stone surface by applying preservative treatments. Although one particular method may not be sufficient to attain the desired results, combinations of methods give better results. Routine preventive measures that manage humidity and reduce causes of dampness in buildings, help reduce the extent of development of biological growth and slow its development. Measures such as the repair of roofs, gutters, and other water-shedding systems; improvement of drainage systems, and installation of water proofing to control rising dampness all aid in drying the stone masonry, at least partially (Ashurst and Ashurst 1988, Ortega-Calvo et al. 1993).

Vegetation could also be used as a preventive and protective measure in archaeological sites or outdoor environments. Vegetation for landscaping around monuments and sites help modify the microclimate and thereby affecting biological colonization of masonry structures. However, selection of plants should be critically done to optimize results and minimize risks related to destructive effects of their root systems. Suitably chosen vegetation may lower the water level, minimize evaporation, reduce air salinity, pollution and erosion (Fosberg 1980, De Marco et al. 1990).

PERIODIC CLEANING OF DIRT AND DUST, SPORES AND SEEDS

Periodic cleaning as a preventive conservation measure is the major way to prevent and control biological attack. Dust, various organic substance deposits, bird droppings and unsuitable restoration materials which are sources of nutrition for organisms are removed from surfaces. It has also been found useful in controlling the early establishment of mosses, lichens, fungi, algae, and higher plants by discouraging the accumulation of wind-borne spores and seeds of plants and their successive germination (Hale 1980, Shah and Shah 1992-93). For cleaning paintings, besides removing the dust and grime, the varnish layer is carefully removed by suitable solvents such as acetone, toluene, ether or ethanol. Care must be taken in order to avoid removing some of the pigment with the varnish, which gives the painting a 'skinned' appearance (Batchelor et al. 1978)

Lasers

Lasers could be used to clean stone. It does not need any physical contact with the stone and thus is ideal for cleaning very delicate surfaces. A laser beam impacts on the surface, and the energy of the beam is dissipated by vaporization of the dirt. The light absorbs the dirty stone and cleaning proceeds. Once the dirt has been removed, the light is reflected by the clean surface, and no more material is removed (Cooper et al. 1993, Orial and Riboulet 1993).

Desalination

Desalination is removal of soluble salts, which are a potential contributor to decay. Clay poultices could extract salts from masonry through repeated use, however for long run source of further salt are to be eliminated (Bowley 1975). Poultices may consist of clay, paper pulp or cellulose ethers. For the removal of less soluble material like calcium sulfate, additional materials may be added in order to increase its solubility. The solution could be drawn out by the use of a vacuum (Friese 1992). Bacteria may also be used in desalination. Gauri et al. (1992) used sulfur-reducing bacteria to eliminate the black crust and Gabrielli (1991) gave an anecdotal account of the use of cow dung to convert nitrates into elemental nitrogen. Microwaves can also be used in desalination (Minder et al. 1994). Other methods of desalination may invoke the movement of ions and of solutions under the influence of electric fields (Skibinski 1985).

BIOLOGICAL CLEANING

Biological cleaning can be followed for the treatment. This application is based on the presence of parasitic or antagonistic organisms of the biodeteriogens. Bacteria, insects and phage might be used, but up to now this kind of intervention has been applied chiefly in the agricultural sector. Antagonistic or predators species have been introduced for pull away the birds from cities. Hempel (1978) was the first to suggest the possibility. Biological cleaning has been little researched. Gauri et al. (1992), however, reported the use of the anaerobic sulfur-reducing bacterium *Desulfovibrio desulfuricans* in removing the black crust on marble. The bacterium converts calcium sulfate back to the calcium carbonate from which it was originally formed (Atlas et al. 1988).

WATER REPELLENT AND CONSOLIDANT TREATMENTS

Consolidation

Weakened stone can be restored by consolidation to some strength and might also resist further decay, thereby prolonging survival. The treatment is cheap, easy to apply, and safe to handle and remains effective for decades at a time. In order to avoid internal stresses the treated stone must have the same moisture expansion, thermal expansion, and elastic modulus as the untreated stone. Ideally, treatment should work equally well on any type of stone, regardless of the cause of decay. The consolidant must have ability to penetrate the stone. It requires a low viscosity and a low contact angle and also needs to stiffen or set once it is in place in order to strengthen the stone. These requirements can be fulfilled by three ways: First, one can apply wax at elevated temperatures that stiffens as it cools down. In practice, it is hard to get a low enough viscosity without excessive heat and the wax tends to be sticky and to pick up dirt. The second approach is to use a consolidant dissolved in a solvent. The third approach uses a low-viscosity system that undergoes a chemical reaction in situ to give a solid product.

Brush, spray, pipette or immersion can be used for application of consolidants and are drawn into the stone by capillarity as 'pocket system' (Domaslowski 1969) that hold the consolidant against the stone, and a system of bottles for maintaining a steady supply of the consolidant at a large number of points (Mirowski, 1988) or a low-pressure application technique can be followed that maximizes capillary absorption (Schoonbrood 1993). Vacuum systems may also be used to facilitate penetration into movable objects (Antonelli 1979). Organic polymers are principally used as consolidants, other than inorganic materials, such as calcium hydroxide (slaked lime) and barium hydroxide. Synthetic polymers and resins can also be used as protective coatings and consolidants in tropical environments (Kumar and Sharma 1992). Water repellents may be used to increase the efficacy of a biocidal treatment by reducing moisture retention and inhibit growth on clean surfaces. These treatments work more effectively when applied to new stone or to clean stone that has already been treated with a biocide (Sharma et al. 1989, Kumar and Sharma 1992).

INDIRECT METHOD

These include control of the environmental surroundings though control of weather variables is practically inapplicable for monumental sites, but the construction of protective structures, preventing the water run-off of surfaces, can successfully lessen biological growth (Valentin 1990). In fact, periodical simple cleaning of exposed surfaces eliminates the 'soiling' effect due to the deposit of environmental particles which can favor the development of reproductive bodies. Consolidants can also increase water repellence and cohesion thus preventing or retarding the recolonization of the substrate, but in critical environmental conditions, they could favor the development of specialized microflora. The temperature and humidity for long periods of time favors the development of the biological reproductive bodies, these should be conditioned at values of temperature ranging between 16°C-18°C and relative humidity level below 60%. This humidity value can be a limiting factor for the development of fungi, while the temperature for the insects' eggs.

REMEDIAL METHODS

Cleaning of Stone Surfaces

Microbes related with the visual disfigurement of monuments can be removed mechanically by dry or wet scrubbing or brushing, and washing with water (Sadirin 1988, Ashurst 1994). Partial removal of biological growth prior to the application of biocidal agents is to be done, especially for stone structures that are encrusted with growth of mosses, heavy lichen, and algal growth over an extended period of time. These procedures endorse biocidal activity by allowing better penetration of biocide (Garg et al. 1988). Biological material can also be physically removed by hand or with tools such as a stiff bristle or nonferrous soft-wire brushes, scalpels, spatulas, scrapers, sickles, pick axes or hoes.

Brushing and washing by water was effective for some algae, spermatophytes and pteridophytes, but mosses and crustose lichens remain mostly ineffective. For trees and creepers cutting the length of the main stem at a convenient height above ground level is suitable. The plant may then be left in this state to die on its own accord or a toxic material may be applied to hasten its destruction. Although these methods do not have long-lasting results, as elimination of superficial mycelium or cutting of vegetation alone does not completely seize vegetative activity of these organisms (Mishra et al. 1995). For complete removal of biological growth, the process has to be repeated from time to time though such methods may damage stone surfaces.

The effectiveness of all cleaning methods involving large quantities of water must be carefully evaluated in relation to microbial load; such treatments reduce infestation only for a short time (Warscheid et al. 1988). To overcome this problem biocides should be applied after cleaning the surface with water. For the removal of lichens on stone, prior application of dilute ammonia has often been recommended as it assists in the swelling and softening of the thalli (Sneyers and Henau 1968). Aqueous ammonia (2-5%) has been found to be very effective in cleaning stone monuments in India which were covered with mosses, lichens, algae and fungi without creating any side effects. This cleaning is usually followed by a biocidal treatment to inhibit biological growth and a water-repellent or preservative treatment to act as a water barrier (Sengupta 1979, Sharma et al. 1985, Kumar 1989).

Algae can be cleaned by use of diluted solutions of hydrogen peroxide and sodium hypochlorite on stone surfaces (Garg et al. 1988, Caneva et al. 1991).Though care has to be taken while applying these agents because of their bleaching action (Nugari et al. 1993). Evaluation of chemical cleaning processes usually indicates a six- to eight-month delay in reappearance of microbial growth in the absence of subsequent application of any biocidal or preservative treatment.

BIOCIDAL TREATMENTS

Biocides are those used to eliminate and inhibit biological growth such as, bactericides, fungicides, algicides and herbicides. Biocides affect the metabolic activity of organisms, thus cause severe damage and even death of the organism (Denyer 1990). Prior to their use on monuments, their effectiveness against target organisms, resistance of target organisms, toxicity to humans, risks of environmental pollution, compatibility with stone, and effects of interactions with other chemical conservation treatments, is to be assayed (Krumbein et al. 1993). Various biocides as polybor (a mixture of polyborates and boric acid), borax and clorox could be used to control microbial growth. Surface-active quaternary compounds have strong biocidal activity (Richardson 1988). Quaternary ammonium compounds are reported to effectively inhibit microbial growth on sandstone monuments (Sadirin 1988). Lichens can be temporarily inhibited by aqueous solutions of benzalkonium chloride

(20%), sodium hypochlorite (13%), and formaldehyde (5%), using soaked cotton strips for about 16 hours, followed by scrubbing with a brush and water (Nishiura and Ebisawa 1992). The treatment can kill all organisms and microorganisms but does not give any kind of protection for future recolonization, thus biocides are to be applied regularly to control the biological growth.

APPLICATION METHODS

Commonly biocides are applied over surfaces by spraying and brushing when the component material is in poor condition. However in the case of organic materials, such as paintings on paper, prints, parchment or old books, spraying or brushing biocides is not always possible since some liquid solutions can dissolve pigments and inks. Poultices can also be applied to increase the contact time and use the dissolving action of water itself. For any of the above mentioned treatments, it is prudent to remove biocidal residues to circumvent possible secondary reactions (due to the degradation of products) or toxicological difficulties (high-persistence products on premises open to the public or for operators). Biocide can also be supplemented to a gelatinous solvent paste used for cleaning stone and mural paintings. In some cases, the removal of biological or chemical incrustations on stone is obtained through non-selective absorbent clays (e.g., sepiolite or attapulgite). Pesticides can be injected to enhance their diffusion for infested wooden structures. In the case of subterranean termites, injection is performed under pressure into the soil. Fumigation methods are widely employed for its rapid effectiveness and reach deep penetration inside the object but need to be applied carefully owing to high toxicity. As an alternative to conventional fumigation techniques, inert gases, such as nitrogen, together with low RH can be employed for controlling museum pests. The time necessary for a substratum to be reinfested by biodeteriogens depends on the preventive measures taken for its conservation — it can be a long (several years) if the restored objects are maintained or else, it is short i.e 1-2 years (Nugari et al. 1987, Gilbert 1989).

Soft and Hard Nanomaterials

Nanomaterials could be applied to the field of cultural heritage preservation. Nanoparticles of calcium and magnesium hydroxide and carbonate are reported to restore and protect wall paints and to de-acidify paper and wood. Nanodispersions of solids, micelle solutions, gels and microemulsions present novel and trustworthy ways to restore and preserve works of art by merging together the main features and properties of soft matter and hard matter systems, allowing the synthesis of systems specifically adapted for the works of art to fight the deterioration processes which menace the conservation of the world cultural heritage (Baglioni and Giorgi 2006).

BIOREMEDIATION

Innovative systems including biological methods can be used for cleaning of cultural objects in order to reduce any possible risk for the materials treated. Enzymes, such as lipase, have been effectively used to eliminate aged acrylic resin coatings in paintings, while sulfate, nitrates and organic matter from artistic stone works can be removed by microbial cultures (Heselmeyer et al. 1991, Bellucci et al. 1999). It has been reported that on a laboratory scale bacterial strains (such as *Pseudomonas stutzeri, P. aeruginosa, Desulfovibrio vulgaris, D. desulfuricans*) were not only able to denitrify and desulfuricate harmful masonry salts such as nitrate and sulfate but also mineralize organic residues or pollutants like carbohydrates, waxes or hydrocarbons that commonly occur in crusts on stonework (Ranalli et al. 2003, 2005). Other microorganisms, including *Bacillus cereus,* have been shown to protect the exposed mineral surfaces, by the formation of sacrificial layers of calcite. These may be dissolved in polluted environments but they can be renewed when necessary (Tiano et al. 1999, Urzì et al. 1999).

REFERENCES

Abbate Edlmann, M.L. 1967. Primo contributo allo studio delle alterazioni da Teredini in vari legni immersi nel Mare Ligure. Contributi Scientifico-Pratici per una Migliore Conoscenza e Utilizzazione del Legno. VII, No. 10. CNR. 9-35. Rome.

Abbate Edlmann, M.L., A. Gambetta, G. Giachi, and E. Orlandi. 1989. Studio del deterioramento di alcune specie legnose appartenenti ad un relitto navale del VII secolo. A.C., effettuato con il microscopio elettronico a scansione. Il restauro del legno. Atti del 21 Congresso Nazionale Restauro del Legno. Nardini Editore. Florence. pp. 121-127.

Allsopp, D. and K.J. Seal. 1986. Introduction to Biodeterioration. Edward Arnold. London.

Andreoli, C., N. Rascio, L. Garlet, S. Leznicka, and A. Strzelczyk. 1988. Interrelationships between algae and fungi overgrowing stone works in natural habitats. In 6th International Congress on Deterioration and Conservation of Stone, vol. 2, comp. J. Ciabach, 324-27. Torun, Poland: Nicholas Copernicus University Press Department.

Antonelli, V. 1979. Il restauro della Porta della Carta in Venezia. Deterioramento e conservazione della pietra: AW del 3 Congresso internazionale, 629-44. Università degli Studi di Padova. Padova, Italy.

Arai, H., N. Matsui, and H. Murakita. 1988. Biochemical investigations on the formation mechanisms of foxing. *In:* S. Mills, P. Smith and K. Yamasaki [eds.]. The Conservation of Far Eastern Art. Preprints of the contributions to the Kyoto Congress, 19-23 September. London: 11C. pp. 11-14.

Ashurst, J. and N. Ashurst. 1988. Stone Masonry. Vol. 1 of Practical Building Conservation. Aldershot. Gower Technical Press, England.

Atlas, R.M., A.N. Chowdhury, and K.L. Gauri. 1988. Microbial calcification of gypsum-rock and sulfated marble. Studies in Conservation 33: 149-153.

Baglioni, P. and R. Giorgi. 2006. Soft and hard nanomaterials for restoration and conservation of cultural heritage. The Royal Society of Chemistry pp. 293-303.

Bassi, M., N. Barbieri, and R. Bonecchi. 1984. St. Christopher's church in Milan: Biological investigations. Arte Lombarda, 68/69: 117-121.

Bell, D. 1984. The role of algae in the weathering of Hawkesbury sandstone: Some implications for rock art conservation in the Sydney area. Institute for the Conservation of Cultural Material Bulletin, Canberra 10: 5-12.

Bell, D. 1997. Guide to International Conservation Charters. The Historic Scotland, Edinburg. ISBN 1-900168-24-3.

Bellucci, R., P. Cremonesi, and G. Pignagnoli. 1999. A preliminary note on the use of enzymes in conservation: The removal of aged acrylic resin coatings with lipase. Studies in Conservation 44: 278-281.

Bhatnagar, P. 2008. Geomicrobiological Study of Gwalior fort with special reference to fungi and restoration measures. Ph.D. thesis, Jiwaji University, India.

Bowley, M.J. 1975. Desalination of stone: A case study. *In:* Building Research Establishment Current Paper CP46/75. Building Research Establishment. Watford, England.

Bravery, A.F., R.W. Berry, L. Carey, and D.E. Cooper. 1987. Recording wood rot and insect damage in buildings. *In:* W.F. Clapp and R. Kenk [eds.]. Building Research Establishment Report.

Brill, T.B. 1980. Light: Its Interaction with Art and Antiquities. Plenum Press, New York.

Burford, P.E., M. Kierans, and G.M. Gadd. 2003. Geomycology: fungi in mineral substrata, Mycologist 17: 98-107 Cambridge University Press.

Caneva, G., A. Danin, S. Ricci, and C. Conti. 1994. The pitting of Trajan's column, Rome: An ecological model of its origin. Conservazione delpatrimonio culturale, 78-102. Contributi del Centro linceo inter-disciplinare "Beniamino Segre," no. 88. Rome: Accademia Nazionale dei Lincei.

Caneva, G. and O. Salvadori. 1988. Biodeterioration of stone. *In:* Deterioration and Conservation of Stone, ed. Lorenzo Lazzarini and Richard Pieper, 182-234. Studies and Documents on the Cultural Heritage, no. 16. Paris: Unesco.

Caneva, G., M.P. Nugari, and O. Salvadori. 1991. Biology in the Conservation of Works of Art. Rome: ICCROM.

Charola, A.E. 1984. "Understanding Stone Decay Through Chemistry" The pHilter, 16, 1.

Clapp, W.F. and R. Kenk. 1963. Marine borers. An annotated bibliography, Office of Naval Research. Dpt. of the Navy, Washington D.C., pp. 10-1134.

Cooper, M.I., D.C. Emmony, and J.H. Larson. 1993. Laser cleaning of limestone sculpture. *In:* N.H. Tennent [ed.]. Conservation Science in the U.K. James and James, London. pp. 29-32.

De Marco, G., G. Caneva, and A. Dinelli. 1990. Geobotanical foundation for a protection project in the Moenjodaro archaeological area. Prospezione Archeologiche 1: 115-120.

Desmarais, G. 2000. "Impact of added insulation on the hygrothermal performance of leaky exterior wall assemblies", MASc, Concordia University (Canada).

Dhawan, S. 1986. Microbial deterioration of paper material—A literature review. M.M. Khan [ed.]. Government of India, Department of Culture, National Research Laboratory for Conservation of Cultural Property. Lucknow, India. pp. 1-18.

Dhawan, S. and O.P. Agrawal. 1986. Fungal flora of miniature paintings and lithographs. Int. Biodet. Bull. 22(2): 95-99.

Diakumaku, E. 1996. Investigations on the role of black fungi and their pigments in the deterioration of monuments. Ph.D. diss., University of Oldenburg, Germany. Division of Building Research, CSIRO.

Domaslowski, W. 1969. Consolidation of stone objects with epoxy resins. Monumentum 4: 51-64.

Dornieden, T., A.A. Gorbushina, and W.E. Krumbein. 2000. Biodecay of mural paintings and stone monuments as a space/time related ecological situation —an evaluation of a series of studies. Int. Biodeterioration and Biodegradation 46: 261-270.

Eckhardt, F.E.W. 1985. Mechanisms of the microbial degradation of minerals in sandstone monuments, medieval frescoes and plaster. *In:* P.G. Felix (comp) Proceedings of the 5th International Congress on Deterioration and Conservation of Stone. Vol. 2, 25-27 September Presses polytechniques romandes. Lausanne, Switzerland. pp. 643-652.

Feilden, B.M. 1982. Insects and other pests as causes of decay. Conservation of Historic Buildings. Butterworth Scientific, London. pp. 131-151.

Fitzner, B., K. Heinrichs, and R. Kownatzki, 1997. Weathering forms at natural stone monuments—classification, mapping and evaluation. International Journal for Restoration of Buildings and Monuments. Vol. 3, No. 2, Aedificatio Verlag/Fraunhofer IRB Verlag, Stuttgart. pp. 105-124.

Fitzner, B., K. Heinrichs and D. La Bouchardiere. 2002. Damage index for stone monuments. *In:* E. Galan and F. Zezza [eds.]. Protection and Conservation of the Cultural Heritage of the Mediterranean Cities. Proceedings of the 5th International Symposium on the Conservation of Monuments in the Mediterranean Basin, Sevilla, Spain, 5-8 April 2000. Swets and Zeitlinger, Lisse, The Netherlands. pp. 315-326.

Florian, M. 1988. Deterioration of organic materials other than wood. *In:* C. Pearson [ed.]. Conservation of Marine Archaeological Objects. Butterworths, London. pp. 21-54.

Fosberg, F.R. 1980. The Plant Ecosystem for Moenjodaro. Unesco Technical Report no. RP/1977-78/4.121.6, FMR/CC/CH/80/189. Unesco. Paris

Friese, P. 1992. Desalination of brickwork. NATO-CCMS pilot study. Conservation of historic brick structures. *In:* Proceedings of the 5th Expert Meeting, Berlin, 1991, 186-194.

Gabrielli, N. 1991. Mural paintings in Science, Technology and European Cultural Heritage. *In:* N.S. Baer, C. Sabbioni, and A.I. Sors [eds.]. Butterworth-Heinemann, Oxford. pp. 168-179.

Gallo, F. 1985. Biological Factors in Deterioration of Paper. ICCROM. Rome.

Garg, K.L., S. Dhawan, and O.P. Agrawal. 1988. Deterioration of Stone and Building Materials by Algae and Lichens: A Review. National Research Laboratory for Conservation of Cultural Property, Lucknow, India.

Gauri, L.K., L. Parks, J. Jaynes, and R. Atlas. 1992. Removal of sulphated-crust from marble using sulphate-reducing bacteria. *In:* G.M. Robin [ed.]. Stone Cleaning and the Nature, Soiling and Decay Mechanisms of Stone. Proceedings of the International Conference, 14 to 16 April, Donhead Publishing Ltd., Webster, Edinburgh, United Kingdom. pp. 160-165.

Gehrmann, C.K., K. Petersen and W.E. Krumbein. 1988. Silicole and calcicole lichens on jewish tombstones—interactions with the environment and biocorrosion. *In:* VIth International Congress on Deterioration and Conservation of Stone, edited by Nicholas Copernicus University, Torun, pp. 33-38.

Gilbert, M. 1989. Inert atmosphere fumigation of museum objects. Studies in Conservation 34: 80-84.

Gill, W.R. and G.H. Bolt. 1955. Pfeffer's studies of the root growth pressures exerted by plants. Agronomy Journal 47: 166-168.

Gorbushina, A.A., K.A. Palinska, K. Sterflinger, and W.E. Krumbein. 1998. Biotechnologische Verfahren zum Schutz von Kulturgütern. BioForum, Forschung und Entwicklung 21, 274-277.

Gorbushina, A.A. and W.E. Krumbein. 1999. Poikilotroph response of micro-organisms to shifting alkalinity, salinity, temperature and water potential. *In:* A. Oren [ed.]. Microbiology and Biogeochemistry of Hypersaline Environments, CRC Press, LLC, Boca Raton. pp. 75-86.

Gorbushina, A. and W.E. Krumbein. 2000. Subaerial Microbial Mats and Their Effects on Soil and Rock. *In:* R. Riding and S.M. Awramik [eds.]. Microbial Sediments. Springer, Heidelberg. pp. 161-170.

Griffin, P.S., N. Indicator, and R.J. Koestler. 1991. The biodeterioration of stone: a review of deterioration mechanisms, conservation case, histories and treatment. International Biodeterioration 28, 187-208.

Guillitte, O. 1995. Bioreceptivity: A new concept for building ecology studies. The Science of the Total Environment 167: 215-220.

Guillitte, O. and R. Dreesen. 1995. Laboratory chamber studies and petrographical analysis as bioreceptivity assessement tool of building materials. The Science of the Total Environment 167: 365-374.

Hale, M.E. 1980. Control of biological growths on Mayan archaeological ruins in Guatemala and Honduras. *In:* National Geographic Research Reports, 1975 Projects. National Geographic Society, Washington, D.C. pp. 305-321.

Hempel, K. 1976. An improved method for the vacuum impregnation of stone. Studies in Conservation 21: 40-43.

Heselmeyer, K., U. Fischer, K.E. Krumbein, and T. Warscheid. 1991. Application of *Desulfovibrio vulgaris* for the bioconversion of rock gypsum crusts into calcite. BIOforum 1/2: 89.

Honeyborne, D.B. 1990. Weathering and decay of masonry. *In:* J. Ashurst and F. G. Dimes [eds.]. Conservation of Building and Decorative Stone, Vol. 1, Butterworth- Heinemann, London. pp. 153-184.

Hyvert, G. 1972. The Conservation of Borobudur Temple. Unesco document no. RMO. RD/2646/CLP. Paris: Unesco.

Jain, K.K., A.K. Mishra, and T. Singh. 1993. Biodeterioration of stone: A review of mechanism involved. *In:* Recent Advances in Biodeterioration and Biodegradation, vol. 1, ed. K.L. Garg, Neelima Garg, and K.G. Mukerji, 323-54. Calcutta: Nay a Prokash.

Jones, D. and M.J. Wilson. 1985. Chemical activity of lichens on mineral surfaces: A review. International Biodeterioration 21: 99-104.

Jones, B. and G. Pemberton. 1987. Experimental formation of spiky calcite through organically mediated dissolution. Journal of Sedimentary Petrology 57: 687-694.

Jones, D., M.J. Wilson, and W.J. McHardy. 1981. Lichen weathering of rock-forming minerals: Application of scanning electron microscopy and microprobe analysis. Journal of Microscopy 124: 95-104.

Kirk, K.T. and M. Shimada. 1985. Lignin biodegradation: The microorganisms involved, and the physiology and biochemistry of degradation by white-rot fungi. *In:* T. Higuchi [ed.]. Biosynthesis and Biodegradation of Wood Components. Academic Press, London. pp. 579-605.

Koestler, R.J. and O. Salvadori. 1996. Methods of evaluating biocides for the conservation of porous building materials. Science and Technology for Cultural Heritage 5(1); 63-68.

Korpi A., A-L. Pasanen, and P. Pasanen, August. 1998. Volatile Compounds Originating from Mixed Microbial Cultures on Building Materials under Various Humidity Conditions, Applied and Environmental Microbiology, Vol. 64, No. 8, 2914-2919.

Kowalik, R. 1980. Microbiodecomposition of basic organic library materials. Microbiodeterioration of library materials. Part 2. Restaurator, 4(34): 135-219.

Krumbein, W.E. 1988. Biotransfer in Monuments—a sociobiological study: Durability of Building Materials 5: 359-382.

Krumbein, W.E., S. Diakumaku, K. Petersen, T. Warscheid, and C. Urzi. 1993. Interactions of microbes with consolidants and biocides used in the conservation of rocks and mural paintings. *In:* M.-J. Thiel, E. London, F.N. Spon, E. Batchelor, K. Mulvaney-Buente, and G.T. Nightwine [eds.]. 1978, Art Conservation: The Race Against Destruction; Cincinnati Art Museum: Cincinnati, Ohio, 589-596.

Krumbein, W.E. and E. Diakumaku. 1996. The role of fungi in the deterioration of stones. *In:* M. deCleene [ed.]. Interactive physical weathering and bioreceptivity study on building stones, monitored by computerized X-ray tomography (CT) as a potential non-destructive research tool. Protection and Conservation of the European Cultural Heritage Research Report Nr. 2. Science Information Office, Gent, 286. pp. 140-170.

Krumbein, W.E. and A.A. Gorbushina. 1996. Organic pollution and rock decay. pp. 277-284. *In:* R. Pancella [ed.], Preservation and restoration of cultural heritage. Proceedings of the 1995 LPC Congress. EPFL, Lausanne, 773p.

Krumbein, W.E. 2003. Patina and cultural heritage – a geomicrobiologist's perspective. pp. 39-47. *In:* R. Kozlowski, [ed.], Proceedings of the Vth EC Conference Cultural Heritage Research: A Pan-European Challenge. EC and ISC, Krakow, 451p.

Kumar, R. 1989. Science branch's physical progress reports. Internal Report, Archaeological Survey of India, Bhubaneswar, India.

Kumar, R. and R.K. Sharma. 1992. Conservation of deul of the Lord Jagannath temple, Puri (Orissa), India: A case study. *In:* J. Delgado Rodrigues, F. Henriques, and F.T. Jeremias [eds.]. Proceedings of the 7th International Congress on Deterioration and Conservation of Stone. Held in Lisbon, Portugal, 15-18 June. Laboratório Nacional de Engenharia Civil. Lisbon. pp. 1471-1480.

Lepidi, A.A. and G. Schippa. 1973. Some aspects of the growth of chemotrophic and heterotrophic microorganisms on calcareous surfaces. *In:* Colloque international sur la détérioration des pierres en oeuvre, 1er, La Rochelle, 11-16 septembre 1972, 143-48. Chambéry, France: Les imprimeries réunies de Chambéry.

Makies, F. 1981. Enzymatic consolidation of paintings ICOM Committee for Conservation. 6th Triennal Meeting. Ottawa, 21-25 September Preprints. ICOM. 81/2/7- 1 - 7n. Paris.

Makies, F. 1984. Enzymatic removal of lining paste from painting. *In:* ICOM Committee for Conservation. 7th Triennial Meeting. Copenhagen, 10-14 September, Preprints. ICOM. 84.2. 26-2.30. Paris.

May, E., F.J. Lewis, S. Pereira, S. Tayler, M.R.D. Seaward, and D. Allsopp. 1993. Microbial deterioration of building stone: A review. Biodeterioration Abstracts 7(2): 109-123.

McGlinchy, C. 1994. "Color and Light in the Museum Environment" *In:* The Changing Image: Studies in Paintings Conservation; The Metropolitan Museum: New York, pp. 44-52.

Mirowski, R. 1988. A new method of impregnation of stone historical objects. 6th International Congress on Deterioration and Conservation of Stone. Nicholas Copernicus University. Torun, Poland. pp. 633-640.

Mishra, A.K., K.K. Jain, and K.L. Garg. 1995. Role of higher plants in the deterioration of historic buildings. The Science of the Total Environment 167: 375-392.

Nasim, A. and A.P. James. 1978. Life under condition of high irradiation. In D.J. Kushner [ed.]. Microbial life in extreme environments. Academic press, New York, 409-439.

Nishiura, T. and T. Ebisawa. 1992. Conservation of carved natural stone under extremely severe conditions on the top of an high mountain. *In:* K. Toishi, H. Arai, T. Kenjo, and K. Yamano [eds.]. Proceedings of the 2nd International Conference on Biodeterioration of Cultural Property, October 5-8. Held at Pacifico Yokohama, International Communications Specialists. Tokyo. pp. 506-511.

Nugari, M.P., G.F. Priori, D. Mate' and F. Scala. 1987. Fungicides for use on textiles employed during the restoration of works of art. Int. Biodet. Bull., 23: 295-306.

Orial, G. and G. Riboulet. 1993. Technique de nettoyage de la statuaire monumentale par désincrustation photonique: Réalisation d'un prototype mobile. *In:* M.-J. Thiel [ed.]. Conservation of Stone and Other Materials, E and F N Spon. London. pp. 542-549.

Orial G., S. Castanier, G. Le Métayer, and J.F. Loubiere. 1993. The biomineralisation: A new process to protect calcareous stone applied to historic monuments. *In:* K. Toishi, H. Arai, T. Kenjo, and K. Yamano [eds.]. Proceedings of 2nd International Conference on Biodeterioration of Cultural Property, Yokohama. Japan, pp. 98-116.

Ortega-Calvo, J.J., M. Hernandez-Marine, and C. Saiz-Jimenez. 1993. Cyanobacteria and algae on historic building and monuments. *In:* K.L. Garg, Neelima Garg, and K.G. Mukerji [eds.]. Récent Advances in Biodeterioration and Biodégradation, Vol. 1, Naya Prokash. Calcutta. pp. 173-203.

Pasanen, A.L., S. Lappalainen, and P. Pasanen. 1996. Volatile organic metabolites associated with some toxic fungi and their mycotoxins. Analyst 121: 1949-1953.

Price. C.A. 1996. Stone Conservation, an Overview of Current Research. J. Paul Getty Trust Publications, Santa Monica, California.

Purvis, O.W. 1984. The occurrence of copper oxalate in lichens growing on sulphide-bearing rocks in Scandinavia. Lichenologist 16: 197-204.

Ranalli, G., E. Zanardini, P. Pasini, and A. Roda. 2003. Rapid biodeteriogen and biocide diagnosis on artwork: A bioluminescent low-light imaging technique. Annals of Microbiology 53(1): 1-13.

Ranalli, G., G. Alfano, C. Belli, G. Lustrato, M.P. Colombini, I. Bonaduce, E. Zanardini, P. Abbruscato, F. Cappitelli, and C. Sorlini. 2005. Biotechnology applied to cultural heritage: Biorestoration of frescoes using viable bacterial cells and enzymes. Journal of Applied Microbiology 98(1): 73-83.

Realini, M., C. Sorlini, and M. Bassi. 1985. The Certosa of Pavia, a case of biodeterioration. *In:* G. Felix (comp), 5th International Congress on Deterioration and Conservation of Stone, Proceedings. Lausanne, 25-27 September, Vol. 2, Presses polytechniques romandes. Lausanne, Switzerland. pp. 627-629.

Richardson, B.A. 1988. Control of microbial growth on stone and concrete. *In:* D.R. Houghton, R.N. Smith, and H.O.W. Eggins [eds.]. Biodeterioration 7:

Selected Papers Presented at the Seventh International Biodeterioration Symposium, Cambridge, U.K., 6-11 September 1987, Elsevier Applied Science, New York. pp. 101-106.

Riederer, J. 1981. The preservation of historical monuments in Sri Lanka. *In:* Conservation of Stone 2: Preprints of the Contributions to the International symposium, Bologna, 27-30 October 1981, ed. Raffaella Rossi-Manaresi, 737-58. Bologna: Centro per la Conservazione delle sculture all'aperto.

Sadirin, H. 1988. The deterioration and conservation of stone historical monuments in Indonesia. *In:* J. Ciabach (comp), 6th International Congress on Deterioration and Conservation of Stone. Vol. 1, Nicholas Copernicus University Press Department, Torun, Poland. pp. 722-731.

Saiz-Jimenez, C. 1993. Deposition of airborne organic pollutants on historic buildings. Atmos Environ 27B: 77-85.

Schaffer, R.J. 1967. Causes of deterioration of building materials. I: Chemical and physical causes. Chemistry and Industry 23: 1584-1586.

Schaffer, R.J. 1972. The Weathering of Natural Building Stones. Building Research Special Report, no. 18. London: Her Majesty's Stationery Office.

Schoonbrood, J.W.M. 1993. Low pressure application technique for stone preservatives. *In:* M.-J. Thiel [ed.]. Conservation of Stone and Other Materials. E and F N Spon. London. pp. 512-518.

Shah, R.P. and N.R. Shah. 1992-93. Growth of plants on monuments. Studies in Museology 26: 29-34.

Sharma, B.R.N., K. Chaturvedi, N.K. Samadhia, and P.N. Tailor. 1985. Biological growth removal and comparative effectiveness of fungicides from central India temples for a decade in-situ. *In:* G. Félix (comp), Proceedings of the 5th International Congress on Deterioration and Conservation of Stone. Lausanne, 25-27 September, Vol. 2, Presses polytechniques romandes. Lausanne, Switzerland. pp. 675-683.

Singh, A. and G.P. Sinha. 1993. Corrosion of natural and monument stone with special reference to lichen activity. Recent Advances in Biodeterioration and Biodégradation 1: 355-577.

Skibinski, S. 1985. Salt removal from stone historical objects by means of membrane electrodialysis. 5th International Congress on Deterioration and Conservation of Stone, Ed. G. Felix, 959-65, Lausanne: Presses polytechniques Romandes.

Sneyers, R.V. and P.J. Henau. 1968. The conservation of stone. *In:* The Conservation of Cultural Property with Special Reference to Tropical Conditions. Unesco, Paris. pp. 209-235.

Sterflinger, K. and W.E. Krumbein. 1997. Dematiaceous fungi as the main agent of biopitting on mediterranean marbles and limestones. Geomicrobiology Journal 22: 219-231.

Strzelczyk, A.B. 1981. Painting and sculpture. *In:* A.H. Rose [ed.]. Microbial Biodeterioration. Economic Microbiology Vol. 6. Academic Press, London. pp. 203-234.

Syers, J.K. and I.K. Iskandar. 1973. Pedogenic significance of lichens. *In:* V. Ahmadjian and M.E. Hale [eds.]. The Lichens. Academic Press, New York. pp. 9-29.

Tiano, P. 1987. Biological deterioration of exposed works of art made of stone. In the Biodeterioration of Constructional Materials: Proceedings of the Summer Meeting of the Biodeterioration Society Held at TNO Division of Technology for Society, Delft, the Netherlands, 18-19 September, 1986, ed. L. H. G. Morton, 37-44. Kew, England: Biodeterioration Society.

Tiano, P., L. Addadi, and S. Weiner. 1992. Stone reinforcement by induction of calcite crystals using organic matrix macromolecules, feasibility study. *In:* Proceeding of 7th International Congress on Deterioration and Conservation of Stone, Vol. 2. Lisbon. pp. 1317-1326.

Tiano, P., L. Biagiotti, and G. Mastromei. 1999. Bacterial Bio-mediated Calcite Precipitation for Monumental Stone Conservation: Methods of Evaluation. Journal of Microbiological Methods 36: 139-145.

Urzì, C., M.T. Garcia-Valles, M. Vendrell, and A. Pernice. 1999. Biomineralisation processes of the rock surfaces observed in field and in laboratory. Geomicrobiology J. 16: 39-54.

Valentin, N., M. Lidstrom, and F. Preusser. 1990. Microbial control by low oxygen and low relative humidity environment. Studies in Conservation 35: 222-230.

Varonina, L.L., O.N. Nazarova, U.P. Petushkova, and N.L. Rebrikova. 1981. Damage of parchment and leather caused by microbes. ICOM Committee for Conservation, 6th Triennial Meeting. Ottawa, 21-25 September, Preprints. ICOM. 1913.1-19/3.11.Paris.

Viitanen, H., A. Hanhijärvi, A. Hukka, and K. Koskela. 2000. Modelling mould growth and decay damages. Healthy Buildings, Espoo, August 6-10, 3: 341-346.

Von Endt, D.W. and W.C. Jessup. 1986. The deterioration of protein materials in museums. *In:* S. Barry and D.R. Houghton [eds.]. Biodeterioration 6. Proceedings of the Sixth International Biodeterioration Symposium, Cab International, Great Britain. pp. 332-337.

Voute, C. 1969. Indonesia: Geological and Hydrological Problems Involved in the Preservation of the Monuments of the Borobudur. Unesco Document No. 1241/BMF-RD/CLT. Unesco.Paris.

Warscheid, T. and W.E. Krumbein. 1996. General aspects and selected cases. *In:* Heitz et al. [eds.]. Microbially Induced Corrosion of Materials. Springer, Berlin. pp. 274-295.

Warscheid, T., K. Petersen, and W.E. Krumbein. 1988. Effect of cleaning on the distribution of microorganisms on rock surfaces. *In:* D.R. Houghton, R.N. Smith, and H.O.W. Eggings [eds.]. Biodeterioration 7. Elsevier Applied Science Publ. pp. 455-460.

Wee, Y.C. and K.B. Lee. 1980. Proliferation of algae on surfaces of buildings in Singapore. International Biodeterioration Bulletin 16: 113-117.

Werner, A. 1981. " Synthetic Materials in Art Conservation" J. Chem. Educ. 58, 321.

Wolters, B., W. Sand, B. Ahlers, F. Sameluck, M. Meincke, C. Meyer, T. Krause-Kupsch, and E. Bock. 1988. Nitrification—the main source for nitrate deposition in building stones. In 6th International Congress on Deterioration and Conservation of Stone, vol. 1, comp. J. Ciabach, 24-31. Torun, Poland: Nicholas Copernicus University Press Department.

Index

Color Plate Section

Chapter 6

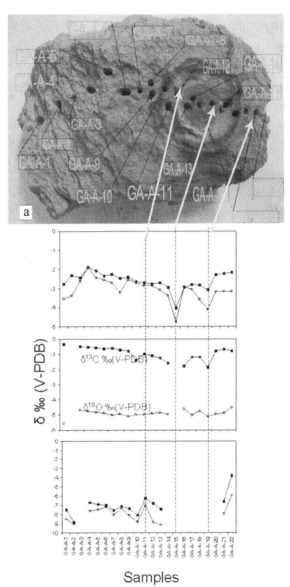

FIG. 6.1 Extinct low-temperature hydrothermal chimney from the Jaroso Mars Analog (SE Spain) (Martínez-Frías et al. 2006). Sampling and stable isotope variations from the central orifice to the outer rim. Note that siderite is not present where ankerite (more stable phase) occurs. The paleoenvironmental interpretation carried out from the isotopic data emphasizes the importance of the stable isotopes as fluid geomarkers.

Chapter 7

Live *Spirulina* sp. (LSP)

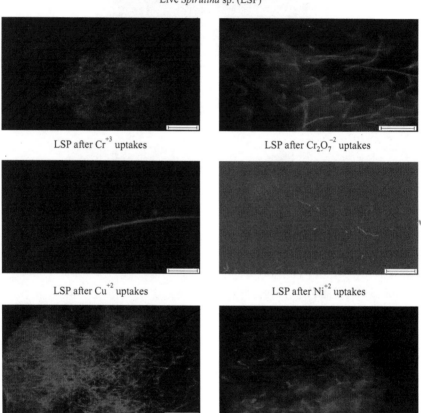

LSP after Cr^{+3} uptakes

LSP after $Cr_2O_7^{-2}$ uptakes

LSP after Cu^{+2} uptakes

LSP after Ni^{+2} uptakes

LSP after As_4O^{-3} uptakes

LSP after Cd^{+2} uptakes

Plate 7.3 Fluorescence microscopic photographs (⊢⊣ : 40 μ) of live Spirulina sp. (LSP) and treated ones.